AGING OF ORGANS AND SYSTEMS

BIOLOGY OF AGING AND ITS MODULATION

VOLUME 3

AGING OF ORGANS AND SYSTEMS

Edited by

RICHARD ASPINALL

Department of Immunology
Imperial College of Science Technology and Medicine
Chelsea and Westminster Hospital
London, UK

KLUWER ACADEMIC PUBLISHERS
DORDRECHT / BOSTON / LONDON

Library of Congress Cataloging-in-Publication Data is available.

ISBN 1-4020-1743-X

Published by Kluwer Academic Publishers,
PO Box 17, 3300 AA Dordrecht, The Netherlands.

Sold and distributed in North, Central and South America
by Kluwer Academic Publishers,
101 Philip Drive, Norwell, MA 02018, USA

In all other countries, sold and distributed
by Kluwer Academic Publishers, Distribution Center,
PO Box 322, 3300 AH Dordrecht, The Netherlands

Printed on acid-free paper

Printed and bound in Great Britain by Antony Rowe Limited.

Contents

About the series "Biology of aging and its modulation"

During the last 40 years, the study of the biological basis of aging has progressed tremendously, and it has now become an independent and respectable field of study and research. Several universities, medical institutes and research centers throughout the world now offer full-fledged courses on biogerontology. The interest of students taking such courses, followed by undertaking research projects for MSc and PhD studies, has also increased significantly. Cosmetic, cosmeceutical and pharmaceutical industry's ever increasing interest in aging research and therapy is also obvious. Moreover, increased financial support by the national and international financial agencies to biogerontological research has given much impetus to its further development.

This five-volume series titled "Biology of Aging and its Modulation" fulfills the demand for books on the biology of aging, which can provide critical and comprehensive overview of the wide range of topics, including the descriptive, conceptual and interventive aspects of biogerontology. The titles of the books in this series and the names of their respective editors are:

1.   Aging at the molecular level (Thomas von Zglinicki, UK)

2.   Aging of cells in and outside the body (S. Kaul and R. Wadhwa, Japan)

3.   Aging of organs and systems (R. Aspinall, UK)

4.   Aging of organisms (H. D. Osiewacz, Germany)

5.   Modulating aging and longevity (S. Rattan, Denmark)

The target readership is both the undergraduate and graduate students in the universities, medical and nursing colleges, and the post-graduates taking up research projects on different aspects of biogerontology. We hope that these books will be an important series for the college, university and state libraries maintaining a good database in biology, medical and biomedical sciences. Furthermore, these books will also be of much interest to pharmaceutical, cosmaceutical, nutraceutical and health-care industry for an easy access to accurate and reliable information in the field of aging research and intervention.

Suresh I.S. Rattan, Ph.D., D.Sc.
Series Editor and Editor-in-Chief, Biogerontology
Danish Centre for Molecular Gerontology, Department of Molecular Biology,
University of Aarhus, Denmark

Preface

"Biological scientists working on aging should study the aging process and not the diseases often cited as the cause of death amongst those over the age of 65" was one side of an old argument put to me by someone in the field of aging research who proposed that diseases such as influenza, pneumonia and heart disease can kill individuals throughout a lifespan and are not purely diseases of the elderly. The other side of the argument is that diseases which result in death in an elderly population are indicators of how the process of aging is affecting the mechanisms of the body, and if one is to start looking at the process of aging, what better place to begin than with the pathology of these diseases? As a scientist working on aging, it is difficult to remain as a disinterested third party in this debate.

This volume on *Aging of Organs and Systems* is therefore an attempt to bring some understanding to both the aging process and the disease processes of old age. Through hypothesis-led scientific work we may understand the aging process enough to extend both healthspan and lifespan, although thumbing through the history of aging would suggest that this has already been achieved. From where I sit in my laboratory it is only a short ride by tube to Westminster Abbey where Thomas Parr is buried in the South Transept. The inscription on his white marble gravestone reads:

THO: PARR OF YE COUNTY OF SALLOP. BORNE
IN AD: L483. HE LIVED IN YE REIGNES OF TEN
PRINCES VIZ: K.EDW.4. K.ED.5. K.RICH.3.
K.HEN.7. K.HEN.8. K.EDW.6. Q.MA. Q.ELIZ.
K.JA. & K.CHARLES. AGED L52 YEARES.
& WAS BURYED HERE NOVEMB. L5. L635.

So, according to this information, Thomas Parr died at the age of 152 years and 9 months, and from other tales his was an active life. He married Jane Taylor when aged about 80, with whom he had a son and daughter, and when aged 100 his penance for being unfaithful to his wife by having an affair with Kathleen Milton (who bore his illegitimate child) was to stand draped in a white sheet in the parish church. Some time after Jane's death he married Jane Lloyd, a marriage which produced no children. This prolonged lifespan (and obvious healthspan) he attributed to a diet of green cheese, onions, coarse bread, buttermilk or mild ale. He was brought to the Court of Charles I by the Earl of Arundel but seems to have failed to adhere to his own advice, since he died within a few weeks of arriving in London, a death ascribed to the change in diet (the rich wines), and the pollution of the City.

Perhaps we would not all like a life as exciting as Old Parr's, but, at least for some, the understanding of the loss of activity with age and whether this can be reversed would be beneficial, and for others the means of understanding the aging process in order to prevent the onset of specific diseases would also prove rewarding.

Richard Aspinall
London 2003

The Relationship Between Cell Turnover and Tissue Aging

Richard G.A. Faragher

School of Pharmacy and Biomolecular Sciences, University of Brighton, Brighton, East Sussex, BN2 4GJ, UK

Introduction

We live in an aging world. By 2050 between 40–50% of the population of the western world and the Pacific Rim will be over the age of 60 [1]. Such increasing life expectancies are a testament to the capacity of progress in science and technology to deliver real benefits to ordinary people. But it should not be forgotten that these increases in life expectancy have resulted almost exclusively through improvements in our ability to deal with childhood mortality and infectious disease, the great killers of the 19th and early 20th centuries. Unfortunately solving the most important problems of one age inevitably brings those lurking in the wings firmly to the centre of the stage. If it becomes possible to prevent a population dying like flies early on in their lives it becomes necessary to deal with the different spectrum of problems that same population may experience in later life. Advancing chronological age is associated with an increased susceptibility to a wide range of degenerative conditions, which range in seriousness from the trivial to the fatal. All of these illnesses cost money and reduce the quality of life of those who suffer from them [2]. The challenge posed by the aging population in the 21st century is thus simple: how to avoid having to spend more money to keep more people more miserable than ever before. This has put efforts to understand the biological basis of the process in the forefront of political vision in a way unique in history.

Research into the nature of the aging process is making steady progress but the subject is conceptually rather difficult, the perspectives of researchers working with the various model systems are often different and consequently the scientific literature on aging is almost as likely to confuse newcomers to the field as it is to inform them. This chapter sets out to review current thinking on one proposed mechanism of mitotic tissue aging *the cell theory of aging*. This postulates that clonal exhaustion and replicative senescence (the permanent exit from the cell cycle of cells

1

R. Aspinall (ed.), Aging of Organs and Systems, 1–28.
© 2003 *Kluwer Academic Publishers. Printed in Great Britain.*

that would normally be able to undertake division) as a consequence of ongoing cell turnover throughout the life of the organism plays a causal role in the aging of mitotic tissue. This theory is sometimes presented as standing in opposition to other mechanistic theories (for example the theory that damage arising from reactive oxygen or nitrogen species plays a primary causal role in organismal aging). However it is in fact complimentary. The rest of this chapter will attempt to demonstrate this by placing the cell hypothesis in the wider context of gerontology.

What is aging and why do organisms do it?

Aging has been defined as *a failure to maintain homeostasis under conditions of physiological stress*. This is accurate as far as it goes but the definition is somewhat loose. For example it would also fit a human's response to being shot, a rabbit's response to ending up under the wheels of the human's car and a bird's response to unintentional collision with a plate glass window because of all the commotion. Strehlar's definition of aging is rather better. According to Strehlar aging is:

- **Universal**. All members of a species show aging.

- **Progressive**. Aging is a continuous series of incremental changes.

- **Intrinsic**. The changes would take place even in a "perfect" environment.

- **Degenerative**. The changes compromise physiological effectiveness in the organism leading to eventual death.

Only this last of these requirements prevents this definition of aging from also being a workable definition of development. It is perhaps also worth stressing that even though the word "aging" is frequently applied to tissue components (as in "cell aging") true aging is something that only happens to entire organisms. It does not happen to proteins (they become adducted or damaged as a function of chronological dwell time in more or less damaging environments) and it does not happen to cells (unless the cell in question also happens to be an independent organism in its own right). The cells of multi-cellular (metazoan) organisms certainly undergo a range of changes over time and these changes almost certainly play causal roles in the aging of the organism but the cells themselves (whether mitotic or post mitotic) are not aging in Strehlar's sense of the term. Organismal aging is a *consequence* of these intrinsic changes. An obvious question is why are these changes permitted to happen at all since they clearly compromise organismal survival. What, in short, is the evolutionary rationale for aging?

Aging is not universal across all species, or even among all metazoans [5], but it is extremely common and thus presumably provides some form of selective advantage to organisms that show it compared to those that do not. Modern explanations for "why aging happens" have as a central tenet the observation that the force of natural selection declines with age [6, 7]. This means that, even in a population of immortal organisms, there are always far fewer chronologically old ones than young ones (because the longer a given organism has been around the more likely it is to have been eaten, met with an accident, etc). Thus, despite the fact that the reproductive ability of "old" and "new" immortal organisms is the same, the "old" organisms contribute fewer offspring to the next generation than the "new" organisms simply because there are fewer of them (Figure 1). Thus, any mutation that favors early life fecundity will be selected for even if it results in deleterious effects later on in the lifetime (a type of gene action termed antagonistic pleiotropy). This view of aging argues against the operation of a "clock" controlling the aging of individuals but suggests that aging will result from an accumulation of faults at different rates in different tissues. Thus aging is not a programmed process, in the sense that no genes are known to have evolved specifically to cause it. It resembles a car breaking down rather than a bomb going off.

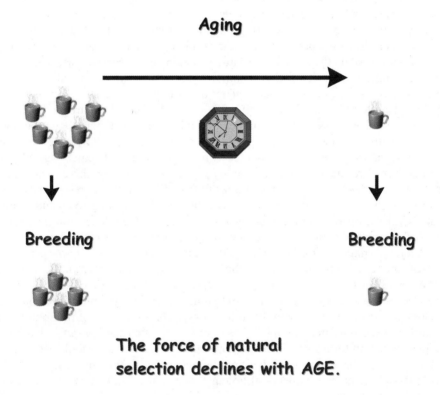

Figure 1. Simple illustration of antagonistic pleiotropy using a population of non-aging "organisms" (i.e., coffee cups). Over time, even organisms which do not show aging have a measurable rate of mortality as a result of accidents, predation, etc. (In this example, coffee cups get broken or stolen). Evolution measures reproductive success, thus if the coffee cups are allowed to have baby coffee cups then the contribution of offspring to the next generation from the chronologically "younger" population is always greater than that from the chronologically "older" population even though coffee cups do not age. Thus, there is potent selective pressure for any mutation that increases early life fecundity, even if it is associated with deleterious effects later on in the lifespan.

The multicausal nature of the aging process: universal versus primary causes

Biologists frequently attempt to unravel the working of processes operating in humans through the use of model systems. These are organisms that are studied as a substitute for the human original because they offer obvious advantages such as a short life cycle, simple growth conditions, a limited number of genes or a readiness to engage in genetic crosses on demand. Enormous progress has been made with the various model organisms in elucidating the operation of a very wide variety of fundamental processes (such as DNA replication and repair). In large measure this approach has been so successful because the molecular machinery necessary to carry out the process being studied has frequently been tightly conserved (i.e., it is a safe bet that a DNA polymerase from *E.coli* will be substantially similar to that found in *H.sapiens* because both species ultimately derive from a common ancestor and the functional requirements of replicating human and bacterial DNA are very similar indeed). There has thus been an expectation in some quarters of close similarities between species when the mechanistic causes of aging have been investigated (i.e., the study of aging in yeast will inform on human aging). To an extent this expectation has been justified. A series of mutants in genes affecting the insulin-IGF axis have been shown to lengthen lifespan in very different species [8–10]. There are also processes (such as oxidative stress) that are practically universal and frequently damaging so would logically be expected to play some role in the aging of all aerobic organisms [11]. However inter-species similarities can be overstated. Because aging is the unprogrammed result of selection for early reproductive success there is no *a priori* reason why the deleterious changes that result from that selection pressure have to be universal across species (by the same token there is also nothing to exclude similar changes if they impact a common pathway). In fact there is evidence that the primary driver of organismal aging can sometimes be quite different between different species. For example, a major cause of aging and death in female *D. melanogastor* is the toxic effect of compounds present in the seminal fluid products secreted from the main cells of the male fruit fly accessory gland [12, 13]. Seminal fluid toxicity is not seriously advanced as a primary cause of mammalian aging. Similarly replicative senescence (the loss of divisional capacity in the mitotic tissue compartments of the soma) is not seriously advanced as an aging mechanism for species such as *C. elegans*, that have a completely post-mitotic soma. Table 1 shows some of the universal and specific aging mechanisms that have been identified in different species (sometimes called "public" and "private" aging mechanisms although the terms are subject to rather elastic usage). In summary an aging mechanism which is "universal" is not necessarily "primary" in all species and a mechanism which is "primary" in one species is not necessarily "universal" in all the others.

The theoretical relationship between replicative senescence and aging

Replicative senescence may be briefly defined as a permanent block to further replication in cells from the mitotic tissue compartments of a metazoan organism. This block usually occurs in the G_1 phase of the cell cycle and is mediated by one or

Table 1. *Sample universal and species-specific aging mechanisms*

Public aging mechanism	Reference	Private aging mechanism	Reference
Oxidative damage	11	Toxic effects of male sperm (*D. melanogaster*)	13
AGE formation	14	Extreme cuticle thickness (*C. elegans*)	15
Insulin-IGF axis	12	Extrachromosomal ribosomal DNA circles (*S. cerevisiae*)	16
Alterations in energy balance	17	Replicative senescence (mammals)	See text

more cyclin dependent kinase inhibitors [18]. The result is a living, viable and metabolically active, but reproductively sterile cell. Such senescent cells are responsible for the decline in growth potential of cell populations which have undergone substantive turnover. They display many biochemical features that are distinct from their growing counterparts as a result of widespread changes in the transciptome (see Table 2 for some sample changes). A series of microarray studies together with earlier subtractive and differential cDNA analyses have shown that the onset of replicative senescence is coupled to highly selective changes in gene expression. The expression of some genes goes up (as a consequence of both increased transcription and mRNA stabilization), the expression of others goes down and the expression of still more is unaffected [19–21]. These changes give senescent cells a radically altered phenotype from their growing counterparts [22].

Failure to divide in response to a mitotic stimulus is the hallmark of senescent cells. However not all cells that fail to divide are senescent. Senescence is distinct from *quiescence*, a transiently growth arrested state (sometimes also called contact inhibition or density dependent inhibition of growth). Quiescence can be reversed *in vitro* by passaging the cells or by the readdition of serum. Again confusingly, both quiescence and senescence are referred to as the G_O phase of the cell cycle (sometimes distinguished as G_O^Q or G_O^S respectively). Also in systems where it is possible to experimentally separate the two, senescence has also been shown to be distinct from terminal differentiation. The best example of such a separation of replicative senescence and terminal differentiation is probably provided by human keratinocytes growing *in vitro*. In this system it is possible to prevent differentiation by growing the cells in medium containing greatly reduced levels of calcium. Nonetheless cultures of such cells still enter senescence and such "senescent" keratinocytes can be induced to differentiate by increasing the calcium concentration [23]. Also keratinocyte cultures are often overtaken by a variant cell type known as ndk (non-differentiating keratinocyte). The emergence of the ndk phenotype is associated with a specific deletion on chromosome 1 and a subsequent expansion of this mutant clone [24]. These are primary keratinocytes which have lost the ability to differentiate and express genes for a number of variant proteins (e.g., scatter factor) [25]. Nevertheless ndk cells still show senescence *in vitro*. This behavior is consistent with senescence and differentiation being parallel, but separate, processes.

The relationship between cell senescence and mitotic tissue aging is shown in (Figure 2a). This flow diagram implies that replicative senescence has the potential to contribute to aging in at least two distinct ways (i) through loss of proliferative capacity and (ii) through subtle alterations in the tissue microenvironment as a result of the accumulation of cells with an altered phenotype [26]. Cutler [27] has proposed a conceptually similar mechanism by which oxidative stress causes aging in the cellular component of a tissue (Figure 2b). The principle difference between the two models is that replicative senescence invokes cell turnover to act as a plausible mechanism but oxidative stress does not.

A priori it is possible to raise two reasonable objections to the idea that senescent cells play a role in the aging process. These may be summarized as:

(i)   Replicative senescence does not exist *in vivo*. It is simply an artefact of "bad tissue culture" or at least results from tissue culture conditions so different from the normal situation *in vivo* that the mechanisms by which growth arrest occur are unlikely to occur in the normal animal and are thus irrelevant to the real aging process.

(ii)   If replicative senescence does occur *in vivo* then the contribution to the aging process resulting from cells becoming senescent is so small that it can effectively be discounted. This is based on the observation that the mitotic fraction of normal tissues is usually extremely low and that the levels of cell turnover undergone by tissues in an organism's lifespan would thus be equally small.

At the center of the first of these objections is the nature of the mechanism by which senescent cells occur, at the centre of the second the frequency with which they occur. Accordingly the section on "Molecular mechanisms controlling entry into senescence and maintenance of the senescent state" will discuss how replicative senescence itself operates at the level of mass culture population dynamics and at the molecular level within individual cells (principally fibroblasts) *in vitro*. The section on "Tissue turnover, senescence and *in vivo* evidence consistent with the cell theory of aging" will then examine some of the available data on tissue kinetics together with the evidence that senescent cells exist *in vivo*. However before doing so it is necessary to introduce some terms that are routinely used within the field of tissue culture.

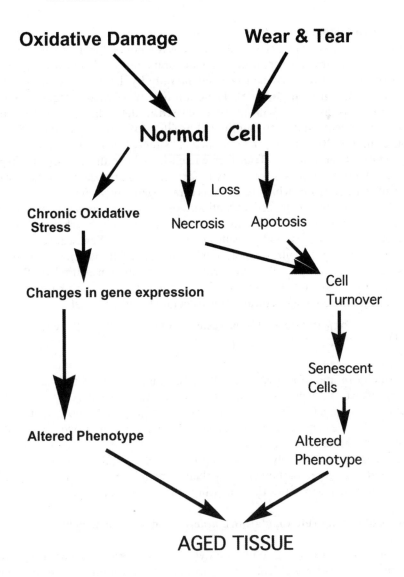

Figure 2. *Simple schematic comparing (a) the cell senescence hypothesis of aging (right hand flow) with (b) the dysdifferentiation hypothesis of aging (left hand flow). The cell senescence hypothesis postulates that in the normal course of life there is cell loss. That loss is balanced by cell division which is actively monitored. One or more "replicometers" act to trigger permanent cell cycle exit (senescence) in individual cells (see text). Cell cycle exit is associated with a broad alteration in gene expression leading to an altered phenotype that affects the microenvironment in which the cell resides and ultimately the entire tissue. In the dysdifferentiation model chronic oxidative stress leads to a regearing of gene expression generating an altered cellular phenotype which contributes to tissue aging. The two models have many essential similarities.*

A note on terminology

Cell cultures can be of three types: *primary* cells are derived from normal tissue and display a limited lifespan in culture. They are sometimes referred to as mortal cell cultures or as cell strains. By contrast *cell lines* are by definition composed of cells with an unlimited growth potential. These fall into two broad categories; *immortal cells* are, for reasons which will not be discussed in detail here, almost exclusively derived from rodents and posses unlimited proliferative capacity without the ability to induce tumors. By contrast, *transformed cells* represent the vast majority of cell lines produced from humans. Transformed cell lines have the unlimited lifespan of immortal cell lines but posses the ability to form tumors in nude mice together with a number of characteristics which reflect this tumorogenicity *in vitro*.

With some highly specialized exceptions, only cell strains show senescence in culture. The term senescence in used in two distinct ways by workers culturing cell strains and this can occasionally result in some confusion. Senescence of the *entire culture*, is a failure of the culture to proliferate under conditions which had previously allowed sustained cell growth. This growth is generally measured as Population Doublings (PDs) calculated as the number of times the cell population doubles in number during the course of culture. It is calculated by the formula:

$$PD = \frac{\log_{10} (\text{number cells harvested}) - \log_{10} (\text{number cells seeded})}{\log_{10} 2}$$

As a rule of thumb most primary cultures of human cells go through 30–60 PD. However this can be significantly less in some cell types (such as the 4–6 PD often reported for human lens epithelial cells) and can be significantly more in others (some human T cell clones and human fibroblast cultures have been grown to over 80–100 population doublings). For comparison rodent fibroblasts cultures routinely enter senescence at approximately 15 population doublings [28] and cell types from animals that are larger or longer lived than humans frequently have significantly greater proliferative capacities than their human equivalent [29].

Dynamics of normal cell populations: kinetic models of their behavior

The majority of data relating to culture dynamics has been obtained using primary human fibroblasts. Although cultures of fibroblasts are a useful system in their own right, because the cells are so easy to grow, the reason underlying the choice of this cell type is principally historical. In an extensive (and now classic) series of studies carried out in the early 1960s Hayflick and Moorhead demonstrated that normal human fibroblasts would only proliferate for a finite number of passages in culture. Latent infectious agents, composition of the medium and depletion of key metabolites were all shown not to be responsible for this failure to grow [30, 31]. Hayflick also observed that fibroblast cultures initiated from adults grew substantially better than those derived from embryos and defined three distinct stages of *in vitro* growth. Phase I, the primary culture, was considered to terminate with the formation of the first confluent sheet of cells. Phase II was characterized by vigorous growth requiring repeated subculture. Phase III was a decline phase in which the cells showed cessation of mitosis, accumulation of cell debris and cell death. These three phases during cell growth can be observed in many different primary cell types.

Although this model identified the central point (that a primary culture of normal cells inevitably ceases to expand) and adequately explained Hayflick's original data it carried two intrinsic assumptions. Firstly, that the cultures studied were composed of homogeneous (unimodal) populations of cells which were either all growing (in Phase I and II) or all non-growing (Phase III). Secondly, that the decline in proliferative ability seen during "Phase III" was a direct consequence of cell death. As we have seen, senescent cells are now known to remain viable for long periods. This data became available relatively early on (at least for fibroblasts) when necrotic cell death as a cause of Phase III was excluded by work which compared RNA synthesis in Phase II and Phase III cultures [32]. This showed that RNA synthesis (as measured by incorporation of tritiated uridine) occurred in all fibroblasts regardless of age. The second assumption of the Hayflick model, unimodality, was disproved somewhat later. Smith and Hayflick [33] and Smith and Whitney [34] demonstrated that primary fibroblast cultures are composed of a mixture of clones with very variable growth potentials. This behavior was shown not to be due to fixed clone character-istics because recloning of a clone with a long lifespan gave a set of subclones with a range of division potentials. The thrust of these experiments is exemplified a series of studies in which the two daughter cells resulting from a single mitotic event were separated immediately following cytokinesis and their replicative capacity followed for a number of population doublings. The daughter cells did not display identical proliferative capacities, indeed they differed by as much as 8 PD [35, 36]. These cloning experiments demonstrated that the reproductive ability of fibroblasts was to some degree determined by a chance (stochastic) process and that the dispersion of reproductive potential gradually moved towards smaller clones as the culture aged.

Work on bulk cultures complemented these single cell analyses. Cristofalo and Scharf [37] carried out an analysis of the cell kinetics of embryonic fibroblasts. Long pulse-labeling experiments with 3H-thymidine were carried out at every passage throughout the lifespan of a culture. The fraction of cells which had entered S phase at each time point was then estimated by autoradiography. It was observed that unlabeled (senescent) cells were present even in very young cultures and that a few labeled (dividing) cells were present even in "senescent" cultures. It was also shown that the fraction of unlabeled cells increased smoothly with serial passage of the culture. These findings were inconsistent with unimodal kinetics, unless the decline in labeling index reflected the behavior of an asynchronous unimodal culture of cells within which the cell cycle duration progressively increased with serial subculture. However, additional experiments excluded such a lengthening of the cell cycle as a significant cause of the decline in labeling index [38]. The only remaining explanation for this behavior is that the fraction of growing cells in the culture declines with population doubling level. The decline in the labeling index thus reflects the independent stochastic "exit decisions" of thousands of clones. To summarise fibroblast cultures have been shown to be bimodal mixtures of senescent and growing cells, the proportions of which alter as the cultures age (Figure 3).

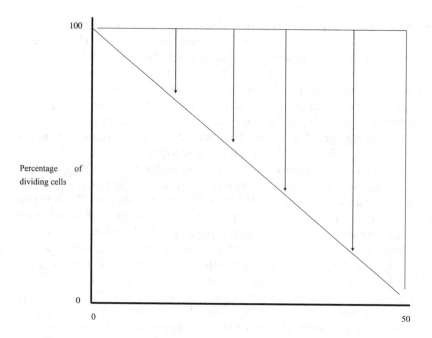

Figure 3. *Simple schematic of the difference in divisional fraction between populations displaying unimodal (dashed line) and bimodal kinetics (dotted line). A unimodal population would display a fixed fraction of dividing cells until the very last population doubling when they would enter senescence simultaneously. By contrast a bimodal population has senescent cells present from the start. Analyses of gene expression within the culture such as northern blots or microarrays thus give an average for the population and can be misleading unless the fraction of dividing cells is known.*

Colony size analysis (and related techniques such as Ponten mini-cloning) [39] are extremely time consuming to undertake and accordingly their application has been limited. However bulk culture kinetic analysis of the Cristofalo–Scharf type are more rapid and have been applied a little more widely. The original Cristofalo–Scharf protocol is unsuited for the accurate estimation of labeling index. This is because labeling times substantially greater than the cell cycle duration give rise to a significant over estimation of the labeling index. Accordingly, modern bulk culture kinetic studies either use short labeling times (~1 hour) or rely on immunocyto-chemical; detection of endogenous proteins (such as proliferation cell nuclear antigen or pKi67) that vary in amount or conformation depending on whether the cell is cycling or not. These studies [40, 41] have shown that the rate at which the division competent fraction of a cell culture is lost (measured as decline in the labeling index per population doublings) varies significantly between different cell types. Thus for a given amount of cell turnover (measured in population doublings) many cell types throw off more senescent cells than a comparable culture of fibroblasts. This has important implications for the effects of cell turnover *in vivo* on the number of senescent cells generated.

Due to the early exclusion of necrosis as a cause of senescence, death was not considered as a factor in the population dynamics given above. However subsequent interest in apoptosis prompted an evaluation of the role of programmed cell death in cell kinetics. Bayreuther and co-workers proposed a rigid seven stage lineage for fibroblasts aging in culture with the last stage (post-mitotic stage VIIa) being apoptosis [42]. This model predicted that senescent fibroblasts are inherently apoptosis prone. By contrast Wang has proposed that senescent fibroblasts represent a highly apoptosis resistant "survivor" population [43]. Thus on these models a kinetic study should produce a rate of apoptosis which either increases as the number of senescent cells in the culture goes up (the Bayreuther model) or goes down as the number of "survivor" senescent cells increases (the Wang model).

At least two studies published to date have undertaken the simultaneous measurement of apoptotic, proliferating and senescent cells throughout the replicative lifespan of a culture. One such study used human keratinocytes and found no change in the rate of apoptosis over the culture lifetime) [23]. The other used human vascular endothelial cells. It combined measurement of the dividing fraction by immunostaining for pKi67 and the senescent fraction by detection of senescence associated β galactosidase. The apoptotic fraction was determined using TUNEL and direct cinematography. This study also demonstrated that the baseline apoptoic rate within a culture of vascular endothelium remained constant over most of the culture lifespan [41]. Apoptosis did increase in the last few population doublings; but this was due to declining cell densities of the entire culture not senescence. Thus different rates of senescence in different cell types principally represent real differences in the likelihood of senescent cells being produced, not elevated cell turnover as a consequence of elevated apoptosis.

That being said, these kinetic measurements of the baseline apoptosis rate also showed that it can differ significantly between one cell type and another. For example approximately 3% of vascular endothelial cells can be shown to be apoptotic by TUNEL staining compared with less than 0.1% of normal fibroblasts (Faragher, unpublished observations). This difference in apoptotic rates probably explains the variability of survival times reported for senescent cultures derived from different tissues; which range from years in the case of fibroblasts [44] to days or weeks in T lymphocytes or vascular endothelium [45].

Several theoretical models have been proposed to account for the senescence of serially passaged cell cultures of which only three will be considered here. Kirkwood and Holliday [46] proposed the "commitment" theory of cell kinetics. This two parameter model postulated that within any culture a population of progenitor cells exists. Each has a unlimited replicative potential but may "commit" to a form in which the replicative potential is limited. Studies of the aggregate growth kinetics of fibroblast cultures suggested this probability of commitment (P_m) to be a fixed 0.275 per cell per generation leading to committed clones with a fixed growth potential of a least 42 generations. On this model senescence in a fibroblast culture results from a combination of a dilution effect by which the progenitor cells are lost (represented by an additional parameter the population size N) and senescence of the residual committed population. This model implies that replicative senescence is essentially artefactual (resulting from the worker's inability to keep all the cells generated). By contrast Shall and Stein [47] showed that the same data was consistent with the idea that the "committed" cells did not divide at all provided that the probability of a cell irreversibly exiting the cell cycle and becoming senescent was allowed to increase in simple one parameter dependency on the number of times the cell had divided before.

The Shall–Stein relationship is:

$$P_m = t \,/\, \gamma + t$$

where P_m is the probability of senescence, where t is the time (in units of generation) and γ is the time in generations at which $P_m = 0.5$.

The Shall–Stein model is independent of population size, requires no intrinsically immortal progenitor population, and thus postulates that senescence is an intrinsic property of normal cells. Both it and the Kirkwood–Holliday modal provided a very good fit for the bulk growth curve of cells in culture. However Prothero and Gallant [48], using the data from the Smith and Whitney experiments discussed above [34], demonstrated that the Kirkwood–Holliday model was a very poor fit to the clone size distribution data for fibroblasts. The Shall–Stein model, although better than the Kirkwood–Holliday, also did not completely fit the Smith–Whitney data (in particular it was unable to explain the emergence of significant numbers of senescent cells at very early passages). Instead Prothero and Gallant proposed a two parameter model in which P_m increases gradually during culture and is followed by a "committed state" in which the cells are able to divide up to seven times. This clonal attenuation model is conceptually very similar to the Shall–Stein (in particular it is independent of population size) but is rather more complex. The best models suggested that senescence was intrinsic to cell populations but could say no more in the absence of knowledge of the molecular mechanisms controlling replicative senescence. It is to these molecular mechanisms that we must now turn.

Molecular mechanisms controlling entry into senescence and maintenance of the senescent state

Having described the kinetic process by which a population of cells enters senescence we can now turn to the more detailed question of the mechanism of senescence in individual cells. Since senescence was first reported in the 1960s a wide variety of possible causes have been put forward as candidates for this role. It is now safe to say that in many human cell types the causal mechanism(s) of senescence have been provisionally established. It is nonetheless helpful to provide a historical summary of the experiments that led to the elucidation of the mechanisms of human replicative senescence so that our current understanding of the processes involved can be placed in context. A primary goal of these studies was to determine if senescence resulted from (i) the accumulation of macromolecular damage or (ii) from the operation of a system capable of counting divisions. This latter concept was receiving support from the kinetic experiments described above (which were being undertaken at the same time).

Some of the most compeling early data came from cell fusion experiments. Early work attempted hybridizations between senescent and growing fibroblast cultures and between senescent human fibroblasts derived from different donors [49]. These experiments had two separate goals. Firstly, in senescent to growing fusions, to determine if the senescent phenotype behaved as a dominant or recessive characteristic. Secondly, in senescent to senescent fusions, to determine whether the senescent phenotype could be overcome by complementation between pairs of cells from different donors, as would be predicted if senescence was due to the accumulation of random damage or error. No proliferating clones were recovered in either case.

These data showed that the senescence mechanism was common between cell strains and, more importantly, that senescence was actively dominant in fusions between growing and senescent cells, suggesting active gene involvement rather than random damage. Further evidence against a causal role for damage was provided by the work of Schneider [50]. This work tested the hypothesis that senescence was due to errors in the transcription or translation machinery by using viruses as probes of macromolecular synthesis. Infection of both growing and senescent fibroblasts with vesicular stomatitis virus (VSV) gave effectively identical numbers of virions. These studies demonstrated that the protein synthesis machinery of senescent fibroblasts functioned adequately. They also showed that the DNA replication machinery of senescent cells remained equally functional. It was simply not being used.

Further cell fusion work demonstrated that if a senescent cell was fused to a growing primary cell to produce a heterokaryon (dual nucleus cell) DNA synthesis in the senescent nucleus was not induced and synthesis in the young nucleus was repressed. Pretreatment of the senescent partner with cycloheximide or puromycin abolished this effect, suggesting that the inhibitory activity was a diffusable protein [51–54]. Experiments with quiescent cells also demonstrated the presence of similar activities capable of repressing DNA synthesis. By 1992 these studies had led to the identification of the molecule that causes growth arrest in senescent human cells. The molecule is question was a 21 Kd protein named $p21^{sdi}$ (senescent cell derived inhibitor), now more commonly known as $p21^{waf}$ [55]. $P21^{waf}$ is a broad-spectrum inhibitor of members of the cyclin dependent kinases. These are a family of enzymes which comprise a catalytic subunit (the cyclin dependent kinase or CDK) and activating subunit (the cyclin). The co-ordinated assembly and activation of these holoenzymes is central to cell cycle progression. Inhibition of CDK activity by $p21^{waf}$ effectively arrests cells at the G_1-S phase boundary [56]. Other cyclin dependent kinase inhibitors were subsequently identified of which the most important with regard to senescence is $p16^{ink4a}$ [57]. This is a selective inhibitor of cyclin D-CDK4 and cyclin D-CDK6 kinase pairs. The presence of active $p16^{INK4a}$ prevents phosphorylation of members of the retinoblastoma (Rb) family of proteins causing arrest at the G_1-S phase transition. p21 was formally demonstrated to be the primary molecule responsible for senescence in human fibroblasts by a series of studies in which antibodies to p21 were microinjected into senescent fibroblasts in order to prevent association of the molecule with CDKs. This removal of $p21^{waf}$ allowed "senescent" fibroblasts to re-enter the cell cycle [58]. Related studies demonstrated an identical effect using antibodies to the tumor supressor protein p53 [59]. Since p53 is a transcription factor for p21^{waf} (among many other things) these studies demonstrated that senescence in human fibroblasts arose from a p21^{waf} mediated arrest as a resulting from activation of p53. The factor responsible for p53 activation remained unclear. However this was also resolved.

As early as 1973 Olovnikov proposed that cells might count divisions through the progressive shortening of chromosome ends (telomeres) [60]. Telomeres distinguish natural chromosome ends from simple double-strand breaks which can lead to fusion and recombination between chromosomes and the activation of genome damage-monitoring systems [61]. Conventional DNA replication faces a major problem with the ends of a linear DNA molecule, the so-called end replication problem. The consequence of this is that a small amount of terminal DNA is not duplicated in each S phase, resulting in an inexorable loss of terminal DNA with repeated cell division. Olovnikov's theory was based on his recognition of this "end replication loss." In mammals this loss can be compensated for by a specialized reverse transcriptase (telomere reverse or telomerase, sometimes abbreviated to hTERT) which is absent in the majority of somatic tissues. By contrast telomerase activity is present in human germ-line cells, tumors and immortal human cell lines [62]. Telomeric shortening in telomerase negative cells has been demonstrated in a variety of mortal cell types which have undergone turnover both *in vitro* and *in vivo* [63]. These data were, of course, correlative and did not indicate a causal role for telomere shortening in the induction of replicative senescence. However cloning of the catalytic subunit of telomerase allowed interventional testing of the hypothesis that telomeric attrition causes replicative senescence [64]. Forced expression of telomerase halts telomere shortening and prevents senescence in a significant number of human cell types (including fibroblasts, vascular endothelial cells and retinal pigmented epithelial cells) conferring apparent immortality [65]. Figure 4a summarizes the telomere-dependent model of replicative senescence.

Telomeric attrition is a particularly attractive candidate for the control of replicative senescence in human fibroblasts because it offers the capacity to explain their population dynamics. This is because telomere shortening is inherently asymmetrical with each round of DNA replication producing W and C strands of differing lengths on each chromosome. These different length strands are then partitioned essentially randomly between the two daughter nuclei. A simple computer model of this process by Levy *et al.* showed that heterogeneity arises spontaneously in such a system [67]. Rubelj and Vondracek took this modeling approach further and demonstrated that this pattern of telomere loss can produce a predicted colony size distribution pattern for human fibroblasts that is very similar to that observed experimentally by Smith and Whitney [68]. The telomere "replicometer" also requires no energy to run and appears to be malleable over short evolutionary timescales [69]. It also provides one answer to the objection that replicative senescence results from poor tissue culture conditions. End replication loss is so fundamental a process it must operate both *in vitro* and *in vivo*. However, telomere-dependent senescence does not appear to be the sole mechanism by which cells become senescent *in vitro*.

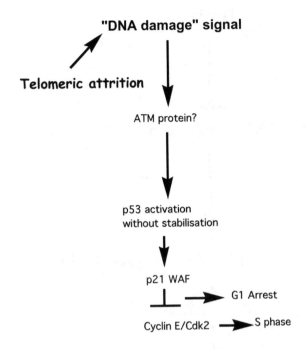

Figure 4a. *Telomere-driven senescence. Progressive telomeric attrition leads to the generation of one or more "critically short" telomeres that signal cell cycle arrest via the activation of the p53 tumor supressor gene product. p53 is a transcription factor for a number of genes, among them the cyclin dependent kinase inhibitor p21^{waf}. Expression of p21 leads to cell cycle arrest at the G1-S phase boundary. ATM is the Ataxia telangiectasia mutated gene product. Adapted from ref. 66.*

Recently there have been a series of papers showing that the ectopic expression of telomerase is insufficient on its own to produce immortalization of several primary human cell types even though telomerase activity is restored and telomere length stabilized. These "telomere-independent" cell types include pancreatic β cells and keratinocytes and seem to also require interventions that principally bypass the activity of the p16^{INK4a} and RB axis of the cell cycle (e.g., overexpression of CDK4 or introduction of the HPV E7 gene product) [70–72]. Also rodent fibroblasts appear to enter senescence in a telomere-independent manner as a consequence of p53 activation *via* the p19^{ARF} gene product) . These two mechanisms are illustrated in Figure 4b and 4c.

Opinion is currently heavily divided as to the exact significance of p16^{INK4a} mediated senescence in humans. Some workers view it exclusively as a tissue culture artefact. Such "artefactual" growth arrest may result from the abnormally high partial oxygen pressures under which cells are routinely cultured. Alternatively it may result from the absence of as yet poorly-defined extrinsic factors that repress

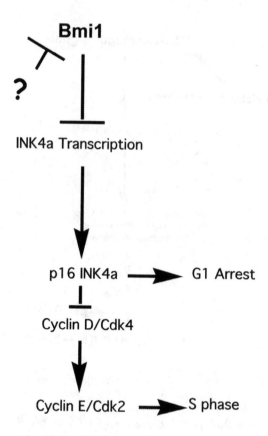

Figure 4b. *Simple schematic telomere-indendent induction of the senescent state. The product of the polycomb gene Bmi-1 repressed the transcription of the cyclin dependent kinase inhibitor $p16^{INK4a}$. Derepression at this locus leads to a build up of p16 protein and inhibition of the activity of cyclin D-Cdk4 and cyclin E-Cdk2 kinase pairs. This leads to a failure to phosphorylate the retinoblastoma gene product and a failure to pass the G1-S phase boundary. Adapted from ref. 66.*

$p16^{INK4a}$ induction. This latter view has been strengthened by the successful production of immortal keratinocyte lines using a combination of hTERT and feeder cell layers *versus* earlier studies showing that hTERT and E7 were required in combination when keratinocytes were grown on plastic [76]. Unfortunately, this immortalization of keratinocytes by hTERT alone in the presence of feeder cells has yet to be replicated by others, who instead show the existence of a telomere independent proliferative lifespan barrier that can be abrogated by forced expression of a $p16^{INK4a}$-insensitive allele of CDK4 [71]. These data seem to suggest, at least, that the situation is rather more complicated than it currently appears.

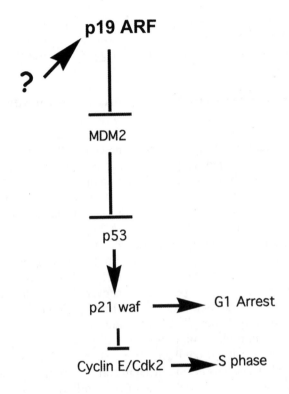

Figure 4c. *Simplified simple pathway by which rodent fibroblasts may enter replicative senescence. Upregulation occurs at the ARF (alternate reading frame) locus in response to a variety of upstream signals including v-Ras and v-Abl. ARF neutralizes the ability of Mdm2 to promote p53 degradation, leading to the stabilization and accumulation of p53. Ectopic expression of ARF causes a senescence-like growth arrest. This pathway is p53 dependent but telomere independent. Adapted from refs. 66 and 75.*

When senescence in human fibroblasts was first reported in the 1960s some in the aging field dismissed it (essentially out of hand) as an artefact of tissue culture resulting from the addition of antibiotics to the medium [77]. Writing in 2003 with the benefit of hindsight it is fair to say that we now know better and that those people were wrong. Accordingly my own view is that we should take care before we dismiss any results obtained in tissue culture. It is also my view that an important, but sometimes overlooked, scientific point lies at the heart of the debate on telomere independent senescence. For a growth arrested state such as p16[INK4a] induced senescence to be truly artefactual it must be demonstrated that such cells do not appear in an organism *in vivo*. If one such cell can be shown to be present *in vivo* then telomere-independent senescence cannot be an *in vitro* artefact. As the next section will show in contrast to the wealth of detail now available on the mechanisms by which cells enter senescence the demonstration of senescent cells *in vivo* is something of an area of honorable poverty. Since the *in vivo* relevance of senescence is the key area of interest that the phenomenon has for gerontologists it is a poverty that I personally deplore.

Tissue turnover, senescence and *in vivo* evidence consistent with the cell theory of aging

Attempts to demonstrate the presence of senescent cells in tissue began relatively early in the history of the field. Hayflick and Moorhead themselves provided the key observation that fetal fibroblasts grew better than adult ones [30]. This is now a commonplace of cell biology but at the time it was surprising. This was because the work of the Nobel Laureate Alexis Carrel suggested that all normal cells had an indefinite proliferate capacity. On a first pass application of this logic fetal and normal cells should not have shown senescence at all and they certainly should not have shown a difference in replicative potential[1]. The implication was that the cultures derived from adult tissue contained cells that has "used up" some of their replicative potential as a result of tissue turnover during life. A more extensive analysis was performed by Schneider and Mitsui [78] on the *in vitro* growth of fibroblasts derived from "young" (21–36 years old) and "old" (63–92 years old) donors. Declines in fibroblast migration rate, *in vitro* lifespan, population growth rate, and saturation density at confluence were shown to be statistically significant. Colony forming efficiency was also reduced in cultures from old donors compared to those from young donors. However, none of these differences was as great as those found between "young" (< 20 PD) and "old" (> 40 PD) embryonic human fibroblasts. These workers also demonstrated that differences in growth potential between fibroblast cultures from old and young donors could not be explained by differences in the "cellularity" (that is, the number of cells per unit tissue) of old and young skin but was consistent with higher numbers of senescent cells in the older tissue biopsies.

Studies being carried out at the time in the area of tissue kinetics might also have suggested that the amounts of turnover in normal tissues were likely to be far from negligible [79]. To calculate an approximate estimate of turnover time, workers in the field of tissue kinetics employed variations on the formula:

$$L.I. = \lambda \, (T_s) \, / \, (T)$$

where L.I. is the absolute (not percentile) labeling index at one hour following injection of tritiated thymidine. T_s is the duration of S phase (typically assumed to be between 7–9 hours). T is the turnover time in hours and λ is a factor that for slowly proliferating populations has a value close to $Log_e 2$.

---

[1]   Although to be fair to Carrel he was aware that a reduction in the growth capacity of cells from adult donors compared to embryos existed. However he suspected this was due to changes in the "properties of the humors" made by the animal as it aged. See Carrel (1923). Measurement of the inherent growth energy of tissues. *J Exp Med.* 521–7.

One example may serve as an illustration of how low labeling indices could indicate significant turnover. One study (Tannock and Hayashi) determined the labeling index of unstimulated endothelium in the skin of mice to be $\sim 3.55 \times 10^{-3}$ (two labeled cells out of 562). This labeling index (0.4%) is extremely small. However it equates to an average turnover (population doubling) time of approximately 8 weeks. If constant that would suggest something of the order of 13 population doublings over the two-year lifespan of a mouse. Since rodent fibroblasts only have a replicative lifespan of ~ 15 population doublings this suggests that some senescent cells at least should be present. Meaningful estimation of lifetime cell turnover by this approach is beset with pitfalls (not least the inability to determine accurately such low L.I.s, the assumption that cell turnover rates are constant and difficulties arising from the relative size of nucleotide pools), but it serves as an interesting "back of the envelope calculation" [80].

A cohort study showing inverse correlation between donor age and the number of population doublings achieved *in vitro* was also reported relatively early [81]. In this study the authors used fibroblast cultures derived from 100 subjects with an age range from fetal to 90 years and obtained a regression line with a slope of 0.20 PD per year of donor life. Work by Goldstein and colleagues [82] produced a similar result but also showed that it was possible to break the experimental subjects down into two subpopulations. Those who were either diabetic or pre-diabetic who showed a very sharp decline with age in the replicative capacity of their fibroblasts and those who appeared to be healthy who showed rates of decline in proliferative capacity that were of questionable significance. A similar study conducted more recently used fibroblast cultures derived from a cohort of the Baltimore Longitudinal Survey on aging [83]. This also demonstrated no statistically significant decline in the replicative potential of fibroblasts derived from healthy donors. It is possible to invoke reasons of experimental technique (e.g., selection of biopsy area) as a reason for this failure to observe a decline however to my mind these studies demonstrate something more fundamental. Implicit in the cell theory of aging is the notion that senescent cells are one of the *causal agents* of aging and disease. Thus if a cohort is selected on the basis of an absence of disease it is also selected on the basis of the absence of the causal agents of that disease. If replicative senescence is a causal agent of age-related degenerative disease then it is possible that selection for the healthy elderly is akin to selection against people with significant numbers of senescent cells in their tissue. It would be close to direct disproof of the theory if the mitotic tissues of the elderly were full of senescent cells but showed no diminution in physiological function. However none of these studies allowed the visualization of senescent cells in tissue.

In a landmark study, Campisi and co-workers [84] adapted the assay conditions for endogenous mammalian β-galactosidase such that it could only be demonstrated histochemically in large cells. In tissue culture this technique was shown to provide highly selective staining for senescent cells, as determined by their failure to incorporate tritiated thymidine. In tissue sections derived from donors of increasing age, increasing numbers of β-galactosidase positve cells were detected. A more detailed study was carried out by Wolf and co-workers [85]. Using the lens epithelium of mice as a target tissue these researchers elegantly combined long bromodeoxyuridine labeling studies *in vivo* with Smith–Whitney colony size analysis *in vitro* and demonstrated (i) that the number of mitotic cells in lens epithelium declined smoothly with the age of the animal (ii) that this decline in mitotic index was associated with an increase in the number of senescent cells as measured by colony size analysis and (iii) that calorie restriction both lengthened the lifespans of the animals and reduced the rate at which senescent cells appeared. This study, to my mind, represents something close to the best demonstration of senescent cells in tissue that it will be possible to make until antisera capable of distinguishing senescent cells from their quiescent counterparts become available and are used in comparative morphometric studies.

It may therefore be tentatively concluded that senescent cells do increase in number in the connective tissue of normal mammals with age. It is thus possible that, to an increasing degree, the physiological responses of this aged tissue may be determined by the phenotype of the senescent cells which compose it. Increasing knowledge of the phenotypic changes that take place in senescent cells (Table 2) provide some justification for this statement. However a more direct demonstration that senescent cells can exert significant "aging" effects was recently provided by studies in which human dermal populations aged *in vitro* were incorporated into reconstituted human skin equivalents [86]. Increasing number of senescent cells in the skin produced material showing increased fragility and subepidermal blistering. This was not observed in the early passage controls.

In my view some of the best evidence that replicative senescence of primary cells can play a causal role in the aging of the organism from which they are taken derives from studies on humans with Werner's syndrome [100]. Fibroblast cultures from patients with this genetic disease have an extremely poor replicative capacity (90% of cultures undergo less than 20 population doublings). The disease also has body wide manifestations that are strongly reminiscent of accelerated aging. These include atherosclerosis, arteriosclerosis, greying of the hair, hypermelanosis, osteoporosis and dermal atrophy [101].

The disease is caused by loss of a member of the RecQ helicase family (*wrn*) which appears to function as a repair helicase during DNA replication [102–104]. Loss of *wrn* results in replication fork stalling and a distinctive mutator phenotype character-ized by large DNA deletions [105]. The occurrence of these deletions at or near the telomere accelerates the rate of telomere-driven senescence in Werner's syndrome fibroblasts and it is tempting on the basis of this alone to suggest that the clinical features of the disease result from the presence of an excess of senescent cells [106]. However, the presence of the mutator phenotype complicates this picture since it could itself play a role in development of the pathology [107]. In addition some mitotic tissues are affected very severely in Werner's syndrome (such as the dermal layer of the skin) and some appear to be clinically very close to normal (such as the immune system). Comparative studies have shown that mass T cell cultures derived from Werner's syndrome patients display no lifespan deficit compared to those taken from normal controls but both they and dermal fibroblasts do display the mutator phenotype [108–110]. The observation of significant markers of global genomic damage in tissues which are both clinically affected and clinically unaffected thus

Table 2. Selected alterations in cell phenotype with the onset of senescence

Phenotypic alteration in senescence.	Cell type	Reference
Repression of c-fos	Fibroblasts, T lymphocytes	87, 88
Repression of cyclin A and B	Fibroblasts	89
G_2 arrest on restimulation without division	Fibroblasts T lymphocytes	90, 91
Elevated collagenase	Fibroblasts	92
Elevated TIMP 2	Fibroblasts, endothelial cells	93, 94
Elevated PAI-1	Fibroblasts, endothelial cells	95
Elevated ceramide	Fibroblasts	118
Transcriptional repression of IGF-1 gene	Fibroblasts	95
Specific induction of Ws3.10 inhibitor of Ca^{2+}-dependent membrane currents	Fibroblasts	96
Elevated IL-1α expression	Fibroblasts	96
Senescence associated β-galactosidase activity	Fibroblasts, keratinocytes, mammary epithelial cells, endothelial cells, neonatal melanocytes	85
Specific induction of SAG gene	Fibroblasts	98
Elevation of cytochrome b and NADH 4/4L subunit	Fibroblasts	99

appears inconsistent with the idea that genomic instability is a primary driver in mitotic tissue aging in Werner's syndrome. In contrast the appearance of premature replicative senescence in an affected tissue (the dermis) but not in an unaffected one (the immune system) is consistent with a causal role for senescent cells in the development of the progeroid pathology seen in the disease. A model that seeks to explain these tissue specific differences in replicative lifespan has recently been published [111, 112].

Thesis, antithesis and synthesis. Oxidative stress, telomeres and "reactive" senescence pathways

At the start of this Chapter it was stated that oxidative stress is sometimes presented as an alternative aging mechanism to replicative senescence. The evolutionary rationale for aging and the limits within which the cell theory of aging are proposed to operate were presented in the hope of dispelling one element of confusion that occasionally arises. Specifically this is the idea that since senescence is the "programmed" operation of a cell division counting system those proposing that it plays a role in aging must also consider organismal aging itself to result from a "programmed" process. This last section presents two areas in which oxidative damage and cell senescence are clearly interrelated processes.

The first of these is the observation (initially by Thomas von Zglinicki but confirmed since in several other laboratories) that by placing cells under mild oxidative stress (mild stress being defined as a stress that still allows some cell proliferation to take place) it is possible to increase the rate at which telomere shortening occurs [113]. Such mild stresses include treatment or hydrogen peroxide or placing the cells under chronic hyperoxia. Since the fraction of cells driven into apoptosis by mild oxidative stress remains extremely low; this increase in telomere loss is unlikely to result from an increase in the cycling frequency of the division competent sub-population. It seems to be a genuine increase in the loss rate of DNA at the telomere at each cell division. There is also some evidence that the rate of telomere shortening *in vitro* is set by the different levels of antioxidant defence enzymes present in the fibroblasts of different donors [114, 115]. The implication of these elegant studies is that localized oxidative stress within tissue has the potential to greatly increase the rate at which senescent cells are generated as a result of turnover compared to the rate that would occur in unstressed tissue. Conversely, protection against oxidative stress may protect against the production of senescent cells and thus against deleterious changes in the tissue microenvironment resulting from their presence. Of course this means of generating senescent cells still requires that tissue turnover takes place.

However, a growing body of work also exists on a phenomenon that I have previously termed "reactive" cell senescence [116]. This is the immediate entry by members of a population of cells into senescence as a result of an external stimulus. Such as treatments include exposure to γ radiation, infection with vectors carrying activated viral oncogenes such as RAS-V12, the use of DNA damaging agents such as mitomycin C and exposure to ceramide [117, 118]. Reactive senescence represents a transcriptional regearing that is so similar to the changes in gene expression proposed by Cutler's dysdifferentiation hypotheses that the differences between them becomes almost semantic. The cell theory of aging postulates that the senescent cell itself (rather than any particular "replicometer" or mode of production) is a primary effector mechanism in mitotic tissue aging. Thus the observation of reactive senescence gives another plausible route by which senescent cells could be generated *in vivo* and adds to the likely relevance of cell senescence to organismal aging [2].

References

1.   Lutz W, Sanderson W, Scherbov S (1997). Doubling of world population unlikely. *Nature* 387: 803–5.
2.   Prophet H, Mills E, *et al.* (1998). *Fit for the Future: The Prevention of Dependency in Later Life.* London: Continuing Care Conference.
3.   Comfort A (1979). *The Biology of Senescence*, 3rd edn. Edinburgh and London: Churchill-Livingstone.
4.   Strehlar BL (1962). *Time, Cells and Aging.* New York and London: Academic Press.
5.   Martinez DE (1998). Mortality patterns suggest lack of senescence in hydra. *Exp Gerontol.* 33: 217–25.
6.   Medawar PB (1952). *An Unsolved Problem of Biology.* London: HK Lewis.
7.   Kirkwood TBL (1996). Human senescence. *BioEssays* 18: 1009–16.
8.   Hsu AL, Murphy CT, Kenyon C (2003). Regulation of aging and age-related disease by DAF-16 and heat-shock factor. *Science* 300(5622):1142–5.
9.   Clancy DJ, Gems D, Hafen E, Leevers SJ, Partridge L (2002). Dietary restriction in long-lived dwarf flies. *Science* 296(5566): 319.
10. Carter CS, Ramsey MM, Ingram RL, *et al.* (2002). Models of growth hormone and IGF-1 deficiency: applications to studies of aging processes and life-span determination. *J Gerontol A Biol Sci Med Sci.* 57(5): B177–88.
11. Beckman KB, Ames BN (1998). The free radical theory of aging matures. *Physiol Rev.* 78(2): 547–81
12. Partridge L, Gems D (2002). The evolution of longevity. *Curr Biol.* 12(16): R544–6.
13. Chapman T, Liddle LF, Kalb JM, Wolfner MF, Partridge L (1995). Cost of mating in Drosophila melanogaster females is mediated by male accessory gland products. *Nature* 373(6511): 241–4.
14. Vlassara H, Brownlee KR, Manogue CA, Dinarello A, Pasagian A (1988). Cachectin/ TNF induced by glucose-modified proteins: role in normal tissue remodelling *Science* 240: 1546–8.
15. Herndon LA, Schmeissner PJ, Dudaronek JM, *et al.* (2002). Stochastic and genetic factors influence tissue-specific decline in ageing *C. elegans. Nature* 419(6909): 808–14.
16. Sinclair DA, Guarente L (1997). Extrachromosomal rDNA circles – a cause of aging in yeast. *Cell* 91(7): 1033–42.
17. Kaeberlein M, Andalis AA, Fink GR, Guarente L (2002). High osmolarity extends life span in *Saccharomyces cerevisiae* by a mechanism related to calorie restriction. *Mol Cell Biol.* 22: 8056–66.
18. Bartkova JB, Lukas J, Bartek J (1997). Aberrations of the G1- and G1/S-regulating genes in human cancer. *Prog Cell Cycle Res.* 3: 211–20.
19. Shelton DN, Chang E, Whittier PS, Choi D, Funk WD (1999). Microarray analysis of replicative senescence. *Curr Biol.* 9(17): 939–45.
20. Doggett DL, Rotenburg MO, Pignolo RJ, Phillips PD, Cristofallo VJ (1992). Differential gene expression between young and quiescent, senescent WI-38 cells. *Mech Age Dev.* 65: 239–55.
21. Lecka-Czernik B, Moerman EJ, Jones RA, Goldstein S (1996). Identification of gene sequences overexpressed in senescent and Werner syndrome human fibroblasts. *Exp Gerontol.* 31(1–2): 159–74.
22. Stanulis-Praeger BM (1987). Cellular senescence revisited: a review. *Mech Age Dev.* 38: 1–48.

23.   Norsgaard H, Clark BFC, Rattan SIS (1996). Distinction between differentiation and senescence and the absence of increased apoptosis in human keratinocytes undergoing cellular aging *in vitro. Exp Gerontol.* 31: 563–70.

24.   Adams JC, Watt FM (1988). An unusual strain of human keratinocytes which do not stratify or undergo terminal differentiation in culture. *J Cell Biol.* 107: 1927–38.

25.   Adams JC, Furlong RA, Watt FM (1991). Production of scatter factor by ndk, a strain of epithelial cells, and inhibition of scatter factor by suramin *J Cell Sci.* 98: 385–94.

26.   Faragher RGA, Kipling D (1988). How might replicative senescence contribute to human ageing? *Bioessays* 20(12): 985–91.

27.   Zs-Nagy I, Cutler RG, Semsei I (1988). Dysdifferentiation hypothesis of ageing and cancer: a comparison with the membrane hypothesis of aging. *Ann NY Acad Sci.* 521: 215–25.

28.   Bruce SA, Scott FD, Ts'o POP (1986). *In vitro* senescence of Syrian hamster mesenchymal cells of fetal to aged adult origin: inverse relationship between *in vivo* donor age and *in vitro* proliferative capacity. *Mech Age Dev.* 34: 151–73.

29.   Rohme D (1981). Evidence for a relationship between longevity of mammalian species and life spans of normal fibroblasts *in vitro* and erythrocytes *in vivo. Proc Natl Acad Sci USA* 78: 5009–13.

30.   Hayflick L, Moorhead PS (1961). The serial cultivation of human diploid cell strains. *Exp Cell Res.* 25 585–621.

31.   Hayflick L (1965). The limited *in vitro* lifetime of human diploid cell strains. *Exp Cell Res.* 37: 617–36.

32.   Macieira-Coelho A, Ponten J, Philipson L (1966). The division cycle and RNA synthesis in diploid human cells at different passage levels *in vitro. Exp Cell Res.* 42: 673–84.

33.   Smith JR, Hayflick L (1974). Variation in the lifespan of clones derived from human diploid cell strains *J Cell Biol.* 62: 48–53.

34.   Smith, JR, Whitney RG (1980). Intraclonal variation in proliferative potential of human diploid fibroblasts: stochastic mechanism for cellular ageing. *Science* 207: 82–4.

35.   Jones RB, Whitney RG, Smith JR (1985). Intramitotic variation in proliferative potential: stochastic events in cellular aging. *Mech Aging Dev.* 29(2): 143–9.

36.   Smith JR, Whitney RG (1980). Intraclonal variation in proliferative potential of human diploid fibroblasts: stochastic mechanism for cellular aging. *Science* 207(4426): 82–4.

37.   Cristofalo VJ, Scharf BB (1973). Cellular senescence and DNA synthesis. *Exp Cell Res.* 76: 419–27.

38.   Grove GL, Cristofalo VJ (1977). Characterisation of the cell cycle of cultured human diploid cells: effects of ageing and hydrocortisone, *J Cell Physiol.* 90: 415–22.

39.   Ponten J, Shall S, Stein WD (1983). A quantitative analysis of the ageing of human glial cells in culture, *J Cell Physiol.* 117: 342–52.

40.   Thomas E, Al-Baker E, Dropcova S, *et al.* (1997). Different kinetics of senescence in human fibroblasts and mesothelial cells. *Exp Cell Res.* 246: 355–8.

41.   Kalashnik L, Bridgeman CJ, King AR, *et al.* (2000). A cell kinetic analysis of human umbilical vein endothelial cells. *Mech Age Dev.* 120: 23–33.

42.   Brenneisen P, Gogol J, Bayreuther K (1993). Regulation of DNA synthesis in mitotic and postmitotic W138 fibroblasts in the fibroblast stem cell system *J Cell Biochem.* 17D: 152.

43.   Wang E, Lee M, Pandey S (1994). Control of fibroblast senescence and activation of programmed cell death *J Cell Biochem.* 54: 432–9.

44.   Smith JR, Lincoln DW II (1984). Aging of cells in culture. *Int Rev Cytol.* 89: 151–77.

45. Effros RB, Pawelec G (1997). Replicative senescence of T cells: does the Hayflick Limit lead to immune exhaustion? *Immunol Today* 18(9): 450–4.
46. Kirkwood TBL, Holliday R (1975). Commitment to senescence: a model for the finite and infinite growth of diploid and transformed human fibroblasts in culture *J Therot Biol.* 53: 481–96.
47. Shall S, Stein WD (1979). A mortalisation theory for the control of cell proliferation and for the origin of immortal cell lines *J Theoret Biol.* 76: 219–31.
48. Protero J, Gallant JA (1981). A model of clonal attenuation *Proc Natl Acad Sci USA* 78: 333–7.
49. Littlefield JW (1973). Attempted hybridizations with senescent human fibroblasts *J Cell Physiol.* 82: 129–32.
50. Schneider EL (1979). Aging and cultured skin fibroblasts *J Invest Dermatol.* 73: 15-1-8.
51. Pereira-Smith OM, Smith JR (1982). Phenotype of low proliferative potential is dominant in hybrids of normal human fibroblasts. *Somatic Cell Genet.* 8(6): 731–42.
52. Drescher-Lincoln CK, Smith JR (1983). Inhibition of DNA synthesis in proliferating human diploid fibroblasts by fusion with senescent cytoplasts. *Exp Cell Res.* 144(2): 455–62.
53. Pereira-Smith OM, Smith JR (1988). Genetic analysis of indefinite division in human cells: identification of four complementation groups. *Proc Natl Acad Sci USA* 85(16): 6042–6.
54. Pereira-Smith OM, Fisher SF, Smith JR (1985). Senescent and quiescent cell inhibitors of DNA synthesis. Membrane-associated proteins. *Exp Cell Res.* 160(2): 297–306.
55. Noda A, Ning Y, Venable SF, Pereira-Smith OM, Smith JR (1994). Cloning of senescent cell-derived inhibitors of DNA synthesis using an expression screen. *Exp Cell Res.* 211(1): 90–8.
56. Ball K (1997). p21: structure and functions associated with cyclin-CDK binding. *Prog Cell Cycle Res.* 3: 125–34.
57. Ruas M, Peters G (1998). The p16INK4a/CDKN2A tumor suppressor and its relatives. *Biochim Biophys Acta* 1378(2): F115–77.
58. Ma Y, Prigent SA, Born TL, Monell CR, Feramisco JR, Bertolaet BL (1999). Microinjection of anti-p21 antibodies induces senescent Hs68 human fibroblasts to synthesize DNA but not to divide. *Cancer Res.* 59(20): 5341–8.
59. Gire V, Wynford-Thomas D (1998). Reinitiation of DNA synthesis and cell division in senescent human fibroblasts by microinjection of anti-p53 antibodies. *Mol Cell Biol.* 18: 1611–21.
60. Olovnikov AM (1996). Telomeres, telomerase, and aging: origin of the theory. *Exp Gerontol.* 31: 443–8.
61. Kipling D (1995). *The Telomere.* Oxford: Oxford University Press.
62. Harley CB, Sherwood SW (1997). Telomerase, checkpoints and cancer. *Cancer Surv.* 29: 263–84.
63. Vaziri H, Dragowska W, Allsopp RC, Thomas TE, Harley CB, Lansdorp PM (1994). Evidence for a mitotic clock in human hematopoietic stem cells: loss of telomeric DNA with age. *Proc Natl Acad Sci USA* 91(21): 9857–60.
64. Bodnar AG, Ouellette M, Frolkis M, *et al.* Extension of life-span by introduction of telomerase into normal human cells. *Science* 279(5349): 349–52.
65. Chiu CP, Harley CB (1997). Replicative senescence and cell immortality: the role of telomeres and telomerase. *Proc Soc Exp Biol Med.* 214(2): 99–106.
66. Parkinson EK, Munro J, Steeghs K, *et al.* (2000). Replicative senescence as a barrier to human cancer. *Biochem Soc Trans.* 28: 226–33.

67. Levy MZ, Allsopp RC, Futcher AB, Greider CW, Harley CB (1992). Telomere end-replication problem and cell aging. *J Mol Biol.* 225: 951–60.
68. Rubelj I, Vondracek Z (1999). Stochastic mechanism of cellular aging–abrupt telomere shortening as a model for stochastic nature of cellular aging. *J Theor Biol.* 197(4): 425–38.
69. Kipling D, Cooke HJ (1990). Hypervariable ultra-long telomeres in mice. *Nature* 347(6291): 400–2.
70. Kim H, Farris J, Christman SA, *et al.* (2002). Events in the immortalizing process of primaryhuman mammary epithelial cells by the catalytic subunit of human telomerase. *Biochem J.* 365(pt 3); 765–72.
71. Rheinwald JG, Hahn WC, Ramsey MR, *et al.* (2002). A two-stage, p16INK4A- and p53-dependent keratinocyte senescence mechanism that limits replicative potential independent of telomere status. *Mol Cell Biol.* 22(14): 5157–72.
72. Halvorsen TL, Beattie GM, Lopez AD, Hayek A, Levine F (2000). Accelerated telomere shortening and senescence in human pancreatic islet cells stimulated to divide *in vitro.* *J Endocrinol.* 166(1): 103–9.
73. Kamijo T, Zindy F, Roussel MF, *et al.* (1997). Tumor suppression at the mouse INK4a locus mediated by the alternative reading frame product p19ARF. *Cell* 91(5): 649–59.
74. Dirac AM, Bernards R (2003). Reversal of senescence in mouse fibroblasts through lentiviral suppression of p53. *J Biol Chem.* 278(14): 11731–4.
75. Wei W, Hemmer RM, Sedivy JM (2001). Role of p14(ARF) in replicative and induced senescence of human fibroblasts. *Mol Cell Biol.* 21: 6748–57.
76. Ramirez RD, Morales CP, Herbert BS, *et al.* (2001). Putative telomere-independent mechanisms of replicative aging reflect inadequate growth conditions. *Genes Dev.* 15: 398–403.
77. Strehler BL (2000). Understanding aging. In: Barnett YA, Barnett CR, eds. *Aging Methods and Protocols. Methods in Molecular Medicine* 38. New Jersey: Humana Press.
78. Schneider EL, Mitsui Y (1976). The relationship between *in vitro* cellular aging and *in vivo* human age. *Proc Natl Acad Sci USA* 73: 3584–8.
79. Tannock IF, Hayashi S (1972). The proliferation of capillary endothelial cells. *Cancer Res.* 32: 77–82.
80. Baserga R (1999). Introduction to the cell cycle. In: Stein GS, Baserga R, Giordano A, Denhardt DT, eds. *The Molecular Basis of Cell Cycle and Growth Control.* New York: Wiley-Liss, pp. 1–14.
81. Martin GM, Sprague CA, Epstein CJ (1970). Replicative lifespan of cultivated human cells. Effects of donor's age, tissue and genotype. *Lab Invest.* 23: 86–92.
82. Goldstein S, Moerman EJ, Soeldner JS, Gleason RE, Barnett DM (1978). Chronologic and physiologic age affect replicative life-span of fibroblasts from diabetic, prediabetic, and normal donors. *Science* 199(4330): 781–2.
83. Cristofalo VJ, Allen RG, Pignolo RJ, Martin BG, Beck JC (1998). Relationship between donor age and the replicative lifespan of human cells in culture: a reevaluation. *Proc Natl Acad Sci USA* 95(18): 10614–19.
84. Dimri G, Lee X, Basile G, *et al.* (1995). A biomarker that identifies senescent human cells in culture and in aging skin *in vivo.* *Proc Natl Acad Sci USA* 92: 9362–7.
85. Li Y, Yan Q, Wolf NS (1997). Long-term calorie restriction delays age-related decline in proliferation capacity of murine lens epithelial cells *in vitro* and *in vivo.* *Invest Ophthmol Vis Sci.* 38: 100–8.

86. Funk WD, Wang CK, Shelton DN, Harley CB, Pagon GD, Hoeffler WK (2000). Telomerase expression restores dermal integrity to *in vitro*-aged fibroblasts in a reconstituted skin model. *Exp Cell Res.* 258(2): 270–8.

87. Sikora E, Kaminska B, Radziszewska E, Kaczmarek L (1992). Loss of transcription factor AP1 DNA binding activity during lymphocyte aging *in vivo*. *FEBS Lett.* 312: 179–82.

88. Sheshadri T, Campisi J (1990). Repression of *c-fos* and an altered genetic programme in senescent human fibroblasts. *Science* 247: 205–9.

89. Tein GH, Dulic V (1995). Origins of G1 arrest in senescent human fibroblasts. *BioEssays* 17: 537–43.

90. Kill, IR, Shall S (1990). Senescent human diploid fibroblasts are able to support DNA synthesis and to express markers associated with proliferation. *J Cell Sci.* 97: 473–8.

91. Perillo NL, Naeim F, Walford RL, Effros RB (1993). *In vitro* cellular aging in T-lymphocyte cultures: analysis of DNA content and cell size. *Exp Cell Res.* 207: 131–5.

92. Millis AJT, Sottile J, Hoyle M, Mann DM, Diemer V (1989). Collagenase production by early and late passage cultures of human fibroblasts. *Exp Gerontol.* 24: 559–75.

93. Zeng G, Millis AJ (1994). Expression of 72Kd gelatinase and TIMP2 in early and late passage human fibroblasts. *Exp Cell Res.* 213: 148–55.

94. West MD, Shay JW, Wright WE, Linskens MHK (1996). Altered expression of plasminogen activator and plasminogen activator inhibitor during cellular senescence. *Exp Gerontol.* 31: 175.

95. Ferber A, Chang C, Sell C, *et al.* (1993). Failure of senescent human fibroblasts to express the insulin-like growth factor 1 gene. *J Biol Chem.* 268: 17883–8.

96. Liu S, Thweatt R, Lumpkin CK, Goldstein S (1994). Suppression of calcium dependent membrane currents in human fibroblasts by replicative senescence and forced expression of a gene sequence encoding a putative calcium-binding protein. *Proc Natl Acad Sci USA* 91: 2186–90.

97. Zeng G, Millis AJT (1996). Differential regulation of collagenase and stromelysin mRNA in late passage cultures of human fibroblasts. *Exp Cell Res.* 222: 150–6.

98. Wistrom, C, Villeponteau B (1992). Cloning and expression of SAG: a novel marker of cellular senescence. *Exp Cell Res.* 199: 355–62.

99. Kodama S, Yamada H, Annab L, Barrett JC (1995). Elevated expression of mitochondrial cytochrome b and NADH dehydrogenase subunit 4/4L in senescent human cells. *Exp Cell Res.* 219: 82–6.

100. Salk D (1982). Werner's syndrome: a review of recent research with an analysis of connective tissue metabolism, growth control of cultured cells, and chromosomal aberrations. *Hum Genet.* 62(1): 1–5.

101. Tollefsbol TO, Cohen HJ (1984). Werner's syndrome: an underdiagnosed disorder resembling premature aging. *Age* 7: 75–88.

102. Shen J, Loeb LA (2001). Unwinding the molecular basis of the Werner syndrome. *Mech Age Dev.* 122(9): 921–44.

103. Rodriguez-Lopez AM, Jackson DA, Nehlin JO, Iborra F, Warren AV, Cox LS (2003). Characterisation of the interaction between WRN, the helicase/exonuclease defective in progeroid Werner's syndrome, and an essential replication factor, PCNA. *Mech Age Dev.* 124(2): 167–74.

104. Rodríguez-López AM, Jackson DA, Iborra F, Cox LS (2002). Asymmetry of DNA replication fork progression in Werner's syndrome. *Ageing Cell* 1: 30–9.

105. Fukuchi K, Martin GM, Monnat RJ Jr (1989). Mutator phenotype of Werner syndrome is characterized by extensive deletions. *Proc Natl Acad Sci USA* 86(15): 5893–7.

106. Wyllie FS, Jones CJ, Skinner JW, *et al.* (2000) Telomerase prevents the accelerated cell ageing of Werner syndrome fibroblasts. *Nat Genet.* 24: 16–17.

107. Murata K, Hatamochi A, Shinkai H, Ishikawa Y, Kawaguchi N, Goto M (1999). A case of Werner's syndrome associated with osteosarcoma. *J Dermatol.* 26(10): 682–6.

108. James SE, Faragher RG, Burke JF, Shall S, Mayne LV (2000). Werner's syndrome T lymphocytes display a normal *in vitro* life-span. *Mech Ageing Dev.* 121:139–49.

109. Fukuchi K, Martin GM, Monnat RJ Jr (1989). Mutator phenotype of Werner syndrome is characterized by extensive deletions. *Proc Natl Acad Sci USA* 86: 5893–7.

110. Fukuchi K, Tanaka K, Kumahara Y, *et al.* (1990). Increased frequency of 6-thioguanine-resistant peripheral blood lymphocytes in Werner syndrome patients. *Hum Genet.* 84: 249–52.

111. Ostler EL, Wallis CV, Sheerin AN, Faragher RG (2002). A model for the phenotypic presentation of Werner's syndrome. *Exp Gerontol.* 37: 285–92.

112. Johnson FB, Marciniak RA, McVey M, Stewart SA, Hahn WC, Guarente L (2001). The *Saccharomyces cerevisiae* WRN homolog Sgs1p participates in telomere maintenance in cells lacking telomerase. *EMBO J.* 20(4): 905–13.

113. von Zglinicki T, Saretzki G, Docke W, Lotze C (1995). Mild hyperoxia shortens telomeres and inhibits proliferation of fibroblasts: a model for senescence? *Exp Cell Res.* 220: 186–93.

114. von Zglinicki T, Serra V, Lorenz M, *et al.* (2000). Short telomeres in patients with vascular dementia: an indicator of low antioxidative capacity and a possible risk factor? *Lab Invest.* 80(11): 1739–47.

115. Lorenz M, Saretzki G, Sitte N, Metzkow S, von Zglinicki T (2001). BJ fibroblasts display high antioxidant capacity and slow telomere shortening independent of hTERT transfection. *Free Radic Biol Med.* 1(6): 824–31.

116. Faragher RGA, Mulholland B, Tuft SJ, Sandeman S, Khaw PT (1997). Aging and the cornea. *Br J Ophthalmol.* 81: 814–7.

117. Serrano M, Lin AW, McCurrach ME, Beach D, Lowe SW (1997). Oncogenic *ras* provokes premature cell senescence associated with accumulation of p53 and p16INK4a. *Cell* 88: 593–602.

118. Venable ME, Lee JY, Smyth MJ, Bielawska A, Obeid LM (1995). Role of ceramide in cell senescence. *J Biol Chem.* 270: 30701–8.

Aging of the Skin

P. Stephens

Department of Oral Surgery, Medicine and Pathology, University of Wales College of Medicine, Cardiff, Wales, UK

Introduction

The skin, as the largest organ of the body, has numerous protective roles to play throughout the span of our lives. In due course it is subject to the effects of both intrinsic (innate) and extrinsic (photo) aging. Therefore, the purpose of this chapter is to consider how the skin ages, examining the effects of aging on the organ and how this relates to changes in its function. Finally, consideration will be made of some of the many potential treatments and therapies which are currently available and which future approaches may be beneficial to ameliorate or even reverse the effects of aging.

The structure of normal skin

The skin is the largest organ of the body which in the average adult exceeds $2\ m^2$ in area and is in most places no more than 2 mm in thickness. This complex organ has many important functions including acting as a mechanical barrier, participating in the process of thermoregulation, initiating immunological functions, communicating external stimuli to the body and protecting against the effects of ultraviolet (UV) light. To fulfil all these functions the skin contains many different structures and resident cell types, which exhibit a wide range of properties [1].

The skin is composed of three major tissue layers. The outermost layer is the epidermis, a thin stratified epithelium which varies relatively little in thickness over most of the body surface, except on the palms of the hands and the soles of the feet. Underlying the epidermis is the dermis, a dense fibro-elastic connective tissue forming the mass of the skin. Found within this structure are extensive vascular and neural networks, excretory and secretory glands and keratinized appendages such as hair and nail. The dermis merges into the third definable layer, the subcutis or hypodermis, without a clear boundary. Composed of loose connective tissue and

29

R. Aspinall (ed.), Aging of Organs and Systems, 29–71.
© 2003 *Kluwer Academic Publishers. Printed in Great Britain.*

fatty tissue, the functions of the subcutis are to anchor the skin to the underlying structures, to act as a mechanical cushion and to provide insulation against heat loss.

The epidermis

The epidermis is a stratified squamous epithelium composed predominantly of keratinocytes with much smaller numbers of melanocytes, Langerhans cells and Merkel cells also present. The epidermis extends into the dermis as broad folds, or rete ridges, giving rise to a broadly undulating interface between the two layers.

Keratinocytes

Keratinocytes are the principal cell type within the epidermis and are so named because of the filamentous keratin proteins that are formed as the cells undergo epidermal differentiation. Ultimately these cells are responsible for the production and maintenance of the tough, protective stratum corneum. The epidermis is in a constant state of turnover and self-replacement. Keratinocytes within the basal layer divide and differentiate giving rise ultimately to squames. These are large, flattened, interconnected, dead cells packed with keratin filaments and surrounded by an insoluble cross-linked cell envelope. They form the tough, protective stratum corneum and as they are lost from the outer surface of the skin, so they are replaced by successive generations of terminally differentiated cells.

Four distinct layers of epidermal cells can be distinguished moving upwards through the epidermis. The stratum basale or stratum germinatum (basal layer) contains keratinocytes which are mitotically active and which give rise to the suprabasal keratinocytes. Arranged in a single layer, these cells have a columnar or cuboidal shape. The first of the suprabasal layers is the stratum spinosum (squamous or prickle cell layer) which comprises several layers of cells with a polyhedral morphology, most of which still have biosynthetic activity. Depending upon the anatomical site in which this layer is situated, its thickness varies (e.g., in the skin of the palms and soles it is thicker). Above this layer is the stratum granulosum (granular layer) which consists of one to three layers of flattened cells containing keratohyalin granules and within which, through the action of the enzyme epidermal transglutaminase, takes place a cross-linking process which initiates the formation of the cornified cell envelope. The keratohyalin granules are the precursors of the protein filaggrin which promotes the aggregation of keratin filaments in the outermost and most superficial of all the epidermal layers, the stratum corneum (cornified layer). This relatively thick, compact layer is composed of multiple layers of polyhedral cells that are the most differentiated cells of the keratinizing system. They are keratin-rich, anucleate cells that form the tough outer protective layer of the skin and which are eventually shed from the surface of the skin as part of everyday dynamic epidermal turnover. A fifth epidermal layer, termed the stratum lucidium, has also been identified as a layer of cells above the stratum granulosum, but is only really clearly defined in the thicker skin of the palms and soles. The entire epidermis is renewed approximately every 27 days.

Melanocytes
Melanocytes are dendritic cells that localize in the basal layer of the epidermis and extend their dendritic processes in all directions. They produce and secrete melanin, whose most important function is to protect against the damaging effects of non-ionizing UV radiation [2]. The synthesis of melanin takes place in melanosomes which are eventually transferred to the neighbouring basal keratinocytes. The color of skin is determined by the number and size of melanosomes present in the keratinocytes and not by the number of melanocytes (which in normal skin of all races is at the relatively constant ratio of one melanocyte to every 4–10 basal keratinocytes [3]).

Langerhans cells
Langerhans cells are dendritic cells located in the upper part of the squamous layer. They are the immunologic cells of the skin and are needed to induce proliferative and cytotoxic T-cell responses. They act as antigen-presenting cells to T lymphocytes.

Merkel cells
Within the epidermis the non-dendritic Merkel cell is present in smaller numbers than any of the other epidermal cells. Attached to adjacent keratinocytes via the action of desmosomes, Merkel cells come into contact with neurites and are believed to function as slow-adapting mechanoreceptors [4].

The basement membrane
A basement membrane (the basal lamina) separates the basal layer of the epidermis from the dermis [5]. The basal cells are attached to the basal lamina by hemidesmosomes and to adjacent keratinocytes by desmosomes. Ultrastructurally the basal lamina is defined as consisting of the following: the plasma membrane of the basal cells containing the hemidesmosomes and the anchoring filaments; the lamina lucida which is composed of laminin and bullous pemphigoid antigen; the lamina densa which is composed of type IV collagen; and the sublamina densa containing the anchoring fibrils (type VII collagen) that extend into the dermis.

The dermis
The dermis is composed of connective tissue, consisting mainly of collagen fibers and to a lesser degree, elastin, with a hydrophilic matrix of glycosaminoglycans (GAGs) and proteoglycans (PGs). It projects into the epidermis in finger-like projections called dermal papillae. Adnexal structures (e.g., sweat glands, hair follicles), muscle, blood and lymph vessels and nerves are also situated in the dermis.

The dermis is generally divided into two zones. The first is the papillary dermis which is the material directly beneath the epidermis and around the adnexal structures. It is composed of a highly woven network of type I collagen mixed with some type III collagen, elastin fibers and PGs. Located within this meshwork are capillaries and the main cell type found within the dermis, the fibroblast. Beneath this is the thick reticular dermis which is composed of thick bundles of type I collagen, thick elastin fibers, PGs and other populations of fibroblasts.

Fibroblasts
The fibroblast is the principal cell type of the dermis but other cells such as histocytes, dermal dendrocytes, lymphocytes and mast cells may also be present. Fibroblasts are responsible for the synthesis of all the dermal matrix components (collagens, fibronectin, elastin, GAGs and PGs) and also for a range of enzymes that are able to degrade all these extracellular matrix (ECM) proteins, a process usually described as remodeling. These biosynthetic and degradatory activities are highly regulated so that the appropriate dermal architecture is maintained. Fibroblasts play a crucial role during wound healing when they are responsible for the production and remodeling of large amounts of granulation tissue within the site of injury. The wound healing stimulus provokes fibroblasts to divide and to migrate into the wound site and subsequently to secrete all the relevant matrix molecules.

The hypodermis
The hypodermis, or subcutaneous tissue, is arranged into lobules of mature adipocytes which are separated by thin bands of dermal connective tissue that constitute the interlobular space. An extensive network of arteries, veins, capillaries, nerves and lymphatics extend through the hypodermis and into the dermis where they supply nutrients and remove waste products. This layer also acts as a mechanical cushion and provides insulation against heat loss.

Age associated changes in normal skin

As a society we are living longer, healthier lives but one of the inescapable signs of aging are changes associated with our skin. The typical signs of aging such as wrinkling and sagging of the skin are all too permanent reminders of the encroachment of time or of too much time spent in the sun without the appropriate protection. Interestingly however, the spectrum of changes in the skin (ranging from wrinkling to the development of malignant lesions) are irrelevant to some but for others can lead to major disfigurement and psychological problems. So, even though aging means different things to different people what exactly is aging and what are the typical characteristics of aging skin?

What is aging?
Aging is a basic biological process characteristic of all living organisms [6]. Inevitably it leads to reductions in maximal function and reserve capacity in all organ systems rendering the individual more susceptible to injury, disease and eventually death. Intrinsic or innate skin aging refers to the slow, but irreversible, degeneration of the skin's structure and function [7]. Extrinsic or "photo" aging is the result of exposure to outdoor elements, primarily UV light [8]. For obvious reasons, extrinsic aging is more prominent on the hands, neck and the face and it has been suggested that as much as 80% of facial aging is attributable to sun exposure [9]. Although it affects many layers of the skin, the predominant effects are observed within the connective tissue of the dermis (for an overall summary see Tables 1 and 2).

Table 1. General effects of aging on the cell types resident within the skin

Cell type	Effect of aging
Keratinocytes	↓ proliferation, ↓ differentiation
Melanocytes	↓ density, ↓ proliferation, ↓ biochemical activity
Epidermal lymphocytes	↓ antigen presentation, ↓ response to activating factors
Fibroblasts	↓ proliferation, ↓ ECM production, ↑ ECM turnover
Endothelial cells	↓ proliferation, ↓ response to vasodilators
Inflammatory cells	↓ proliferation, ↓ response to mitogens

Table 2. General effects of aging on the function of individual components of the skin

Skin structure	General effects of aging
Epidermis	Little change in overall structure and function
Basement membrane	Flattening (loss of rete ridges)
Dermis	↓ thickness, ↑ stiffness
Vasculature	↓ numbers of blood vessels, ↓ blood flow
Sebaceous glands	↓ secretion of sebum in women
Hypodermis	↓ or ↑ dependent on body location
Hair	Greying, hair loss
Nails	↓ growth and change in appearance

The clinical and histological signs of aging

The clinical signs of aging can be attributed to both intrinsic and extrinsic factors (Figure 1A). Clinically, intrinsically aged skin is thin, pale and finely wrinkled. Histological analyzes demonstrate a flattening of the dermo-epidermal junction (Figures 1B,C). The effects of extrinsic aging are often more pronounced. Extrinsically aged skin can acquire two phenotypes [10]. The first phenotype is characterized by deep wrinkles, laxity and a leathery appearance, increased fragility, blister formation and poor wound healing. The second phenotype is an atrophic, teleangiectatic one which demonstrates less wrinkling but is still distinct from sunprotected areas. Extrinsically aged skin can have epithelium which is hyperplastic or severly atrophic with evidence of cytologic atypia and disorderly maturation of keratinocytes [11]. Actinic keratosis is sometimes observed which is seen as a rough, scaly, often erythematous patch caused by pronounced nuclear atypis and loss of polarity within the epidermis [12]. "Age spots" or "liver spots" often appear on sun-exposed skin due to increased numbers and activity of melanocytes as can total

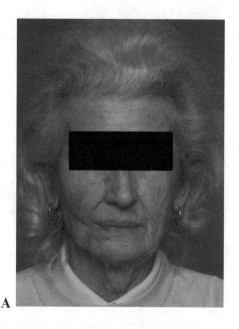

A

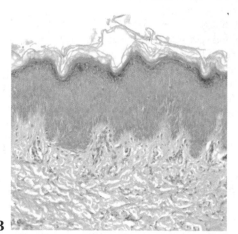

B

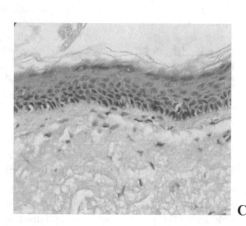

C

Figure 1. *Histological and clinical appearance of aged skin. (A) The clinical signs of aging include a thinning and paleness of the skin with evidence of wrinkling and laxity. Greying of the hair is also apparent. Histological analysis of (B) young and (C) old skin. Notice the lack of rete ridges within the old skin indicative of age-related dermo-epidermal flattening. (Images courtesy of Dr Cath Adams, UWCM, Cardiff [A] and Dr John Potts, UWCM, Cardiff [B,C]).*

depigmentation in fair skinned individuals [6]. Naevi or moles are local proliferations at the dermo-epidermal junction of melanocytes, which more often occur in sun-exposed sites.

Aging of the epidermis

Keratinocytes: Aged skin demonstrates a reduced keratinocyte proliferative capacity [13], an inability to properly terminally differentiate in order to form the stratum corneum and an inability to produce the appropriate cell signals in response to environmental stimuli [14]. Indeed, reduced proliferation within the rete ridges may be one of the reasons which contributes to the flattening of the dermal-epidermal junction [6]. However, despite these differences the structure and function of the stratum corneum are relatively normal [6] and the number of epidermal cell layers remains unaltered during aging [15].

Melanocytes: With the onset of intrinsic aging, despite a decrease in the density of melanocytes during intrinsic aging, there are minimal changes in the skin pigmentary system [16]. Whilst this decrease in melanocyte density may contribute to the pale appearance of older skin, this is also related to a decrease in skin vascularity [13]. As the skin ages, loss of melanocytes affects hair color leading to "greying" or total depigmentation of scalp hairs [17]. This may be explained by decreases in their tyrosinase activity, melanosomal transfer, proliferation and migration to an area close to the dermal papilla [18]. During extrinsic aging sun-exposure alters melano-cyte number and function, starting in childhood with the onset of freckling (characterized by enlarged, overactive melanocytes with a slight increase in overall density [19]) and in some situations leading to chronic hyper-pigmentation (char-acterized by increased melanocyte density, increased epidermal melanin and in-creased numbers of dermal melanophages [20]).

Epidermal lymphocytes: Little evidence is available concerning the effects on epidermal lymphocyte number or function in aging except that the antigen-present-ing lymphocytes within the epidermis decrease with age within non-sun exposed skin but to a greater degree within sun exposed skin [21–23]. Other age-related changes in epidermal immune function include morphologic, functional and positional changes in Langerhans cells and decreased production and response to epidermal thymocyte-activating factor [21, 24, 25].

Aging of the dermo-epidermal junction

The dermo-epidermal junction flattens with age due to retraction of the epidermal papillae and the micro-projections of the basal cells into the dermis [26]. As a result, cellular turnover within the epithelium is affected. Despite the fact that elastin organization at the dermo-epidermal junction alters [27] little change with respect to the expression of the basement membrane components (types IV and VII collagen and laminins B1 and B2) are evident [28]. All this leads ultimately to the development of a dermo-epidermal junction and a skin structural unit which is less resistant to shear forces than younger skin is.

Aging of the dermis

Fibroblasts: Skin thickness tends to decrease after the 7th decade [29]. In general this is thought to be due to alterations in cellular biosynthetic/degradatory responses and direct alterations of the aging ECM itself. The principal changes which occur in the resident cell populations are suggested by some to result from the accumulation of senescent cells (with a decreased ability to divide in response to damage or cell loss) within the aged dermis [30–32] Whilst a number of these changes may relate to environmental factors (e.g., photo-damage), the inverse relationship between donor-age and replicative lifespan suggests that the accumulation of senescent cells within the dermis, observed histochemically, represents a component of organismal aging [30, 32]. In support of the idea of increased senescence within aged dermal tissue it has been reported that hypo-cellularity is typical of intrinsically aged skin and that in photo-aged skin the fibroblasts adopt a stellate phenotype and demonstrate altered biosynthetic activities [33]. Furthermore, a decrease in the number and size of fibroblasts in aged skin has been reported [34]. However, the precise role of replicative senescence on cellular function within the tissues of aged individuals awaits full characterization.

The vasculature: As the skin ages it has been demonstrated that there are reductions in the total number of papillary loop microvessels, decreased thickness of microvessel basement membranes and decreased numbers of perivascular cells [35]. An obvious consequence of these structural alterations are decreased perfusion and increased capillary fragility. It has been suggested that this gradual loss of vascular cell homeostatic capacity and their resultant phenotypic alterations are characteristic of replicative senescence [36]. These alterations include changes in proteolytic gene expression and protein activation [37–39] and cell structural proteins [40]. Within aging skin, increased vasoconstrictor responses and decreases in both vasodilators and vasoprotective agents have also been demonstrated [41–43]. However, as with fibroblasts, whether or not replicative senescence in endothelial cells plays a direct or indirect role in altered vascular responses remains unclear. Recent reports on the prevention of telomere erosion in endothelial cells using telomerase may suggest a direct role for senescence in aging of the vasculature [36].

Inflammatory and immune cells: Long-term sun exposure of the skin results in chronic inflammation [10]. This is supported by the observations of increased numbers of mast cells, mononuclear cells and neutrophils within photo-aged skin [44, 45]. However, these findings are not universal since others have reported decreases in both the number and size of mast cells in aged skin [34]. Furthermore, within aged individuals T lymphocyte numbers are reduced and their ability to proliferate in response to mitogens (e.g., interleukin (IL)-2 and -4) is impaired [46].

Sebaceous glands: The number of sebaceous glands remains approximately the same throughout life although their size tends to change with age [47, 48]. In aging skin, whilst the sebaceous gland volume increases, no obvious morphological abnormalities have been reported [49, 50]. In aged adults alterations in sebaceous gland activity are sex dependent. In adult men few changes are observed compared to young individuals until the eighth decade [49, 51]. However, in women these sebaceous gland secretions starts to fall gradually after the menopause such that by the sixth/seventh decade decreases of up to 40% have been reported [49]. Mean sebum levels are therefore reduced more in females than in males [47, 49, 52]. Within the aged, sebaceous gland activity is still modulated by androgens but reduced androgen levels associated with aging can lead to decreased cellular turnover in the aged sebaceous glands of the face, often resulting in glandular hyperplasia [49, 50]. Other factors responsible for reduced sebaceous gland activity include reduced levels of the growth hormones normally required for the induction of lipid synthesis [53].

Clinical manisfestations of sebaceous gland aging include skin xerosis, sebaceous gland hyperplasia and benign and malignant sebaceous gland tumors which can be exacerbated by prolonged UV exposure [54]. Xerosis results from a reduced production of sebum combined with a decrease of eccrine sweat output [47]. Hyperplastic lesions are characterized by yellowish or skin-colored papules and nodules and can have large dilated follicular openings where sebum can be squeezed out [55]. Biochemically, development of these lesions has been linked with the overexpression of both Smad7 [56] (which blocks transforming growth factor-β signaling) and parathyroid hormone-related protein [57]. Sebaceous gland tumors most often occur in the seventh to ninth decade of life but have a low incidence (0.2% of all skin malignancies) [55, 58]. They start as a yellowish papule or nodule (similar to hyperplasia) then metastasize early to local lymph nodes and bones.

Extracellular matrix molecules: The characteristic appearance of aging dermis is due to distinct changes in its ECM composition. Extrinsically aged skin is characterized by a loss of mature collagen and a basophilic appearance of the collagen. Levels of pro-collagen type I mRNA and protein in naturally and photo-aged aged human skin *in vivo* are significantly lower than young skin [59, 60]. However, an increase in type III collagen in old skin has been reported [61]. Furthermore, type VII collagen anchoring fibrils are lost with age which leads to an increased susceptibility to blister formation.

Another hallmark of aged skin is an increased deposition of dystrophic elastic material in the deep dermis with disorganized tropoelastin and its associated microfibrillar component fibrillin [62, 63] . However, in the upper layers of extrinsically aged skin there is evidence of truncated and depleted fibrillin [64]. In photo-damaged skin increased elastin gene expression has been observed [65, 66] along with decreased expression of fibrillin-1 but not fibrillin-2 [64].

Due to age-associated changes in PG content, skin also undergoes dramatic changes in its mechanical properties, including changes in tissue hydration and resilience. As skin ages there is a decrease in the total GAG content in adult, relative to fetal, skin and this is evident with respect to hyaluronan and dermatan sulphate [67, 68]. Furthermore, in aged skin there is a decrease in the proportion of large chondroitin sulphate PGs (versican) and an increase in the proportion of a small dermatan sulphate PGs (decorin) [69]. There is also a decrease in chondroitin-4-sulphate and condroitin-6-sulphate (within basal lamina) with aging [70]. Versican differs between young and aged skin with respect to its chondroitin sulphate chains. The decorin of older skin shows a greater polydiversity in both its size and charge to mass ratio. In aged skin there is also evidence of another PG which is similar to, but smaller than decorin [69]. Versican has also been reported to increase, whereas decorin decreases, in areas of skin which demonstrate solar elastosis [71]. PGs are vitally important because they retain large amounts of water and therefore changes in their composition could lead to skin dehydration and loss of function.

As well as alterations in the biosynthesis of ECM molecules there are also alterations in the cross-linking of these molecules with age [72]. This is in part related to the fact that both collagen and elastin turnover slowly and are thus susceptible to age-related changes. The major alteration in collagen and elastin is the formation of intermolecular cross-links which can result in an over stiffening of the fibers when present in excess [73]. Initial stabilization of collagen is through lysyl oxidase which oxidatively deaminates collagen N- and C-terminal lysines to form lysyl aldehyde. In elastin the cross-linking serves to restrain excessive stretching rather than render the protein virtually inextensible as in the case of collagen. Further stiffening of the fibers can then occur through a second process based on the reaction of glucose or its metabolites, giving rise to non-enzymic glycosylation-derived cross-links (glycation). Glycation reactions are enhanced in diabetics due to hyperglycemia. Glycation reactions are complex [74] and result finally in the formation of advance glycation end-products (AGEs). Administration of AGEs to animals results in the acquisition of typical age- and diabetes-related changes [75]. The most damaging AGEs are those which form intermolecular cross-links. With respect to normal ECM/cellular responses glycation modification of lysine or arginine alters the charge profiles of the molecules leading to alterations in cell/collagen interactions and its subsequent remodeling. Glycation of type I collagen by arginine directly affects the arginine-glcyine-aspartic acid (RGD) site and its interactions with cells via the $\alpha 1 \beta 1$ and $\alpha 2 \beta 1$ integrin receptors which in turn affects adhesion and cell spreading [76].

Concurrent with altered biosynthesis and cross-linking of the ECM there is also increased ECM turnover within aging skin. This is brought about by a number of proteolytic enzymes (Figure 2) including serine proteases and matrix metalloproteinases (MMPs) [for a review of MMP structure, classification and function see refs 77, 78]. Indeed the direct observation of senescent fibroblasts in the aging dermis [30] strongly suggests that the tissue balance in aged skin is altered towards a more catabolic state since senescent fibroblasts are known to up-regulate MMPs [79–81]. Furthermore, it has been suggested that the altered tissue remodeling within aged individuals is related to the increased levels of MMP-2, plasminogen activator inhibitor-1 (PAI-1) and tissue inhibitor of metalloproteinases (TIMP)-2 which have previously been demonstrated in senescent, compared to normal, fibroblasts [38, 39, 81, 82]. With respect to photo-aging it has been reported that levels of MMP-1 and activity of MMP-2 were higher in the dermis of photo-aged skin than in naturally aged skin. Furthermore, exposure of fibroblasts to both UVA and UVB increases their production of MMPs, serine and other proteases which in turn degrade numerous components of the ECM including collagen, elastin and PGs [83–87]. UV damage can also act directly to produce an unstable intermediate collagen triple helix

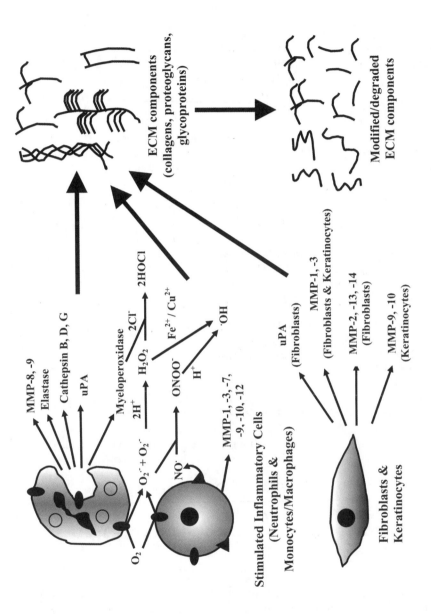

Figure 2. Summary of the major, host-derived enzymic and oxidative mechanisms of extracellular matrix modification and degradation associated with aging and the development of chronic wounds (courtesy of Dr Ryan Moseley, UWCM, Cardiff).

Table 3. *Mechanisms of ROS generation in vivo*

Nature of ROS	Source
Superoxide radical (O_2^-)	*NADPH oxidase:* $NADPH + 2O_2 \longrightarrow NADP^+ + H^+ + 2O_2{}^{-}$
Hydrogen peroxide (H_2O_2)	*Dismutation reaction:*

$$2O_2^- + 2H^+ \quad \overset{\text{Spontaneous or}}{\underset{\text{superoxide dismutase}}{\longrightarrow}} \quad H_2O_2 + O_2$$

Hydroxyl radical (˙OH)	*Haber-Weiss reaction:*

$$O_2^- + H_2O_2 \quad \overset{Fe^{2+}/}{\underset{Cu^{2+}}{\longrightarrow}} \quad ˙OH + OH^- + O_2$$

Fenton-type reaction:
$$Fe^{2+} + H_2O_2 \quad \longrightarrow \quad ˙OH + OH^- + Fe^{3+}$$

Hypochlorous acid (HOCl)	*Myeloperoxidase:*

$$2Cl^- + H_2O_2 \quad \longrightarrow . \quad 2HOCl$$

Nitric oxide (NO˙) and peroxynitrite (ONOO⁻)	*Nitric oxide synthase:*

$$\text{L-arginine} + NADPH + O_2 \longrightarrow N^G\text{-hydroxy-L-arginine} + H_2O + NADP^+ + H^+$$

$$N^G\text{-hydroxy-L-arginine} + \tfrac{1}{2} NADPH + O_2 \longrightarrow \text{L-citrulline} + NO˙ + H_2O$$

$$NO˙ + O_2^- \longrightarrow ONOO^- + H^+ \longrightarrow ˙OH + NO_2$$

which under further irradiation causes fragmentation of the molecule to form a short polypeptide which subsequently leads to helix collapse [88]. It has also been reported that UV treatment of cells affects their population doubling levels driving them towards a senescent phenotype [10]. These effects are not restricted solely to fibroblasts since a senescent phenotype has also been reported for endothelial cells [36].

UV exposure of skin and its constituent cells also stimulates the generation of reactive oxygen and nitrogen species (ROS and RNS respectively [89, 90] and *see below*). These ROS/RNS are formed in aerobic organisms as a result of metabolic activity (Table 3) and cause damage to a number of important cell components such as lipids, proteins and DNA [91]. Antioxidants (e.g., tocopherol, ubiquinone, glutathione, ascorbate and urate) [92] help protect against the damaging effects of these oxidative molecules. Studies suggest a correlation between the aging process and the formation of ROS with increases in oxidative stress observed in some tissues (possibly due to alterations in mitochondrial function) [93, 94]. Furthermore, aging is associated with a reduction of both enzymatic and non-enzymatic antioxidants [95]. The slow build-up of irreversible cell damage leads inevitably to the symptoms of aging. With respect to photo-damaged skin UV light can lead to the formation of ROS giving rise to dose-dependent production of lipid and protein oxidation products [96–98]. Conversely, the increase of elastin in photo-damaged skin may be due to increased tropoelastin mRNA levels by ROS [99]. Direct cellular effects of ROS/RNS include an impairment of migratory, proliferative and ECM biosynthetic properties of both dermal fibroblasts and keratinocytes [100, 101].

Secondary effects of ROS on ECM degradation are that they up-regulate and activate MMPs and decrease TIMPs. Singlet oxygen and H_2O_2 are the major ROS involved in the UVA-dependent activation of MMP-1, -2 and -3 [102–105] whilst hydroxyl radical and intermediates of lipid peroxidation play roles in UVB-induction of MMP-1 and -3 [84]. Furthermore, UV irradiation of the skin activates mitogen-activated protein kinase dependent pathways including c-Jun amino-terminal kinase and p38 resulting in activator protein-1 activation and enhanced expression and activation of MMPs [106]. Therefore, the turnover of the aging matrix arises from an imbalance between the proteolytic species (MMPs, ROS/RNS) and their respective inhibitors (TIMPs/antioxidants). Importantly, such changes will also have a major effect on the biomechanical properties of aged skin [47, 107, 108].

Aging of the hypodermis
Atrophy and hypertrophy of the subcutaneous tissue are common in aged individuals. This is demonstrated by a loss of fat in the hands, feet and facial regions (making these areas more susceptible to injury) and increases in other areas such as the waist and thighs [47]. This has been suggested to be attributed to the fact that aging influences hormone-induced lipolysis [109] and to altered fatty acid handling (resulting in lipid accumulation), dysdifferentiation of mesenchymal precursors (e.g., muscle satellite cells and osteoblast precursors) into a partial adipocyte phenotype, or a combination of these mechanisms [110].

Aging of hair and nails
As mentioned above, the loss of melanocytes within aging skin affects hair color leading to "greying" or total depigmentation of scalp hairs [17]. These alterations are thought to be due to ROS-mediated damage to nuclear and mitochondrial DNA. There is also evidence of reduced hair growth, thickness and density in aged individuals [47, 111–113]. This may go some way to explaining the how both male-

and female-pattern hair loss develops. As well as alterations in hair cycle kinetics there are also age-related alterations in the rate of linear growth of nails [47, 111, 114]. In humans, the rate of growth decreases by up to 50% over the normal life span and there is evidence of a decrease in nail plate thickness and a change in the overall appearance of the nail.

Age associated changes in skin function

Wound healing

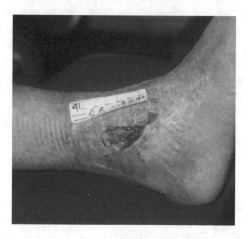

Figure 3. A typical chronic venous leg ulcer characterized by a loss of the epithelial tissues with slough (and sometimes necrotic tissue) within the non-healing ulcer bed. (Image courtesy of Professor Keith Harding, Wound Healing Research Unit, UWCM, Cardiff).

The repair of physical damage is an essential day-to-day function of the skin and as with other tissue function, such as the mounting of an immune reaction against an infectious agent, it involves both cell proliferation and differentiation. Indeed, the ability of an organism to survive crucially depends on its ability to maintain integumental integrity [115]. Wound healing is therefore of paramount importance in mammalian homeostasis. Wound healing is a complex process, involving interactions between resident and migratory cells, the ECM and cytokines/growth factors. In normal wound healing, the regeneration of the epithelium and mesenchymal tissues of the skin is effected by keratinocytes, endothelial cells and fibroblasts and is co-ordinated via complex cell-cell and cell-ECM interactions [116]. These responses are, however, altered in chronic wounds (pressure sores, venous ulcers and diabetic ulcers), normally associated with the aged, and are characterized by prolonged inflammation, defective wound ECM and failure of re-epithelialization [115, 117] (Figure 3; Table 4). Impaired wound healing in the skin has now become a significant worldwide quality of life issue since it affects 3% of the population over 60.

Table 4. *A summary of age associated changes in skin function*

Skin function	General effects of aging
Wound healing:	
– Inflammatory response	Dysfunctional and protracted
– Re-epithelialization	Slowed and sometimes inhibited
– Dermal repair	Impaired granulation tissue formation and remodeling
– Angiogenesis	Reduced
– Overall	Impaired overall response sometimes leading to the formation of chronic wounds
Immunoregulation	Dysfunctional and impaired leading to the generation of neoplasms
Thermoregulation	Impaired ability to perceive the cold, decreased sweat response
Barrier function	Decreased with respect to UV protection (and possibly others)

The financial impact on the UK of these conditions is considerable; chronic skin wounds currently cost the National Health Service an estimated £1 billion per annum. This sum will only increase as the number of aged individuals within our communities increases over the next few decades.

A few theories have been put forward to try to explain why chronic wounds form and ultimately persist. One theory is the fibrin cuff hypothesis [118]. The high ambulatory venous pressure within the calf muscle pump is transmitted causing a widening of endothelial pores, allowing the escape of large molecules such as fibrinogen. Fibrin complexes then form and are not broken down due to inadequate fibrinolytic activity within the blood and tissue fluid. The fibrin complexes prevent the passage of oxygen and other nutrients, which normally sustain the cells of the dermis, the epidermis and the vasculature. This leads directly to cell death and ulceration. Browse and Burnand's theory was refined and expanded by Falanga and Eaglstein [119] in their "trap" hypothesis. They suggested that the macromolecules leaking into the dermis bind or "trap" growth factors and matrix materials, hence making them unavailable for tissue repair and maintenance of tissue integrity. This hostile environment may also inhibit the *de novo* synthesis of ECM molecules by cells in the wound bed. Another theory, put forward by Scott *et al.* [120], suggested that an increased number of white blood cells in the skin of patients with venous disease may play a part in the development of ulcers, due to the build up of free radicals, toxic metabolites and proteolytic enzymes. Free radicals also form the basis of a theory proposed by Cheatle [121] suggesting that the accumulation of iron sets up a reaction leading to the production of hydroxyl radicals causing tissue damage. Despite all these hypotheses it is still not clear exactly how the increased tissue pressure can result in the drastic cellular and matrix changes associated with venous ulceration.

Alterations in aged immune responses and inflammation

The generation of an inflammatory reaction is crucial to successful wound healing but is dysfunctionally regulated in wounds associated with the aged. Within human wounds an age-related increased in the number of mature macrophages has been reported [122, 123], which may seem counter-intuitive since macrophages are crucial for a successful wound healing response. Similar findings have been reported in mice with increased numbers of macrophages (but no change in neutrophils) in wounds made in aged versus young animals [124]. However, a report by Moore et al. [125] suggests that despite their increased numbers the cells actually demonstrate a decreased activation and so may be dysfunctional in their responses. Production of monocyte-derived IL-1 and -6 is altered with aging [126, 127]. This is paralleled by a decline in macrophage production of vascular endothelial growth factor (VEGF) [128] which has important implications for wound healing. Investigations of leukocyte chemoattractant levels would suggest that there is increased monocyte chemoattractant protein (MCP)-1 in the wounds of aged mice but that macrophage inflammatory protein (MIP)-2, MIP-1α and MIP-1β and eotaxin are decreased [124]. Although the exact role of chemokines in aged, dysfunctional healing remains to be fully elucidated it has been demonstrated that mice deficient for the CXC chemokine IP-10 or XCX receptor CXCR2 show delayed healing [129, 130] .

Additionally wound macrophages from aged mice exhibit a reduced phagocytic activity due to decreases in both the number of phagocytic cells and in the number of particles consumed by each cell [124]. Decreases in macrophage phagocytic response may also be due to age-related changes in intracellular signaling as has been described for T cells [131–133]. Also the oxidative burst of both neutrophils and macrophages diminishes with aging [134–136]. A delayed T cell infiltration into wounds in aged mice has also been demonstrated but despite this their number was ultimately greater in aged mice [124]. Alterations in wound T cell content in humans have also been noted in other investigations [122]. Interestingly, although T cell depletion impairs healing [137] wounds in athymic mice demonstrate increased breaking strength [138].

Chronic wounds are characterized by a protracted inflammatory response although the exact initiators of this disease state remain to be delineated. One of the main contributory features of the chronic environment is the formation of the fibrin cuffs which are thought to trap not only growth factors but also neutrophils (see above). This leads to the persistence of these cells with a resulting increase in proteolytic enzyme and ROS release which in turn leads to increased ECM break-down. What is also typical of the chronic wound environment is the imbalance between pro-inflammatory cytokines/growth factors and their inhibitors. For example, it has been demonstrated by comparing chronic and acute wounds that overall levels of tumor necrosis factor (TNF)-α and IL-1β are increased in chronic wounds due to decreased levels of their respective inhibitors; namely p55 (the soluble TNF receptor protein) and the IL-1 receptor antagonist [139–141]. Furthermore, in some chronic non-healing wounds increased levels of pro-inflammatory molecules (interferon-γ, TNF-α, IL-6 and -8) but decreased levels of anti-inflammatory molecules (IL-2 and -10) have been reported [141].

With respect to T lymphocytes there is a decreased ratio of CD4+:CD8+ cells due to increasing numbers of CD8+ T lymphocytes within non-healing wounds [123]. Furthermore, of the CD4+ cells that are present a disproportionate number are of the pro-inflammatory Th1 type. Despite there being very few B lymphocytes within wounds normally, numbers are generally increased with chronic wounds as are numbers of plasma cells [123]. As in normal aging skin, significantly higher numbers of macrophages are present within chronic wounds [123] but these are thought not to be active due to low expression of the activation markers CD16 and CD35 [125]. Persistent inflammation therefore seems to arise due to dysfunctional regulation of both cell numbers and cellular response leading to an imbalance between pro- and anti-inflammatory molecules.

Dysfunctional age-related re-epithelialization
The time taken to heal epidermal wounds is reported to increase during aging. Analysis of the re-epithelialization of split thickness wounds in humans demonstrated that the process occurred more slowly in aged individuals [142]. This has been supported by investigations using aged mice which exhibit significantly delayed wound repair [128]. Age-related decreases in the proliferative capacity of keratinocytes and a failure to terminally differentiate correctly or respond to/produce the appropriate cell signals may contribute to this slow healing of minor injuries, ultimately giving rise to weaker scars and non-healing wounds [6, 143]. This is evidenced by the fact that, with age, the epidermal labeling index decreases, the turnover time increases [143] and aged keratinocytes demonstrate a decreased response to mitogens [144]. It is plausible that these alterations in epithelial renewal are attributable to the senescence of the keratinocyte populations since terminal differentiation of senescent keratinocytes occurs slower than for their growing counterparts [145]. There are also problems in restoration of the barrier function of the stratum corneum after wounding in the aged which is at least in part due to the decreased lipid synthetic capacity associated with aging [146].

Within chronic wounds re-epithelialization is notably retarded, with evidence of "piling up" of the keratinocytes at the edge of the wound and inhibition of cell migration. This may in part be due to aberrant regulation of MMPs at the wound edges [147–149] or to altered cellular responses to the hypoxic wound environment [150]. Whether such failure of re-epithelialization is a direct cause of chronic wounds or merely a consequence of other dysfunctional cellular responses is still open for debate.

Impaired dermal repair in the aged
Impaired healing in the aged dermis has widely been suggested as a potential example whereby replicative senescence impacts on tissue dysfunction. Cell death within the wound and subsequent proliferative "treadmilling" could reduce or even exhaust the

proliferative capacity of the cells in the wound, leading to a local accumulation of senescent cells. The phenotype of these senescent cells may then impact on the poor wound healing response. This is supported by investigations demonstrating impaired formation of granulation tissue in the aged, which was postulated to be due to a decrease in fibroblast number [34] and the fact that wounds in aged mice exhibit significantly delayed collagen synthesis [128]. The well-characterized changes in the steady state ECM of aged individuals [115] demonstrates that the proportions of collagen, elastin and GAGs alter in a manner reminiscent of changes in ECM production seen between senescent and proliferating fibroblasts *in vitro* [151–153]. This, coupled with the direct observation of senescent fibroblasts in the aging dermis [30] strongly suggests that the tissue balance in aged skin is altered towards a more catabolic state. Indeed, tissue remodeling is known to be altered in aged patients and increases in MMP-2, PAI-1 and TIMP-2 production have previously been demonstrated in senescent versus normal fibroblasts [38, 39, 81, 82]. However, counter-intuitively, it has also been reported that fibroblasts aged *in vitro* display increased abilities to reorganize their collagenous environment [154], suggesting a possible increased rate of matrix reorganization *in vivo* and therefore more rapid wound healing.

The healing of dermal wounds is a complex multi-cellular process and whilst one obvious effect of senescence on the fibroblasts which subsequently infiltrate the wound would be their inability to proliferate, secondary effects include a decreased rate of movement [155, 156], increased latent time [157], reduced responsiveness to stimulatory growth factors [115] and reduced overall ability of the population to remodel the ECM components [158] (although this is debated). Various MMPs such as collagenase, stromelysin and elastase are over-expressed by senescent fibroblasts [37, 80, 159]. This may adversely effect matrix re-synthesis and lead to the reduction in tensile strength of the closed wound seen in the aged [160]. Interestingly, microarray studies into replicative aging in different tissues have demonstrated that a number of the changes in gene expression are common to a variety of cell types [161]. The accumulation of senescent fibroblasts with such altered phenotypes within the tissues could, therefore, play an important role in mediating the observed impaired ECM remodeling which is a feature of the delayed wound healing in aged skin. Indeed it has been suggested that the major impact of aging in the skin is evident within the dermal tissues [32] in which the principal cell type is the fibroblast.

A chronic, persistent wound is the end result of delayed dermal healing in the aged. Workers have attempted to characterize the population of senescent fibroblasts within the dermis of chronic wounds to determine if the impaired healing is related to cellular senescent changes [162–168]. The results of these studies, almost exclusively in fibroblasts, have been equivocal. Nevertheless, decreased population doubling levels, proliferation and increased fibroblast senescence have been described in some studies [163–168]. The concept that replicative senescence in wound fibroblasts (induced as a result of excessive proliferation, oxidative stress and DNA damage) results in reduced proliferation and the failure of refractory chronic wounds to respond to treatment has therefore been proposed as an important factor in the tissue phenotype [163–168]. These findings are, however, not universal and contrast with studies in three-dimensional collagen lattice systems [169] and with the active cellular proliferation that is observed within these chronic wounds. Moreover, workers have shown that chronic wound fibroblasts synthesize comparable amounts of ECM to normal fibroblasts; suggesting that the defective wound matrix within chronic wound lesions may be due to differences in ECM remodeling within the wound site, possibly mediated by alterations in local MMP activity [169]. In chronic wound fibroblasts, usually from aged individuals, the precise relationship between the observed alterations in cellular responses with aging and replicative senescence remains to be determined. Although a recent study adds further support to the involvement of senescence in chronic wounds, in that telomerase activity (i.e., the ability to overcome telomere induced senescence; *see below*) is observed within acute dermal wounds but not in chronic wounds [170].

Insufficient angiogenesis in aged skin wounds
In aged tissues the vascularity of the skin is diminished [171] which would have obvious effects on the wound repair process. Age-related declines in angiogenesis during repair have been reported [172] and delayed angiogenesis has been demonstrated in an aged murine wound model [128]. It has been suggested that angiogenesis in wounds sustained within the aging population is different to younger individuals since endothelial cells from older individuals produce different ECM molecules (e.g., thrombospondin) compared to those isolated from younger individuals [173]. Furthermore, they demonstrated increased adhesiveness to leukocytes, increased responsiveness to TNF (which inhibits proliferation and phenotypically alters the cells [174]) and increased production of IL-1 (resulting in decreased proliferation). These phenotypic changes may be due to the acquisition by the endothelial cell of senescent characteristics [36]. Senescence has also been associated with atherosclerosis with evidence of shortened telomeres with age, especially in blood vessels where the endothelium is under hemodynamic stress [175, 176]. Telomere shortening is possibly due to extensive cell division to replace cells lost due this hemodynamic stress, which in turn could contribute to the formation of atherosclerotic plaques [175, 176].

As with the dysfunctional responses of all other cells types, failure of angiogenesis in chronic wounds is also observed. Whilst the presence of fibrin cuffs may have a direct role to play in this process [118], decreased levels of the $\alpha_v\beta_3$ integrin receptors on chronic wound endothelial cells and their incorrect distribution within the tissues have also been suggested to be important [177].

Immunoregulation
Within the skin, dendritic cells are one of the major cell types involved in immunoregulation and defence of the organism. The skin is a site of antigen, allergen and toxin contacts and the dysfunction of dendritic cells is likely to contribute to allergic reactions, infectious diseases and reduced defence against neoplasias [25]. Altered Langerhans cell number and function may favor the development of age- and UV-associated skin diseases, in particular the generation of skin cancers. One postulated theory about the age-associated development of skin cancer is that aged

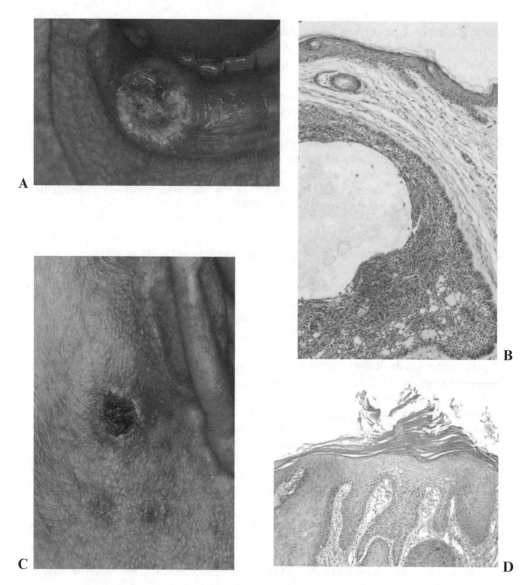

Figure 4. Histological and clinical appearance of skin tumors. (A) Basal cell carcinomas appear as slightly elevated nodules with evidence of central ulceration. (B) Histologically, basal cell carcinomas demonstrate distinct nests of darkly stained tumor cells within the dermis. (C) Squamous cell carcinomas often present clinically as a shallow ulcer surrounded by a wide, elevated and indurated border. (D) Histologically, squamous cell carcinomas consist of irregular masses of epidermal cells that proliferate downward and invade the dermis. (Images courtesy of Dr Cath Adams, UWCM, Cardiff [A,C] and Dr John Potts, UWCM, Cardiff [B,D]).

dendritic cells are impaired in their ability to present antigen and regulate T-cell functions [178]. This, coupled with a decreased immune function through age-associated decreasing telomere length and senescence in T-lymphocytes [179], suggests the existence of a dysfunctional immunoregulatory state within aged skin.

The skin of most aged individuals reveals a variety of benign cellular growths which in some cases can become malignant [6] (Figure 4). Seborrhoeic keratoses are common benign neoplasms which are variable in size and color and begin to appear in the third to fifth decade [180]. They are thought to result from over-proliferation of both keratinocytes and melanocytes in which endothelin-1 has been implicated [181, 182]. These lesions have been suggested to be one of the best biomarkers of intrinsic aging. As the skin ages due to both intrinsic and extrinsic factors, basal cell carcinomas and squamous cell carcinomas (both derived from keratinocyte malignancies) along with melanomas (melanocyte-derived) occur with increasing incidence [183, 184]. Commonly, these are linked to sun-exposure especially in fair-skinned individuals [12, 185]. A vast number of reports exist relating to the biochemical mechanisms responsible for the development of these malignant neoplasms, which implicate UV-induced mutations in the p53 tumor suppressor gene [186, 187], *patched* [188], smoothened [189], sonic hedgehog [190] and the cell cycle regulator p16 [191]. It therefore appears that irreparable DNA damage has profound effects on the development of skin malignancies however, the development of the tumor is probably also supported by alterations in the surrounding ECM environment and the reduced immunosurveillance that occurs with aging.

Thermoregulation
Aging is associated with a reduced intensity of the vasconstriction and shivering responses during exposure to the cold [192]. As aged individuals have a reduced cutaneous thermal sensitivity and a reduced thermal perception during cooling it has been suggested that they require a more intense stimulus to perform protective actions against cold stress [193] (i.e., they have an impaired ability to perceive the cold). This decreased vasoconstrictive response to the cold in aged skin is partly related to a reduced skin vasomotor sensitivity to sympathoneuronal stimuli. Furthermore, it has been postulated that this impaired thermoregulatory response may be a result of decreased norepinephrine release, a decreased vasomotor response for a given amount of norepinephrine at its receptor [192, 194, 195] or a down regulation of α-adrenoceptor numbers during aging [196].

In hot environments the most effective way of dissipating heat is through sweat production and its evaporation from the skin surface. However, in aged individuals there is a decreased sweat response which has been linked to both a decreased skin flow (due to less active vasodilation) [197] and to age-related modifications of peripheral mechanisms involving the sweat glands and modifications of the sensitivity of thermoreceptors within the skin to thermal stimuli [198]. These occur despite the fact that there is no apparent change in the ability of the epithelium to prevent trans-epidermal water loss [15].

Barrier function
As the epidermis ages its ability to act as a permeability barrier changes. This is linked both to a reduction in the lipid content of the stratum corneum and abnormalities in cholesterol synthesis [199]. Evidence for this comes from a report of a decrease rate of barrier formation after tape stripping in the aged which was partly due to a decreased lipid synthetic capacity with aging [146]. These alterations are suggested to arise via alterations in cytokine/growth factor signaling pathways within aged skin, particularly in the IL-1 family [199]. Importantly, recent evidence suggests that dynamic tests of barrier function are more reliable indicators of barrier function in the clinical setting than traditional tests such as basal trans-epidermal water loss [200]. Such tests may eventually provide some evidence as to whether any alterations occur in the ability of the aged skin to provide a barrier to invasion by microorganisms. Whilst both aerobic and anaerobic microorganisms undoubtedly have roles to play in the persistence of chronic wounds (where the open wound allows microbial colonization and migration into the deep tissues [201]) little is yet known about how their interactions with and invasion into aged skin may be altered.

Another vital role of the skin is to act as a barrier to UV bombardment. Within the skin two lines of defence against this insult are via the production of antioxidants and heat shock proteins (HSPs) [202]. As mentioned above, UV light can up-regulate the production of ROS leading to accelerated symptoms of aging [96–98]. Therefore, the decreased levels of antioxidants reported in aged skin [95] may ultimately be linked to a decreased barrier function with respect to UV. The HSPs are associated with increased resistance to UV-induced cell death [203, 204] but have also been demonstrated to be down regulated within aged skin and aged or stress exposed cells [205, 206]. Therefore, the lack of HSP induction in aged individuals may further contribute to a decline in UV protective barrier function.

Mechanisms of aging

One of the most perplexing questions asked of gerontologists is "how do we age?" Such a question is not easily answered since human aging covers a multitude of systems, tissues and biological processes. However, a number of hypotheses have been proposed which attempt to answer this question [for a review see ref. 207] and so some of these will now be considered.

Mechanisms of skin aging
Replicative senescence and telomere loss
The first suggested mechanism is one which involves telomere loss (Figure 5). In was Hayflick [208] who demonstrated that the number of divisions for a normal cell such as a diploid fibroblast is finite. A theory which has been proposed to explain this is the "telomere hypothesis." Telomeres are made out of thousands of hexameric (TTAGGG) repeats [209] and function to protect chromosomes against illegitimate fusions, to prevent the chromosome end from being recognized as a double strand break and to guide the pairing and movement of the chromosomes during mitosis and meiosis. Telomere structure and several associated telomeric binding proteins are

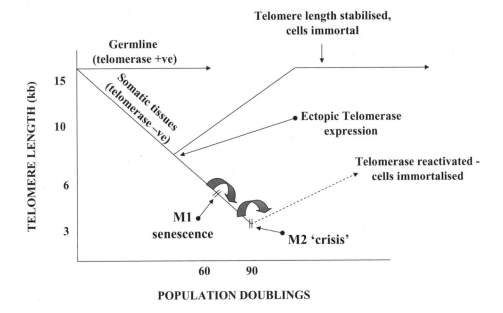

Figure 5. *The telomere hypothesis. Within germ line cells telomere lengths are kept constant due to the presence of telomerase. The telomere lengths in telomerase negative somatic cells gradually decreases with each cell division. M1 signifies the start of replicative senescence but this can be overcome by the expression of viral oncogenes that inactivate p53 and pRb. A second checkpoint is then reached at a critical telomere length called "crisis" (M2) at which point most cells die due to extensive chromosome abnormalities. Cells however, can be rescued form replicative senescence by the ectopic expression of telomerase (courtesy of Dr Duncan Baird, UWCM, Cardiff).*

all necessary for the mediation of telomere function. The telomeric binding proteins act to stabilize the chromosome and ensure its normal function. However, during DNA replication the cells are faced with an "end replication problem" [210] (Figure 6). Due to the nature of the replication process, the far end of each telomere is not replicated and thus the telomeres shorten with each round of cell division. Senescence occurs when a short, critical telomere length is achieved and the cell irreversibly exits the cell cycle. It is hypothesized that the reason why short telomeres generate a senescence response is that they can no longer from the necessary secondary structure and thus generate a DNA damage response [211]. It still remains to be established whether it is always the same chromosome that exhibits the shortest telomeres or whether the process is random. Furthermore, it is unsure exactly what length a "critically short" telomere constitutes since, depending on the source, cells can senesce with different length telomeres [212]. Support for this theory of gradual telomere loss comes from reports of the inverse relationship between donor-age and cellular replicative lifespan [155, 213] however, these findings have been recently questioned [214, 215]. The result of acquisition of the senescent phenotype is that

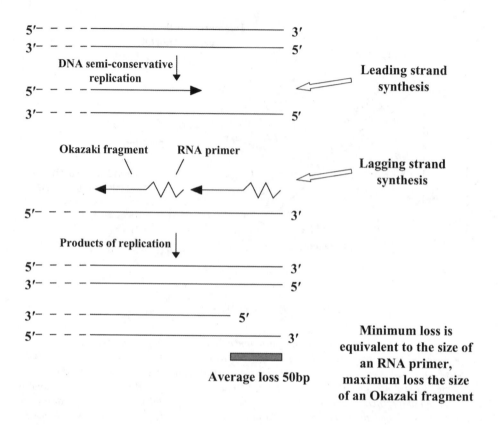

Figure 6. *The "end replication problem." As the DNA is replicated RNA primers serve as starting points for DNA polymerase. These primers can be replaced by DNA everywhere except at the extreme 5' end leading to an average loss of 50 bp of DNA with each round of cell division (courtesy of Dr Duncan Baird, UWCM, Cardiff).*

cells alter their biosynthetic and degradatory capacities and their abilities to migrate and respond to growth factors (*see above*). Most importantly, the senescent cells acquire resistance to apoptosis [216] which leads to an accumulation of senescent cells within the tissue and exacerbation of the age-related tissue changes.

Once the Hayflick limit or M1 stage (Mortality stage 1) is reached, cells undergo senescence (Figure 5). Such cell cycle inhibition is brought about by the induction of inhibitors of cyclin-dependent kinases, notably p21$^{\text{Waf1}}$ and p16$^{\text{INK4A}}$ [217, 218]. Furthermore, the retinoblastoma tumor suppressor protein (pRb) remains constitutively unphosphorylated and so in its growth suppressive form [219]. However, the expression of viral oncogenes that inactivate p53 and pRb [220] can cause the M1 barrier to proliferation to be overcome, leading to further proliferation and subsequent telomere shortening [221]. A second checkpoint is then reached at a critical telomere length called "crisis" (M2) at which point most cells die due to extensive chromosome abnormalities.

There are however, exceptions to this rule of limited cell division through telomere shortening in, for example, germ line cells, cancer cells and stem cells [222]. This is because these cells posses the enzyme telomerase which is able to synthesize telomeric sequence *de novo* and can therefore overcome telomere shortening. Telomerase is a ribonucleoprotein which is consists of two core components; an RNA component (hTR) which acts as an anchor and template for the telomeric DNA and a catalytic subunit (hTERT) which catalyzes the addition of nucleotides to the end of the telomere [223]. Indeed, forced expression of hTERT (the rate-limiting catalytic subunit of telomerase) by transfection is sufficient to confer telomerase activity to a number of cell types, halt telomere erosion with ongoing cell division and prevent replicative senescence. In so doing the cells appear to maintain their normal phenotype in that they retain their cell cycle checkpoints and differentiated state and do not appear to have undergone any other changes associated with the malignant phenotype [224, 225]. However, it should be noted that although telomere maintenance is ultimately necessary for cellular immortality, several cell types do not show lifespan extension following forced expression of telomerase. Before telomere erosion becomes a barrier to cell division, such cells enter a telomere-independent growth arrest state. This has been observed in a number of cell types including thyroid follicular epithelial cells human astrocytes, ovarian surface epithelium and corneal endothelium, islet β cells keratinocytes and mammary epithelium [226–229].

Free radical damage theory
Another theory of aging is the "free radical theory" proposed by Harmann in 1956 [230]. That is, ROS and RNS are formed in aerobic organisms as a result of metabolic activity and that with time these cause irreversible damage to a number of important cell components. Reactive oxygen species is a collective term encompassing oxygen free radicals, including the superoxide radical (O_2^-) and hydroxyl radical ($^.OH$) species, as well as non-radical oxygen derivatives, such as hydrogen peroxide (H_2O_2) and hypochlorous acid (HOCl). Reactive nitrogen species include the nitric oxide radical ($NO^.$) species and peroxynitrite ($ONOO^-$) [91]. To counteract these oxidative agents a number of cellular (enzymic) and extracellular (non-enzymic) antioxidants (e.g., tocopherol, ubiquinone, glutathione, ascorbate and urate) exist which help protect against the damaging effects of the ROS and RNS.

Studies have suggested a correlation between the aging process and the formation of ROS [for review see ref. 231] with increased oxidative stress observed in some tissues (possibly due to alterations in mitochondrial function [93]. Furthermore, aging is associated with a reduction of both enzymatic and non-enzymatic antioxidants [95, 232]. These imbalances in oxidant/antioxidant status lead to an excess of ROS/RNS available to cause indiscriminate damage to cellular constituents such as DNA, lipids and proteins [233]. Furthermore, ROS/RNS can directly and indirectly (via proteolytic activation) modify and degrade ECM components such as collagen, proteoglycans and hyaluronan [91] (Figure 2). Direct cellular effects of ROS/RNS include the impaired migratory, proliferative and ECM synthetic properties of dermal fibroblasts and keratinocytes [100, 101]. Inevitably the slow build up of irreversible cell damage leads to symptoms of aging. In terms of photo-damaged skin, UV light can up-regulate the production of ROS giving rise to dose-dependent increases in lipid and protein oxidation products [98] further accelerating the aging process.

Differential gene expression theory

Another theory to explain the aging phenomenon centers on the concept that as cells age they alter their cellular activities due to programmed or epigenetic changes in gene expression [7, 234]. For example, with respect to aging within the skin, distinct alterations in a variety of ECM and protease genes have been reported which lead to markedly lower ECM biosynthesis and increased ECM turnover in aged individuals compared to fetal tissues [235]. Therefore, it may be the differential gene expression profiles of cells which ultimately determine how readily a tissue will age.

Lessons from genetically altered organisms and aging disease states

The study of gene knock-down or gene ablation in mice and of premature aging syndromes have enabled some of the theories about "how we age" to be examined. For example, the development of the telomerase-null mouse has enabled a controlled analysis of the role that telomere shortening has at the organismal level [236]. Using this model system, although generalized signs of aging were not demonstrated, an increase in hair greying and alopecia were reported and these were correlated with a decrease in telomere length. Furthermore, the mice showed a higher incidence of severe ulcerative skin lesions, delayed wound closure, an increased incidence of spontaneous malignancies and a decreased lifespan. However, the consequences of telomere shortening were only evident after five generations of shortening [237] because the mouse telomere length (40–150 kb) is much longer than human telomeres (5–15 kb). Another suggested model has been the use of the nematode, *Caenorhabditis elegans*, since its telomeres are composed of the replicative TTAGGC repeats and have a length of 4–6 kb (which is comparable to human telomeres) [238]. However, the evidence for the existence of telomerase within this organism is equivocal [238, 239]. Decreases in lifespan and an early aging associated phenotype have also been reported for the senescence-accelerated mouse [240] and mice with defects in the gene *klotho* [241]. Interestingly, a recent report of a p53 mutant mouse reported delayed skin wound healing, osteoporosis and reduced longevity in the mice due to activation of the p53 protein [242].

Support for the importance of the control of ROS on longevity comes from analysis of p66shc-deficient mice which demonstrate an increase in life-span of up to 30% as a result of a higher resistance to ROS and oxidative stress [243]. It is speculated that p66shc is a ROS senor driving an apoptotic pathway in damaged cells [244]. Further support for the ROS theory comes from studies of *Drosophila melanogaster* and *Caenorhabditis elegans* in which overexpression of the antioxidants superoxide dismutase and catalase [245] or the use of superoxide dismutase/catalase mimetics [246] lead to an extension of their lifespan. However, in some cases it has been suggested that these lifespan extensions should be carefully evaluated in the context of the life spans of the control strains [231]. Unfortunately, gene ablation of MnSOD is neonatal lethal [247]. However, the use of conditional (Cre/Lox) knockouts [248] may, in the future, prove useful in delineating the direct effects of ROS on skin aging.

Analysis of hereditary premature aging syndromes provides a genetic link between aging and replicative lifespan [249]. Examples of these syndromes include Werner syndrome, Bloom syndrome, Hutchinson-Gilford progeria and Down syndrome; all of which show alterations in telomere biology. The best studied is Werner syndrome which is caused by a mutation in the WRN gene which encodes a member of the RecQ family of DNA helicases [250]. Although affected individuals live a relatively normal early life, they go on to prematurely develop a number of aging characteristics including hair and dermal thinning, osteoporosis, atherosclerosis and cancer [251]. Although WRN is not a telomere binding protein or a component of the telomerase enzyme the altered phenotype of Werner syndrome fibroblasts can be reversed *in vitro* by the over-expression of hTERT [252]. Further studies of affected individuals will, it is hoped, reveal more information about the reasons behind how and why we age.

Mechanisms of slowing/reversing skin aging
In an ideal world the ultimate aim of all the clinical and scientific knowledge of skin aging is to attempt to reverse or at least slow down the aging process. In the light of the findings that as much as 80% of facial aging is attributable to sun exposure [9], one obvious way to combat aging is to stay out of the sun or at least wear sun cream with a high sun protection factor. However, aside from this idea a vast number of studies are ongoing which aim to "roll back the years"; some scientific and some not quite so. A few of these will now be considered.

Attempts to reduce ROS levels within aged skin have been a major focus of research. Free radical scavenging through the addition of low molecular weight antioxidants (tocopherol and ascorbate) has yielded some interesting results [94]. Beneficial effects include a reduction in lipid peroxidation [253] and a stimulation of the immune system [254]. In animals, tocopherol can reduce UV skin damage [255] but data on the effects in humans is equivocal, although ascorbate has been demonstrated to inhibit erythema in a study involving human subjects [256]. With respect to application to the skin there are often problems with penetration and adsorption of the agents. Agents such as lipoic acid penetrates the skin easily and is believed to contribute to the overall antioxidant properties of the skin [94]. However, this has yet to be fully investigated with respect to skin aging. A more plausible route of delivery may be through a diet rich in fruit or vegetables either orally or via topical administration. Interestingly, as iron has also been implicated in the up-regulation of oxidative stress topical application of iron chelators to hairless mice skin have been demonstrated to significantly delayed photoaging [10]. What is apparent though from the literature is that there is a paucity of long-term randomized controlled trials into the roles of low molecular weight antioxidants for the prevention or treatment of skin aging.

The other potentially damaging species present in increasing amounts within aging skin are the MMPs. Studies targeted within this area have demonstrated that treatment of human skin with retinoic acid forty eight hours prior to UVB irradiation leads to inhibition of MMP expression [257]. The importance of retinoic acid is also seen in its role in the repair of photo-aged dermis [258, 259]. Furthermore, as ROS activity is implicated in activation of MMPs then control of these reactive oxygen species (*see above*) will go some way to controlling MMP activity within the aged tissues. The tetracyclins may also be an important class of MMP inhibitors for use during skin aging since they have been used successfully in the treatment of tissue-destructive diseases such as arthritis and periodontitis [260, 261].

Hormonal approaches have also been explored in an effort to alleviate the signs of aging. Estrogens have been used to try to protect against skin aging at, or after, the menopause but have shown little beneficial effect [262]. However, in other studies estrogen has been used to treat aging sebaceous glands within the facial skin of menopausal females resulting in induced sebaceous gland activity (increased sebum secretion) and reduced skin xerosis [263, 264]. Furthermore, treatment with female hormones such as estrogen can influence stratum corneum sphingolipid composition which may in turn affect barrier function post menopause [265].

Additionally attempts have also been made to reduce the amount of ECM cross-linking observed with the dermis of aging individuals. As a simple treatment option, changes in lifestyle including alterations in diet and smoking can act to reduce the levels of AGEs [266–268]. Aminoguanidine inhibits formation of AGEs and can reduce the damage of tissue [269] but further trials of these are needed. Other reagents including pyridoxamine (which inhibits AGE formation) and isoenzymes derived from *Aspergillus* sp. (which result in enzymic de-glycation [270]) may also prove useful in the future. Interestingly, macrophages and endothelial cells express receptors that bind AGEs and may therefore be involved in their natural clearance [271, 272]. Amelioration of macrophage and endothelial aging phenotypes may therefore be an attractive future therapeutic approach.

The identification of senescent cell populations within aged tissues may suggest that a reversal of this phenotype may be beneficial. As mentioned above this can be achieved by the use of telomerase. However, a degree of caution should be noted here since extension of cellular lifespan with hTERT may create a malignant cellular phenotype. Indeed, recent reports have demonstrated that this can lead to activation of *c-myc* within cells [273] suggesting that anti-telomerase therapy might be beneficial for the treatment of cancers [274]. Prevention of cancer in aged individuals may also arise from the topical use of α-interferon cream which has been demonstrated to increase the number of epidermal dendritic cells in both intrinsically and extrinsically aged skin [275]. Where lifespan extension maybe useful in the future is in the expansion of cells for re-implantation or tissue replacement [276]. Although simple skin based replacements are already available they are sometimes limited because of a lack of available cells and because of their oversimplified nature (i.e., lacking meloncytes, blood vessels etc). Telomerase treatment might help overcome these barriers in the future.

Conclusions

The skin is the major barrier our bodies have to protect us against the environment in which we live. However, as we live longer lives one of the inescapable signs of aging are changes associated with the appearance and function of this barrier. Intrinsic and extrinsic aging causes wrinkling and sagging of the skin and in extreme cases can lead to the formation of malignant neoplasms. As the aging process ensues cellular functions are altered leading to problems with wound healing, immunosurveillance, temperature regulation and general barrier function. Molecular and cell biological research has shed new light on the mechanisms associated with the aging process giving insights into how future efforts may be targeted to ameliorate or even reverse age-related changes/diseases. This research would seem extremely pertinent in the light that we are not only all living longer lives but are expecting a better quality of life throughout these years.

References

1.   Urmacher C (1990). Histology of normal skin. *Am J Surg Pathol.* 14: 671–686.
2.   Quevedo WC, Fitzpatrick TB, Szabo G, Jimbow K (1987). In: Fitzpatrick TB, Eisen AZ, Wolff K, Freedberg IM, Austen KF, eds. *Dermatology in General Medicine.* New York: McGraw-Hill, pp. 224–51.
3.   Nordlund JJ, Sober AJ, Hansen TW (1985). Periodic synopsis on pigmentation. *J Am Acad Dermatol.* 12: 359–63.
4.   Stenn KS, Goldenhersh MA, Trepera RW (1992). In: Weedon D, ed. *The Skin.* London: Churchill Livingston, pp. 1–19.
5.   Holbrook KA, Wolff K (1987). In: Fitzpatrick TB, Eisen AZ, Wolff K, Freedberg IM, Austen KF, eds. *Dermatology in General Medicine.* New York: McGraw-Hill, pp. 93–153.
6.   Yaar M, Gilchrest BA (2001). Ageing and photoageing of keratinocytes and melanocytes. *Clin Exp Dermatol.* 26: 583–91.
7.   Uitto J, Bernstein EF (1998). Molecular mechanisms of cutaneous aging: connective tissue alterations in the dermis. *J Investig Dermatol Symp Proc.* 3: 41–4.
8.   Gilchrest BA (1995). *Photodamage.* Cambridge, MA: Blackwell Science.
9.   Gilchrest BA (1989). Skin aging and photoaging: an overview. *J Am Acad Dermatol.* 21: 610–3.
10. Ma W, Wlaschek M, Tantcheva-Poor I, *et al.* (2001). Chronological ageing and photoageing of the fibroblasts and the dermal connective tissue. *Clin Exp Dermatol.* 26: 592–99.
11. Kligman AM, Kligman LH (1999). Photoaging. In: Freedberg IM, Eisen AZ, Wolff K, eds. *Fitzpatrick's Dermatology in General Medicine.* New York: McGraw-Hill, pp. 1717–23.
12. Schwartz RA, Stoll HLJ (1999). Squamous cell carcinoma. In: Freedberg IM, Eisen AZ, Wolff K, eds. *Fitzpatrick's Dermatology in General Medicine.* New York: McGraw-Hill, pp. 840–56.
13. Yaar M, Gilchrest BA (1999). Aging of the skin. In: Freedberg IM, Eisen AZ, Wolff K, eds. *Fitzpatrick's Dermatology in General Medicine.* New York: McGraw-Hill, pp. 1697–706.

14. Yaar M (1995). Molecular mechanisms of skin aging. *Adv Dermatol.* 10: 63–75.
15. Leyden JJ (1990). Clinical features of ageing skin. *Br J Dermatol.* 122(Suppl 35): 1–3.
16. Gilchrest BA, Blog FB, Szabo G (1979). Effects of aging and chronic sun exposure on melanocytes in human skin. *Trends Biotechnol.* 73: 141–3.
17. Keogh EV, Walsh RJ (1965). Rate of greying of human hair. *Nature* 207: 877–8.
18. Tobin DJ, Paus R (2001). Graying: gerontobiology of the hair follicle pigmentary unit. *Exp Gerontol.* 36: 29–54.
19. Rhodes AR, Albert LS, Barnhill RL, Weinstock MA (1991). Sun-induced freckles in children and young adults. A correlation of clinical and histopathologic features. *Cancer* 67: 1990–2001.
20. Bhawan J, Andersen W, Lee J, Labadie R, Solares G (1995). Photoaging versus intrinsic aging: a morphologic assessment of facial skin. *J Cutan Pathol.* 22: 154–9.
21. Gilchrest BA, Szabo G, Flynn E, Goldwyn RM (1983). Chronologic and actinically induced aging in human facial skin. *Trends Biotechnol.* 80(Suppl): 81s–5s.
22. Thiers BH, Maize JC, Spicer SS, Cantor AB (1984). The effect of aging and chronic sun exposure on human Langerhans cell populations. *Trends Biotechnol.* 82: 223–6.
23. Toyoda M, Bhawan J (1997). Ultrastructural evidence for the participation of Langerhans cells in cutaneous photoaging processes: a quantitative comparative study. *J Dermatol Sci.* 14: 87–100.
24. Sauder DN (1986). Effect of age on epidermal immune function. *Dermatol Clin.* 4: 447–54.
25. Grewe M (2001). Chronological ageing and photoageing of dendritic cells. *Clin Exp Dermatol.* 26: 608–12.
26. Kurban RS, Bhawan J (1990). Histologic changes in skin associated with aging. *J Dermatol Surg Oncol.* 16: 908–14.
27. Lavker RM, Zheng PS, Dong G (1987). Aged skin: a study by light, transmission electron, and scanning electron microscopy. *Trends Biotechnol.* 88: 44s–51s.
28. Olsen DR, Uitto J (1989). Differential expression of type IV procollagen and laminin genes by fetal vs adult skin fibroblasts in culture: determination of subunit mRNA steady-state levels. *Trends Biotechnol.* 93: 127–31.
29. de Rigal J, Escoffier C, Querleux B, Faivre B, Agache P, Leveque JL (1989). Assessment of aging of the human skin by *in vivo* ultrasonic imaging. *Trends Biotechnol.* 93: 621–5.
30. Dimri GP, Lee X, Basile G, *et al.* (1995). A biomarker that identifies senescent human cells in culture and in aging skin *in vivo*. *Proc Natl Acad Sci USA* 92: 9363–7.
31. Campisi J (1996). Replicative senescence: an old lives' tale? *Cell* 84: 497–500.
32. Faragher RG, Kipling D (1998). How might replicative senescence contribute to human ageing? *Bioessays* 20: 985–91.
33. Uitto J (1986). Connective tissue biochemistry of the aging dermis. Age-related alterations in collagen and elastin. *Dermatol Clin.* 4: 433–46.
34. Kligman AM, Lavker RM (1988). Cutaneous ageing: the differences between intrisic aging and photoageing. *J Cutan Aging Cosmetol Dermatol.* 1: 5–12.
35. Braverman IM, Fonferko E (1982). Studies in cutaneous aging: II. The microvasculature. *Trends Biotechnol.* 78: 444–8.
36. Chang E, Yang J, Nagavarapu U, Herron GS (2002). Aging and survival of cutaneous microvasculature. *Trends Biotechnol.* 118: 752–8.
37. Millis AJ, McCue HM, Kumar S, Baglioni C (1992). Metalloproteinase and TIMP-1 gene expression during replicative senescence. *Exp Gerontol.* 27: 425–8.

38. West MD, Shay JW, Wright WE, Linskens MH (1996). Altered expression of plasmino-
    gen activator and plasminogen activator inhibitor during cellular senescence. *Exp
    Gerontol*. 31: 175–93.

39. Ashcroft GS, Horan MA, Herrick SE, Tarnuzzer RW, Schultz GS, Ferguson MW
    (1997). Age-related differences in the temporal and spatial regulation of matrix
    metalloproteinases (MMPs) in normal skin and acute cutaneous wounds of healthy
    humans. *Cell Tissue Res*. 290: 581–91.

40. Vasile E, Tomita Y, Brown LF, Kocher O, Dvorak HF (2001). Differential expression of
    thymosin beta-10 by early passage and senescent vascular endothelium is modulated by
    VPF/VEGF: evidence for senescent endothelial cells *in vivo* at sites of atherosclerosis.
    FASEB J. 15: 458–66.

41. Tokunaga O, Yamada T, Fan JL, Watanabe T (1991). Age-related decline in prostacyclin
    synthesis by human aortic endothelial cells. Qualitative and quantitative analysis. *Am J
    Pathol*. 138: 941–9.

42. Sato I, Morita I, Kaji K, Ikeda M, Nagao M, Murota S (1993). Reduction of nitric
    oxide producing activity associated with *in vitro* aging in cultured human umbilical vein
    endothelial cell. *Biochem Biophys Res Commun*. 195: 1070–6.

43. Cooper LT, Cooke JP, Dzau VJ (1994). The vasculopathy of aging. *J Gerontol*. 49: B191–
    6.

44. Kligman LH, Murphy GF (1996). Ultraviolet B radiation increases hairless mouse mast
    cells in a dose-dependent manner and alters distribution of UV-induced mast cell
    growth factor. *Photochem Photobiol*. 63: 123–7.

45. Lavker RM, Kligman AM (1988). Chronic heliodermatitis: a morphologic evaluation of
    chronic actinic dermal damage with emphasis on the role of mast cells. *Trends
    Biotechnol*. 90: 325–30.

46. Nagelkerken L, Hertogh-Huijbregts A, Dobber R, Drager A (1991). Age-related
    changes in lymphokine production related to a decreased number of CD45RBhi CD4+
    T cells. *Eur J Immunol*. 21: 273–81.

47. Fenske NA, Lober CW (1986). Structural and functional changes of normal aging skin.
    J Am Acad Dermatol. 15: 571–85.

48. Zouboulis CC, Boschnakow A (2001). Chronological ageing and photoageing of the
    human sebaceous gland. *Clin Exp Dermatol*. 26: 600–7.

49. Pochi PE, Strauss JS, Downing DT (1979). Age-related changes in sebaceous gland
    activity. *Trends Biotechnol*. 73: 108–11.

50. Plewig G, Kligman AM (1978). Proliferative activity of the sebaceous glands of the
    aged. *Trends Biotechnol*. 70: 314–17.

51. Pierard GE, Pierard-Franchimont C, Marks R, Paye M, Rogiers V (2000). EEMCO
    guidance for the *in vivo* assessment of skin greasiness. The EEMCO Group. *Skin
    Pharmacol Appl Skin Physiol*. 13: 372–89.

52. Yamamoto A, Serizawa S, Ito M, Sato Y (1987). Effect of aging on sebaceous gland
    activity and on the fatty acid composition of wax esters. *Trends Biotechnol*. 89: 507–12.

53. Deplewski D, Rosenfield RL (1999). Growth hormone and insulin-like growth factors
    have different effects on sebaceous cell growth and differentiation. *Endocrinology* 140:
    4089–94.

54. Lesnik RH, Kligman LH, Kligman AM (1992). Agents that cause enlargement of
    sebaceous glands in hairless mice. II. Ultraviolet radiation. *Arch Dermatol Res*. 284:
    106–8.

55. Leyden J (2001). What is photoaged skin? *Eur J Dermatol*. 11: 165–7.

56.   He W, Li AG, Wang D, *et al.* (2002). Overexpression of Smad7 results in severe
       pathological alterations in multiple epithelial tissues. *EMBO J.* 21: 2580–90.
57.   Foley J, Longely BJ, Wysolmerski JJ, Dreyer BE, Broadus AE, Philbrick WM (1998).
       PTHrP regulates epidermal differentiation in adult mice. *Trends Biotechnol.* 111: 1122–
       8.
58.   Entius MM, Keller JJ, Drillenburg P, Kuypers KC, Giardiello FM, Offerhaus GJ
       (2000). Microsatellite instability and expression of hMLH-1 and hMSH-2 in sebaceous
       gland carcinomas as markers for Muir-Torre syndrome. *Clin Cancer Res.* 6: 1784–9.
59.   Chung JH, Seo JY, Choi HR, *et al.* (2001). Modulation of skin collagen metabolism in
       aged and photoaged human skin *in vivo*. *Trends Biotechnol.* 117: 1218–24.
60.   Trautinger F, Mazzucco K, Knobler RM, Trenz A, Kokoschka EM (1994). UVA- and
       UVB-induced changes in hairless mouse skin collagen. *Arch Dermatol Res.* 286: 490–4.
61.   Lovell CR, Smolenski KA, Duance VC, Light ND, Young S, Dyson M (1987). Type I
       and III collagen content and fibre distribution in normal human skin during ageing. *Br
       J Dermatol.* 117: 419–28.
62.   Mitchell RE (1967). Chronic solar dermatosis: a light and electron microscopic study of
       the dermis. *Trends Biotechnol.* 48: 203–20.
63.   Werth VP, Shi X, Kalathil E, Jaworsky C (1996). Elastic fiber-associated proteins of skin
       in development and photoaging. *Photochem Photobiol.* 63: 308–13.
64.   Watson RE, Griffiths CE, Craven NM, Shuttleworth CA, Kielty CM (1999). Fibrillin-
       rich microfibrils are reduced in photoaged skin. Distribution at the dermal-epidermal
       junction. *Trends Biotechnol.* 112: 782–7.
65.   Bernstein EF, Chen YQ, Tamai K, *et al.* (1994). Enhanced elastin and fibrillin gene
       expression in chronically photodamaged skin. *Trends Biotechnol.* 103: 182–6.
66.   Bernstein EF, Brown DB, Urbach F, *et al.* (1995). Ultraviolet radiation activates the
       human elastin promoter in transgenic mice: a novel *in vivo* and *in vitro* model of
       cutaneous photoaging. *Trends Biotechnol.* 105: 269–73.
67.   Breen M, Johnson RL, Sittig RA, Weinstein HG, Veis A (1972). The acidic glycosami-
       noglycans in human fetal development and adult life: Cornea, sclera and skin. *Connect
       Tissue Res.* 1: 291–303.
68.   Longas MO, Russell CS, He XY (1987). Evidence for structural changes in dermatan
       sulfate and hyaluronic acid with aging. *Carbohydr Res.* 159: 127–36.
69.   Carrino DA, Sorrell JM, Caplan AI (2000). Age-related changes in the proteoglycans of
       human skin. *Arch Biochem Biophys.* 373: 91–101.
70.   Willen MD, Sorrell JM, Lekan CC, Davis BR, Caplan AI (1991). Patterns of
       glycosaminoglycan/proteoglycan immunostaining in human skin during aging. *Trends
       Biotechnol.* 96: 968–74.
71.   Bernstein EF, Fisher LW, Li K, LeBaron RG, Tan EM, Uitto J (1995). Differential
       expression of the versican and decorin genes in photoaged and sun-protected skin.
       Comparison by immunohistochemical and northern analyses. *Lab Invest.* 72: 662–9.
72.   Bentley JP (1979). Aging of collagen. *Trends Biotechnol.* 73: 80–3.
73.   Bailey AJ (2001). Molecular mechanisms of ageing in connective tissues. *Mech Ageing
       Dev.* 122: 735–55.
74.   Bailey AJ, Paul RG, Knott L (1998). Mechanisms of maturation and ageing of collagen.
       Mech Ageing Dev. 106: 1–56.
75.   Vlassara H, Striker LJ, Teichberg S, Fuh H, Li YM, Steffes M (1994). Advanced
       glycation end products induce glomerular sclerosis and albuminuria in normal rats.
       Proc Natl Acad Sci USA 91: 11704–8.

76. Paul RG, Bailey AJ (1999). The effect of advanced glycation end-product formation upon cell-matrix interactions. *Int J Biochem Cell Biol.* 31: 653–60.

77. Barrick B, Campbell EJ, Owen CA (1999). Leukocyte proteinases in wound healing: roles in physiologic and pathologic processes. *Wound Repair Regen.* 7: 410–22.

78. Parks WC (1999). Matrix metalloproteinases in repair. *Wound Repair Regen.* 7: 423–32.

79. West MD, Pereira-Smith OM, Smith JR (1989). Replicative senescence of human skin fibroblasts correlates with a loss of regulation and overexpression of collagenase activity. *Exp Cell Res.* 184: 138–47.

80. Zeng G, Millis AJ (1996). Differential regulation of collagenase and stromelysin mRNA in late passage cultures of human fibroblasts. *Exp Cell Res.* 222: 150–6.

81. Zeng G, Millis AJ (1994). Expression of 72-kDa gelatinase and TIMP-2 in early and late passage human fibroblasts. *Exp Cell Res.* 213: 148–55.

82. Ashcroft GS, Herrick SE, Tarnuzzer RW, Horan MA, Schultz GS, Ferguson MW (1997). Human ageing impairs injury-induced *in vivo* expression of tissue inhibitor of matrix metalloproteinases (TIMP)-1 and -2 proteins and mRNA. *J Pathol.* 183: 169–76.

83. Brenneisen P, Oh J, Wlaschek M, *et al.* (1996). Ultraviolet B wavelength dependence for the regulation of two major matrix-metalloproteinases and their inhibitor TIMP-1 in human dermal fibroblasts. *Photochem Photobiol.* 64: 649–57.

84. Brenneisen P, Wenk J, Klotz LO, *et al.* (1998). Central role of Ferrous/Ferric iron in the ultraviolet B irradiation-mediated signaling pathway leading to increased interstitial collagenase (matrix-degrading metalloprotease (MMP)-1) and stromelysin-1 (MMP-3) mRNA levels in cultured human dermal fibroblasts. *J Biol Chem.* 273: 5279–87.

85. Koivukangas V, Kallioinen M, Autio-Harmainen H, Oikarinen A (1994). UV irradiation induces the expression of gelatinases in human skin *in vivo*. *Acta Derm Venereol.* 74: 279–82.

86. Petersen MJ, Hansen C, Craig S (1992). Ultraviolet A irradiation stimulates collagenase production in cultured human fibroblasts. *Trends Biotechnol.* 99: 440–4.

87. Scharffetter K, Wlaschek M, Hogg A, Bolsen K, Schothorst A, Goerz G, Krieg T, Plewig G (1991). UVA irradiation induces collagenase in human dermal fibroblasts *in vitro* and *in vivo*. *Arch Dermatol Res.* 283: 506–11.

88. Miles CA, Sionkowska A, Hulin SL, Sims TJ, Avery NC, Bailey AJ (2000). Identification of an intermediate state in the helix-coil degradation of collagen by ultraviolet light. *J Biol Chem.* 275: 33014–20.

89. Jurkiewicz BA, Buettner GR (1996). EPR detection of free radicals in UV-irradiated skin: mouse versus human. *Photochem Photobiol.* 64: 918–22.

90. Masaki H, Atsumi T, Sakurai H (1995). Detection of hydrogen peroxide and hydroxyl radicals in murine skin fibroblasts under UVB irradiation. *Biochem Biophys Res Commun.* 206: 474–9.

91. Waddington RJ, Moseley R, Embery G (2000). Reactive oxygen species: a potential role in the pathogenesis of periodontal diseases. *Oral Dis.* 6: 138–51.

92. Shindo Y, Witt E, Han D, Epstein W, Packer L (1994). Enzymic and non-enzymic antioxidants in epidermis and dermis of human skin. *Trends Biotechnol.* 102: 122–4.

93. Kwong LK, Sohal RS (2000). Age-related changes in activities of mitochondrial electron transport complexes in various tissues of the mouse. *Arch Biochem Biophys.* 373: 16–22.

94. Podda M, Grundmann-Kollmann M (2001). Low molecular weight antioxidants and their role in skin ageing. *Clin Exp Dermatol.* 26: 578–82.

95. Beckman KB, Ames BN (1998). The free radical theory of aging matures. *Physiol Rev.* 78: 547–81.

96. Lippman RD (1985). Rapid *in vivo* quantification and comparison of hydroperoxides and oxidized collagen in aging mice, rabbits and man. *Exp Gerontol.* 20: 1–5.

97. Niwa Y, Kasama T, Kawai S, *et al.* (1988). The effect of aging on cutaneous lipid peroxide levels and superoxide dismutase activity in guinea pigs and patients with burns. *Life Sci.* 42: 351–6.

98. Podda M, Traber MG, Weber C, Yan LJ, Packer L (1998). UV-irradiation depletes antioxidants and causes oxidative damage in a model of human skin. *Free Radic Biol Med.* 24: 55–65.

99. Kawaguchi Y, Tanaka H, Okada T, *et al.* (1997). Effect of reactive oxygen species on the elastin mRNA expression in cultured human dermal fibroblasts. *Free Radic Biol Med.* 23: 162–5.

100. O'Toole EA, Goel M, Woodley DT (1996). Hydrogen peroxide inhibits human keratinocyte migration. *Dermatol Surg.* 22: 525–9.

101. Agren UM, Tammi RH, Tammi MI (1997). Reactive oxygen species contribute to epidermal hyaluronan catabolism in human skin organ culture. *Free Radic Biol Med.* 23: 996–1001.

102. Scharffetter-Kochanek K, Wlaschek M, Briviba K, Sies H (1993). Singlet oxygen induces collagenase expression in human skin fibroblasts. *FEBS Lett.* 331: 304–6.

103. Herrmann G, Wlaschek M, Bolsen K, Prenzel K, Goerz G, Scharffetter-Kochanek K (1996). Photosensitization of uroporphyrin augments the ultraviolet A-induced synthesis of matrix metalloproteinases in human dermal fibroblasts. *Trends Biotechnol.* 107: 398–403.

104. Wlaschek M, Briviba K, Stricklin GP, Sies H, Scharffetter-Kochanek K (1995). Singlet oxygen may mediate the ultraviolet A-induced synthesis of interstitial collagenase. *Trends Biotechnol.* 104: 194–8.

105. Wenk J, Brenneisen P, Wlaschek M, *et al.* (1999). Stable overexpression of manganese superoxide dismutase in mitochondria identifies hydrogen peroxide as a major oxidant in the AP-1-mediated induction of matrix-degrading metalloprotease-1. *J Biol Chem.* 274: 25869–76.

106. Fisher GJ, Talwar HS, Lin J, *et al.* (1998). Retinoic acid inhibits induction of c-Jun protein by ultraviolet radiation that occurs subsequent to activation of mitogen-activated protein kinase pathways in human skin *in vivo. J Clin Invest.* 101: 1432–40.

107. Doubal S, Klemera P (1998). Changes in mechanical properties of skin as a marker of biological age. *Sb Lek* 99: 423–8.

108. Pierard GE, Henry F, Castelli D, Ries G (1998). Ageing and rheological properties of facial skin in women. *Gerontology* 44: 159–61.

109. Engfeldt P, Arner P (1988). Lipolysis in human adipocytes, effects of cell size, age and of regional differences. *Horm Metab Res Suppl.* 19: 26–9.

110. Kirkland JL, Tchkonia T, Pirtskhalava T, Han J, Karagiannides I (2002). Adipogenesis and aging: does aging make fat go MAD? *Exp Gerontol.* 37: 757–67.

111. Cerimele D, Celleno L, Serri F (1990). Physiological changes in ageing skin. *Br J Dermatol.* 122(Suppl 35): 13–20.

112. Courtois M, Loussouarn G, Hourseau C, Grollier JF (1995). Ageing and hair cycles. *Br J Dermatol.* 132: 86–93.

113. Birch MP, Messenger JF, Messenger AG (2001). Hair density, hair diameter and the prevalence of female pattern hair loss. *Br J Dermatol.* 144: 297–304.

114. Orentreich N, Markofsky J, Vogelman JH (1979). The effect of aging on the rate of linear nail growth. *Trends Biotechnol.* 73: 126–30.

115. Ashcroft GS, Horan MA, Ferguson MW (1995). The effects of ageing on cutaneous wound healing in mammals. *J Anat.* 187(Pt 1): 1–26.
116. Martin P (1997). Wound healing – aiming for perfect skin regeneration. *Science* 276: 75–81.
117. Herrick SE, Sloan P, McGurk M, Freak L, McCollum CN, Ferguson MW (1992). Sequential changes in histologic pattern and extracellular matrix deposition during the healing of chronic venous ulcers. *Am J Pathol.* 141: 1085–95.
118. Browse NL, Burnand KG (1982). The cause of venous ulceration. *Lancet* 2: 243–5.
119. Falanga V, Eaglstein WH (1993). The "trap" hypothesis of venous ulceration. *Lancet* 341: 1006–8.
120. Scott HJ, Coleridge SP, Scurr JH (1991). Histological study of white blood cells and their association with lipodermatosclerosis and venous ulceration. *Br J Surg.* 78: 210–11.
121. Cheatle T (1991). Venous ulceration and free radicals. *Br J Dermatol.* 124: 508.
122. Ashcroft GS, Horan MA, Ferguson MW (1998). Aging alters the inflammatory and endothelial cell adhesion molecule profiles during human cutaneous wound healing. *Lab Invest.* 78: 47–58.
123. Loots MA, Lamme EN, Zeegelaar J, Mekkes JR, Bos JD, Middelkoop E (1998). Differences in cellular infiltrate and extracellular matrix of chronic diabetic and venous ulcers versus acute wounds. *Trends Biotechnol.* 111: 850–7.
124. Swift ME, Burns AL, Gray KL, DiPietro LA (2001). Age-related alterations in the inflammatory response to dermal injury. *Trends Biotechnol.* 117: 1027–35.
125. Moore K, Ruge F, Harding KG (1997). T lymphocytes and the lack of activated macrophages in wound margin biopsies from chronic leg ulcers. *Br J Dermatol.* 137: 188–94.
126. Bradley SF, Vibhagool A, Kunkel SL, Kauffman CA (1989). Monokine secretion in aging and protein malnutrition. *J Leukoc Biol.* 45: 510–14.
127. Roubenoff R, Harris TB, Abad LW, Wilson PW, Dallal GE, Dinarello CA (1998). Monocyte cytokine production in an elderly population: effect of age and inflammation. *J Gerontol A Biol Sci Med Sci.* 53: M20–6
128. Swift ME, Kleinman HK, DiPietro LA (1999). Impaired wound repair and delayed angiogenesis in aged mice. *Lab Invest.* 79: 1479–87.
129. Luster AD, Cardiff RD, MacLean JA, Crowe K, Granstein RD (1998). Delayed wound healing and disorganized neovascularization in transgenic mice expressing the IP-10 chemokine. *Proc Assoc Am Physicians* 110: 183–96.
130. Devalaraja RM, Nanney LB, Du J, *et al.* (2000). Delayed wound healing in CXCR2 knockout mice. *Trends Biotechnol.* 115: 234–44.
131. Patel HR, Miller RA (1992). Age-associated changes in mitogen-induced protein phosphorylation in murine T lymphocytes. *Eur J Immunol.* 22: 253–60.
132. Shi J, Miller RA (1993). Differential tyrosine-specific protein phosphorylation in mouse T lymphocyte subsets. Effect of age. *J Immunol.* 151: 730–9.
133. Chakravarti B, Abraham GN (1999). Aging and T-cell-mediated immunity. *Mech Ageing Dev.* 108: 183–206.
134. Lipschitz DA, Udupa KB (1986). Influence of aging and protein deficiency on neutrophil function. *J Gerontol.* 41: 690–4.
135. Davila DR, Edwards CK, Arkins S, Simon J, Kelley KW (1990). Interferon-gamma-induced priming for secretion of superoxide anion and tumor necrosis factor-alpha declines in macrophages from aged rats. *FASEB J.* 4: 2906–11.

136. Alvarez E, Santa MC (1996). Influence of the age and sex on respiratory burst of human monocytes. *Mech Ageing Dev.* 90: 157–61.

137. Peterson JM, Barbul A, Breslin RJ, Wasserkrug HL, Efron G (1987). Significance of T-lymphocytes in wound healing. *Surgery* 102: 300–5.

138. Barbul A, Shawe T, Rotter SM, Efron JE, Wasserkrug HL, Badawy SB (1989). Wound healing in nude mice: a study on the regulatory role of lymphocytes in fibroplasia. *Surgery* 105: 764–9.

139. Tarnuzzer RW, Schultz GS (1996). Biochemical analysis of acute and chronic wound environments. *Wound Repair Regen.* 4: 321–5.

140. Wallace HJ, Stacey MC (1998). Levels of tumor necrosis factor-alpha (TNF-alpha) and soluble TNF receptors in chronic venous leg ulcers – correlations to healing status. *Trends Biotechnol.* 110: 292–6.

141. Agren MS, Eaglstein WH, Ferguson MW, *et al.* (2000). Causes and effects of the chronic inflammation in venous leg ulcers. *Acta Derm Venereol Suppl (Stockh)*, 210: 3–17.

142. Holt DR, Kirk SJ, Regan MC, Hurson M, Lindblad WJ, Barbul A (1992). Effect of age on wound healing in healthy human beings. Surgery, 112: 293–7.

143. Gilchrest BA (1983). *In vitro* assessment of keratinocyte aging. *Trends Biotechnol.* 81: 184s–9s.

144. Rattan SI, Derventzi A (1991). Altered cellular responsiveness during ageing. *Bioessays* 13: 601–6.

145. Norsgaard H, Clark BF, Rattan SI (1996). Distinction between differentiation and senescence and the absence of increased apoptosis in human keratinocytes undergoing cellular aging *in vitro*. *Exp Gerontol.* 31: 563–70.

146. Ghadially R, Brown BE, Sequeira-Martin SM, Feingold KR, Elias PM (1995). The aged epidermal permeability barrier. Structural, functional, and lipid biochemical abnormalities in humans and a senescent murine model. *J Clin Invest.* 95: 2281–90.

147. Parks WC (1995). The production, role,and regulation of matrix metalloproteinsases in the healing epidermis. *Wounds: Compendium Clin Res Pract.* 7: 23A–37A.

148. Saarialho-Kere UK (1998). Patterns of matrix metalloproteinase and TIMP expression in chronic ulcers. *Arch Dermatol Res.* 290(Suppl): S47–54.

149. Mirastschijski U, Impola U, Jahkola T, Karlsmark T, Agren MS, Saarialho-Kere U (2002). Ectopic localization of matrix metalloproteinase-9 in chronic cutaneous wounds. *Hum Pathol.* 33: 355–64.

150. Xia YP, Zhao Y, Tyrone JW, Chen A, Mustoe TA (2001). Differential activation of migration by hypoxia in keratinocytes isolated from donors of increasing age: implication for chronic wounds in the elderly. *Trends Biotechnol.* 116: 50–6.

151. Deie M, Marui T, Allen CR, *et al.* (1997). The effects of age on rabbit MCL fibroblast matrix synthesis in response to TGF-beta 1 or EGF. *Mech Ageing Dev.* 97: 121–30.

152. Passi A, Albertini R, Campagnari F, De Luca G (1997). Modifications of proteoglycans extracted from monolayer cultures of young and senescent human skin fibroblasts. *FEBS Lett.* 420: 175–8.

153. Passi A, Albertini R, Campagnari F, De Luca G (1997). Modifications of proteoglycans secreted into the growth medium by young and senescent human skin fibroblasts. *FEBS Lett.* 402: 286–90.

154. Gibson JM, Milam SB, Klebe RJ (1989). Late passage cells display an increase in contractile behavior. *Mech Ageing Dev.* 48: 101–10.

155. Schneider EL, Mitsui Y (1976). The relationship between *in vitro* cellular aging, *in vivo* human age. *Proc Natl Acad Sci USA* 73: 3584–8.

156. Kondo H, Yonezawa Y (1992). Changes in the migratory ability of human lung and skin fibroblasts during *in vitro* aging and *in vivo* cellular senescence. *Mech Ageing Dev.* 63: 223–33.

157. Muggleton-Harris AL, Reisert PS, Burghoff RL (1982). *In vitro* characterization of response to stimulus (wounding) with regard to ageing in human skin fibroblasts. *Mech Ageing Dev.* 19: 37–43.

158. Bell E, Ivarsson B, Merrill C (1979). Production of a tissue-like structure by contraction of collagen lattices by human fibroblasts of different proliferative potential *in vitro*. *Proc Natl Acad Sci USA* 76: 1274–8.

159. Millis AJ, Sottile J, Hoyle M, Mann DM, Diemer V (1989). Collagenase production by early and late passage cultures of human fibroblasts. *Exp Gerontol.* 24: 559–75.

160. Ballas CB, Davidson JM (2001). Delayed wound healing in aged rats is associated with increased collagen gel remodeling and contraction by skin fibroblasts, not with differences in apoptotic or myofibroblast cell populations. *Wound Repair Regen.* 9: 223–37.

161. Shelton DN, Chang E, Whittier PS, Choi D, Funk WD (1999). Microarray analysis of replicative senescence. *Curr Biol.* 9: 939–45.

162. Hasan A, Murata H, Falabella A, *et al.* (1997). Dermal fibroblasts from venous ulcers are unresponsive to the action of transforming growth factor-beta 1. *J Dermatol Sci.* 16: 59–66.

163. Stanley AC, Park HY, Phillips TJ, Russakovsky V, Menzoian JO (1997). Reduced growth of dermal fibroblasts from chronic venous ulcers can be stimulated with growth factors. *J Vasc Surg.* 26: 994–9.

164. Mendez MV, Stanley A, Park HY, Shon K, Phillips T, Menzoian JO (1998). Fibroblasts cultured from venous ulcers display cellular characteristics of senescence. *J Vasc Surg.* 28: 876–83.

165. Mendez MV, Stanley A, Phillips T, Murphy M, Menzoian JO, Park HY (1998). Fibroblasts cultured from distal lower extremities in patients with venous reflux display cellular characteristics of senescence. *J Vasc Surg.* 28: 1040–50.

166. Vande BJ, Rudolph R, Hollan C, Haywood-Reid PL (1998). Fibroblast senescence in pressure ulcers. *Wound Repair Regen.* 6: 38–49.

167. Agren MS, Steenfos HH, Dabelsteen S, Hansen JB, Dabelsteen E (1999). Proliferation and mitogenic response to PDGF-BB of fibroblasts isolated from chronic venous leg ulcers is ulcer-age dependent. *Trends Biotechnol.* 112: 463–9.

168. Vande BJ, Smith PD, Haywood-Reid PL, Munson AB, Soules KA, Robson MC (2001). Dynamic forces in the cell cycle affecting fibroblasts in pressure ulcers. *Wound Repair Regen.* 9: 19–27.

169. Herrick SE, Ireland GW, Simon D, McCollum CN, Ferguson MW (1996). Venous ulcer fibroblasts compared with normal fibroblasts show differences in collagen but not fibronectin production under both normal and hypoxic conditions. *Trends Biotechnol.* 106: 187–93.

170. Osanai M, Tamaki T, Yonekawa M, Kawamura A, Sawada N (2002). Transient increase in telomerase activity of proliferating fibroblasts and endothelial cells in granulation tissue of the human skin. *Wound Repair Regen.* 10: 59–66.

171. Van de Kerkhof PC, Van Bergen B, Spruijt K, Kuiper JP (1994). Age-related changes in wound healing. *Clin Exp Dermatol.* 19: 369–74.

172. Reed MJ, Corsa A, Pendergrass W, Penn P, Sage EH, Abrass IB (1998). Neovascularization in aged mice: delayed angiogenesis is coincident with decreased levels of transforming growth factor beta1 and type I collagen. *Am J Pathol.* 152: 113–23.

66 P. Stephens

173. Kramer RH, Fuh GM, Bensch KG, Karasek MA (1985). Synthesis of extracellular matrix glycoproteins by cultured microvascular endothelial cells isolated from the dermis of neonatal and adult skin. *J Cell Physiol.* 123: 1–9.

174. Gamble JR, Harlan JM, Klebanoff SJ, Vadas MA (1985). Stimulation of the adherence of neutrophils to umbilical vein endothelium by human recombinant tumor necrosis factor. *Proc Natl Acad Sci USA* 82: 8667–71.

175. Chang E, Harley CB (1995). Telomere length and replicative aging in human vascular tissues. *Proc Natl Acad Sci USA* 92: 11190–4.

176. Okuda K, Khan MY, Skurnick J, Kimura M, Aviv H, Aviv A (2000). Telomere attrition of the human abdominal aorta: relationships with age and atherosclerosis. *Atherosclerosis* 152: 391–8.

177. Herouy Y, Mellios P, Bandemir E, *et al.* (2000). Autologous platelet-derived wound healing factor promotes angiogenesis via alphavbeta3-integrin expression in chronic wounds. *Int J Mol Med.* 6: 515–19.

178. Pawelec G, Solana R, Remarque E, Mariani E (1998). Impact of aging on innate immunity. *J Leukoc Biol.* 64: 703–12.

179. Weng NP, Levine BL, June CH, Hodes RJ (1995). Human naive and memory T lymphocytes differ in telomeric length and replicative potential. *Proc Natl Acad Sci USA* 92: 11091–4.

180. Ho VCY (1999). Benign epithelial tumors. In: Freedberg IM, Eisen AZ, Wolff K, eds. *Fitzpatrick's Dermatology in General Medicine.* New York: McGraw-Hill, pp. 873–90.

181. Teraki E, Tajima S, Manaka I, Kawashima M, Miyagishi M, Imokawa G (1996). Role of endothelin-1 in hyperpigmentation in seborrhoeic keratosis. *Br J Dermatol.* 135: 918–23.

182. Imokawa G, Yada Y, Miyagishi M (1992). Endothelins secreted from human keratinocytes are intrinsic mitogens for human melanocytes. *J Biol Chem.* 267: 24675–80.

183. Lin AN, Carter DM, Balin AK (1989). Nonmelanoma skin cancers in the elderly. In: Gilchrest BA, eds. *Clinics in Geriatric Medicine.* Philadelphia: W.B. Saunders Co., pp. 161–70.

184. Morris BT, Sober AJ (1989). Cutaneous malignant melanoma in the older patient. In: Gilchrest BA, eds. *Clinics in Geriatric Medicine.* Philadelphia: W.B. Saunders Co., pp. 171–81.

185. Leffell DJ, Fitzgerald DA (1999). Basal cell carcinoma. In: Freedberg IM, Eisen AZ, Wolff K, eds. *Fitzpatrick's Dermatology in General Medicine.* New York: McGraw-Hill, pp. 857–64.

186. Brash DE, Rudolph JA, Simon JA, *et al.* (1991). A role for sunlight in skin cancer: UV-induced p53 mutations in squamous cell carcinoma. *Proc Natl Acad Sci USA* 88: 10124–8.

187. Ziegler A, Leffell DJ, Kunala S, *et al.* (1993). Mutation hotspots due to sunlight in the p53 gene of nonmelanoma skin cancers. *Proc Natl Acad Sci USA* 90: 4216–20.

188. Johnson RL, Rothman AL, Xie J, *et al.* (1996). Human homolog of patched, a candidate gene for the basal cell nevus syndrome. *Science* 272: 1668–71.

189. Xie J, Murone M, Luoh SM, *et al.* (1998). Activating smoothened mutations in sporadic basal-cell carcinoma. *Nature* 391: 90–2.

190. Dahmane N, Lee J, Robins P, Heller P, Ruiz (1997). Activation of the transcription factor Gli1 and the Sonic hedgehog signalling pathway in skin tumours. *Nature* 389: 876–81.

191. Herbst RA, Gutzmer R, Matiaske F, *et al.* (1997). Further evidence for ultraviolet light induction of CDKN2 (p16INK4) mutations in sporadic melanoma *in vivo. J Invest Dermatol.* 108: 950.

192. Frank SM, Raja SN, Bulcao C, Goldstein DS (2000). Age-related thermoregulatory differences during core cooling in humans. *Am J Physiol Regul Integr Comp Physiol.* 279: R349–54

193. Smolander J (2002). Effect of cold exposure on older humans. *Int J Sports Med.* 23: 86–92.

194. Scott PJ, Reid JL (1982). The effect of age on the responses of human isolated arteries to noradrenaline. *Br J Clin Pharmacol.* 13: 237–9.

195. Nielsen H, Hasenkam JM, Pilegaard HK, Aalkjaer C, Mortensen FV (1992). Age-dependent changes in alpha-adrenoceptor-mediated contractility of isolated human resistance arteries. *Am J Physiol.* 263: H1190–6

196. Docherty JR (1990). Cardiovascular responses in ageing: a review. *Pharmacol Rev.* 42: 103–25.

197. Inoue Y, Shibasaki M, Hirata K, Araki T (1998). Relationship between skin blood flow and sweating rate, and age related regional differences. *Eur J Appl Physiol Occup Physiol.* 79: 17–23.

198. Inoue Y, Shibasaki M, Ueda H, Ishizashi H (1999). Mechanisms underlying the age-related decrement in the human sweating response. *Eur J Appl Physiol Occup Physiol.* 79: 121–6.

199. Elias PM, Ghadially R (2002). The aged epidermal permeability barrier: basis for functional abnormalities. *Clin Geriatr Med.* 18: 103–20.

200. Ghadially R (1998). Aging and the epidermal permeability barrier: implications for contact dermatitis. *Am J Contact Dermat.* 9: 162–9.

201. Davies CE, Wilson MJ, Hill KE, *et al.* (2001). Use of molecular techniques to study microbial diversity in the skin: chronic wounds reevaluated. *Wound Repair Regen.* 9: 332–40.

202. Trautinger F (2001). Heat shock proteins in the photobiology of human skin. *J Photochem Photobiol.* 63: 70–7.

203. Volloch V, Rits S (1999). A natural extracellular factor that induces Hsp72, inhibits apoptosis, and restores stress resistance in aged human cells. *Exp Cell Res.* 253: 483–92.

204. Park KC, Kim DS, Choi HO, *et al.* (2000). Overexpression of HSP70 prevents ultraviolet B-induced apoptosis of a human melanoma cell line. *Arch Dermatol Res.* 292: 482–7.

205. Muramatsu T, Hatoko M, Tada H, Shirai T, Ohnishi T (1996). Age-related decrease in the inductability of heat shock protein 72 in normal human skin. *Br J Dermatol.* 134: 1035–8.

206. Gutsmann-Conrad A, Heydari AR, You S, Richardson A (1998). The expression of heat shock protein 70 decreases with cellular senescence *in vitro* and in cells derived from young and old human subjects. *Exp Cell Res.* 241: 404–13.

207. Finch CE (1990). *Longevity, Senescence and the Genome.* Chicago: University of Chicago Press.

208. Hayflick L (1965). The limited *in vitro* life time of human diploid cell strains. *Exp Cell Res.* 37: 614–36.

209. Griffith JD, Comeau L, Rosenfield S, *et al.* (1999). Mammalian telomeres end in a large duplex loop. *Cell.* 97: 503–14.

210. Olovnikov AM (1973). A theory of marginotomy. The incomplete copying of template margin in enzymic synthesis of polynucleotides and biological significance of the phenomenon. *J Theor Biol.* 41: 181–90.

211. Boukamp P (2001). Ageing mechanisms: the role of telomere loss. *Clin Exp Dermatol.* 26: 562–5.

212. Figueroa R, Lindenmaier H, Hergenhahn M, Nielsen KV, Boukamp P (2000). Telomere erosion varies during *in vitro* aging of normal human fibroblasts from young and adult donors. *Cancer Res.* 60: 2770–4.

213. Morin GB (1997). Telomere control of replicative lifespan. *Exp Gerontol.* 32: 375–82.

214. Cristofalo VJ, Allen RG, Pignolo RJ, Martin BG, Beck JC (1998). Relationship between donor age and the replicative lifespan of human cells in culture: a reevaluation. *Proc Natl Acad Sci USA* 95: 10614–19.

215. Tesco G, Vergelli M, Grassilli E, *et al.* (1998). Growth properties and growth factor responsiveness in skin fibroblasts from centenarians. *Biochem Biophys Res Commun.* 244: 912–16.

216. Wang E (1995). Senescent human fibroblasts resist programmed cell death, and failure to suppress bcl2 is involved. *Cancer Res.* 55: 2284–92.

217. Noda A, Ning Y, Venable SF, Pereira-Smith OM, Smith JR (1994). Cloning of senescent cell-derived inhibitors of DNA synthesis using an expression screen. *Exp Cell Res.* 211: 90–8.

218. Hara E, Smith R, Parry D, Tahara H, Stone S, Peters G (1996). Regulation of p16CDKN2 expression and its implications for cell immortalization and senescence. *Mol Cell Biol.* 16: 859–67.

219. Stein GH, Beeson M, Gordon L (1990). Failure to phosphorylate the retinoblastoma gene product in senescent human fibroblasts. *Science* 249: 666–9.

220. Shay JW, Pereira-Smith OM, Wright WE (1991). A role for both RB and p53 in the regulation of human cellular senescence. *Exp Cell Res.* 196: 33–9.

221. Harley CB (1991). Telomere loss: mitotic clock or genetic time bomb? *Mutat Res.* 256: 271–82.

222. Dhaene K, Van Marck E, Parwaresch R (2000). Telomeres, telomerase and cancer: an up-date. *Virchows Arch.* 437: 1–16.

223. Nugent CI, Lundblad V (1998). The telomerase reverse transcriptase: components and regulation. *Genes Dev.* 12: 1073–85.

224. Jiang XR, Jimenez G, Chang E, *et al.* (1999). Telomerase expression in human somatic cells does not induce changes associated with a transformed phenotype. *Nat Genet.* 21: 111–14.

225. Morales CP, Holt SE, Ouellette M, *et al.* (1999). Absence of cancer-associated changes in human fibroblasts immortalized with telomerase. *Nat Genet.* 21: 115–18.

226. Kiyono T, Foster SA, Koop JI, McDougall JK, Galloway DA, Klingelhutz AJ (1998). Both Rb/p16INK4a inactivation and telomerase activity are required to immortalize human epithelial cells. *Nature* 396: 84–8.

227. Halvorsen TL, Beattie GM, Lopez AD, Hayek A, Levine F (2000). Accelerated telomere shortening and senescence in human pancreatic islet cells stimulated to divide *in vitro*. *J Endocrinol.* 166: 103–9.

228. Jones CJ, Kipling D, Morris M, *et al.* (2000). Evidence for a telomere-independent "clock" limiting RAS oncogene-driven proliferation of human thyroid epithelial cells. *Mol Cell Biol.* 20: 5690–9.

229. Ramirez RD, Morales CP, Herbert BS, *et al.* (2001). Putative telomere-independent mechanisms of replicative aging reflect inadequate growth conditions. *Genes Dev.* 15: 398–403.

230. Harman D (1956). Aging a theory based on free radical and radiation chemistry. *J Gerontol.* 11: 298–300.

231. Sohal RS (2002). Role of oxidative stress and protein oxidation in the aging process. *Free Radic Biol Med.* 33: 37–44.

232. Kohen R, Gati I (2000). Skin low molecular weight antioxidants and their role in aging and in oxidative stress. *Toxicology* 148: 149–57.

233. Halliwell B, Gutteridge JM, Cross CE (1992). Free radicals, antioxidants, and human disease: where are we now? *J Lab Clin Med*. 119: 598–620.

234. Campisi J (1998). The role of cellular senescence in skin aging. *J Investig Dermatol Symp Proc*. 3: 1–5.

235. Uitto J, Fazio MJ, Olsen DR (1989). Molecular mechanisms of cutaneous aging. Age-associated connective tissue alterations in the dermis. *J Am Acad Dermatol*. 21: 614–22.

236. Rudolph KL, Chang S, Lee HW, *et al*. (1999). Longevity, stress response, and cancer in aging telomerase-deficient mice. *Cell* 96: 701–12.

237. Blasco MA, Lee HW, Hande MP, *et al*. (1997). Telomere shortening and tumor formation by mouse cells lacking telomerase RNA. *Cell* 91: 25–34.

238. Wicky C, Villeneuve AM, Lauper N, Codourey L, Tobler H, Muller F (1996). Telomeric repeats (TTAGGC)$_n$ are sufficient for chromosome capping function in *Caenorhabditis elegans*. *Proc Natl Acad Sci USA* 93: 8983–8.

239. Malik HS, Burke WD, Eickbush TH (2000). Putative telomerase catalytic subunits from Giardia lamblia and *Caenorhabditis elegans*. *Gene* 251: 101–8.

240. Mori A, Utsumi K, Liu J, Hosokawa M (1998). Oxidative damage in the senescence-accelerated mouse. *Ann NY Acad Sci*. 854: 239–50.

241. Kuro-o M, Matsumura Y, Aizawa H, *et al*. (1997). Mutation of the mouse klotho gene leads to a syndrome resembling ageing. *Nature* 390: 45–51.

242. Tyner SD, Venkatachalam S, Choi J, *et al*. (2002). p53 mutant mice that display early ageing-associated phenotypes. *Nature* 415: 45–53.

243. Migliaccio E, Giorgio M, Mele S, *et al*. (1999). The p66shc adaptor protein controls oxidative stress response and life span in mammals. *Nature* 402: 309–13.

244. Lithgow GJ, Andersen JK (2000). The real Dorian Gray mouse. *Bioessays* 22: 410–13.

245. Orr WC, Sohal RS (1994). Extension of life-span by overexpression of superoxide dismutase and catalase in *Drosophila melanogaster*. *Science* 263: 1128–30.

246. Melov S, Ravenscroft J, Malik S, *et al*. (2000). Extension of life-span with superoxide dismutase/catalase mimetics. *Science* 289: 1567–9.

247. Huang TT, Carlson EJ, Raineri I, Gillespie AM, Kozy H, Epstein CJ (1999). The use of transgenic and mutant mice to study oxygen free radical metabolism. *Ann NY Acad Sci*. 893: 95–112.

248. Ikegami T, Suzuki Y, Shimizu T, Isono K, Koseki H, Shirasawa T (2002). Model mice for tissue-specific deletion of the manganese superoxide dismutase (MnSOD) gene. *Biochem Biophys Res Commun*. 296: 729–36.

249. Bohr VA (2002). Human premature aging syndromes and genomic instability. *Mech Ageing Dev*. 123: 987–93.

250. Gray MD, Shen JC, Kamath-Loeb AS, *et al*. (1997). The Werner syndrome protein is a DNA helicase. *Nat Genet*. 17: 100–3.

251. Goto M (2001). *From Premature Gray Hair to Helicase – Werner Syndrome: Implications for Aging and Cancer*. Japanese Science Society.

252. Wyllie FS, Jones CJ, Skinner JW, *et al*. (2000). Telomerase prevents the accelerated cell ageing of Werner syndrome fibroblasts. *Nat Genet*. 24: 16–17.

253. Wartanowicz M, Panczenko-Kresowska B, Ziemlanski S, Kowalska M, Okolska G (1984). The effect of alpha-tocopherol and ascorbic acid on the serum lipid peroxide level in elderly people. *Ann Nutr Metab*. 28: 186–91.

254. Penn ND, Purkins L, Kelleher J, Heatley RV, Mascie-Taylor BH, Belfield PW (1991). The effect of dietary supplementation with vitamins A, C and E on cell-mediated

immune function in elderly long-stay patients: a randomized controlled trial. *Age Ageing* 20: 169–74.

255. Dreher F, Maibach H (2001). Protective effects of topical antioxidants in humans. *Curr Probl Dermatol.* 29: 157–64.

256. Dreher F, Gabard B, Schwindt DA, Maibach HI (1998). Topical melatonin in combination with vitamins E and C protects skin from ultraviolet-induced erythema: a human study *in vivo. Br J Dermatol.* 139: 332–9.

257. Fisher GJ, Talwar HS, Lin J, *et al.* (1998). Retinoic acid inhibits induction of c-Jun protein by ultraviolet radiation that occurs subsequent to activation of mitogen-activated protein kinase pathways in human skin *in vivo. J Clin Invest.* 101: 1432–40.

258. Griffiths CE, Russman AN, Majmudar G, Singer RS, Hamilton TA, Voorhees JJ (1993). Restoration of collagen formation in photodamaged human skin by tretinoin (retinoic acid). *N Engl J Med.* 329: 530–5.

259. Ellis CN, Weiss JS, Hamilton TA, Headington JT, Zelickson AS, Voorhees JJ (1990). Sustained improvement with prolonged topical tretinoin (retinoic acid) for photoaged skin. *J Am Acad Dermatol.* 23: 629–37.

260. Greenwald RA, Moak SA, Ramamurthy NS, Golub LM (1992). Tetracyclines suppress matrix metalloproteinase activity in adjuvant arthritis and in combination with flurbiprofen, ameliorate bone damage. *J Rheumatol.* 19: 927–38.

261. Ingman T, Sorsa T, Suomalainen K, *et al.* (1993). Tetracycline inhibition and the cellular source of collagenase in gingival crevicular fluid in different periodontal diseases. A review article. *J Periodontol.* 64: 82–8.

262. Oikarinen A (2000). Systemic estrogens have no conclusive beneficial effect on human skin connective tissue. *Acta Obstet Gynecol Scand.* 79: 250–4.

263. Callens A, Vaillant L, Lecomte P, Berson M, Gall Y, Lorette G (1996). Does hormonal skin aging exist? A study of the influence of different hormone therapy regimens on the skin of postmenopausal women using non-invasive measurement techniques. *Dermatology* 193: 289–94.

264. Dunn LB, Damesyn M, Moore AA, Reuben DB, Greendale GA (1997). Does estrogen prevent skin aging? Results from the First National Health and Nutrition Examination Survey (NHANES I). *Arch Dermatol.* 133: 339–42.

265. Denda M, Koyama J, Hori J, *et al.* (1993). Age- and sex-dependent change in stratum corneum sphingolipids. *Arch Dermatol Res.* 285: 415–17.

266. Reiser KM (1994). Influence of age and long-term dietary restriction on enzymatically mediated crosslinks and nonenzymatic glycation of collagen in mice. *J Gerontol.* 49: B71–9.

267. Sell DR, Lane MA, Johnson WA, *et al.* (1996). Longevity and the genetic determination of collagen glycoxidation kinetics in mammalian senescence. *Proc Natl Acad Sci USA* 93: 485–90.

268. Leung WC, Harvey I (2002). Is skin ageing in the elderly caused by sun exposure or smoking? *Br J Dermatol.* 147: 1187–91.

269. Brownlee M (1994). Lilly Lecture 1993. Glycation and diabetic complications. *Diabetes* 43: 836–41.

270. Takahashi M, Pischetsrieder M, Monnier VM (1997). Molecular cloning and expression of amadoriase isoenzyme (fructosyl amine:oxygen oxidoreductase, EC 1.5.3) from *Aspergillus fumigatus. J Biol Chem.* 272: 12505–7.

271. Schmidt AM, Vianna M, Gerlach M, *et al.* (1992). Isolation and characterization of two binding proteins for advanced glycosylation end products from bovine lung which are present on the endothelial cell surface. *J Biol Chem.* 267: 14987–97.

272. Schmidt AM, Hori O, Cao R, *et al.* (1996). RAGE: a novel cellular receptor for advanced glycation end products. *Diabetes* 45(Suppl 3): S77–80.

273. Wang J, Hannon GJ, Beach DH (2000). Risky immortalization by telomerase. *Nature* 405: 755–6.

274. Hahn WC, Stewart SA, Brooks MW, *et al.* (1999). Inhibition of telomerase limits the growth of human cancer cells. *Nat Med.* 5: 1164–70.

275. Ghersetich I, Lotti T (1994). α-Interferon cream restores decreased levels of Langerhans/indeterminate (CD1a+) cells in aged and PUVA-treated skin. *Skin Pharmacol.* 7: 118–20.

276. Klapper W, Parwaresch R, Krupp G (2001). Telomere biology in human aging and aging syndromes. *Mech Ageing Dev.* 122: 695–712.

Skeletal Muscle Aging

Caroline S. Broome, Aphrodite Vasilaki and Anne McArdle

Department of Medicine, University of Liverpool, Liverpool, L69 3GA, UK

Aging is usually defined as "the progressive loss of function accompanied by decreasing fertility and increasing mortality with advancing age" [1]. It is a complex physiologic process involving morphologic and biochemical changes in single cells and in the whole organism. However, the aging process is as yet poorly understood. During the last few years, the effect of aging on skeletal muscle has been increasingly studied.

Skeletal muscle structure and function

Skeletal muscle comprises 40–50% of the human body. The muscle bulk is composed of several kinds of tissue including muscle tissue, nerves, blood vessels and various types of connective tissue [2]. Skeletal muscles consist of long parallel bundles of multinucleated cells called muscle fibers, which are formed during development by the fusion of numerous precursor cells known as myoblasts. Each muscle fiber is, in turn, composed of parallel bundles of approximately one thousand myofibrils, which are bundles of contractile filaments made up from actin and myosin [3].

Fiber types of skeletal muscle
Skeletal muscle fibers have been grouped into two types according to their performance and biochemical characteristics of the individual muscle cells: type I (or slow-twitch) and type II (intermediate or fast-twitch).

Type I fibers (also called slow oxidative fibers) contain larger numbers of mitochondria and higher concentrations of the red pigment myoglobin. Therefore, muscles containing many type I fibers are also called red muscles due to their darker color. Red muscles contract relatively slowly, have a high capacity for aerobic metabolism and higher resistance to fatigue [2, 4].

Type II fibers are divided into several sub-groups including type IIA and type IIB. Type IIB fibers (also called fast-glycolytic fibers) contain smaller numbers of

R. Aspinall (ed.), Aging of Organs and Systems, 73–99.
© 2003 *Kluwer Academic Publishers. Printed in Great Britain.*

mitochondria, have a limited capacity for aerobic metabolism and less resistance to fatigue. However, type IIB fibers contain large amounts of glycolytic enzymes providing them with a high capacity for anaerobic metabolism [2]. Muscles containing many type IIB fibers are known as white muscles; these muscles have short twitch duration and are specialized for fine, rapid and precise movements [4].

Type IIA fibers (also called fast-oxidative fibers) have physiological properties that fall between type I and type IIB fibers. They are relatively fatigue resistant with intermediate levels of glycolytic activity. The properties of the three main fiber types in skeletal muscle are summarized in Table 1.

Table 1. Classification of fiber types in skeletal muscles (Modified from ref. 2)

Other names	Type I Slow oxidative; red	Type IIB Fast glycolytic; white	Type IIA Fast oxidative
Twitch speed	Slow	Fast	Fast
ATPase activity	Low	High	High
Glycolytic activity	Moderate	High	Moderate to high
Myoglobin content	High	Low	High
Resistance to fatigue	High	Low	Intermediate
Predominant energy system	Aerobic	Anaerobic	Combination

Types of contraction of skeletal muscle
When a single stimulus is given to a muscle, the muscle responds with a simple twitch. This twitch is divided into three phases; a latent period prior to the beginning of contraction, a contraction period and a relaxation period. The duration and the maximum force generated by the twitch varies with the type of the muscle. Thus, type II fibers contract in a shorter time period when stimulated than type I fibers. With rapidly repeated stimulation, individual responses fuse in one continuous contraction prior to any relaxation. This response is known as tetanus, during which maximum force is produced. Muscular contractions that occur during normal body movements are primarily tetanic contractions [4].

Activation of muscle movement involves three types of contraction:

a)   Shortening, concentric or dynamic contractions, where muscle allows shortening of the contractile elements during activation. Shortening contractions are common in most types of exercise.

b)   Isometric or static contractions, where muscle contracts without an appreciable change in the length of the whole muscle. Clenching of fists or tensing of muscles are common examples where muscles undergo isometric contractions, in which pairs of contracting muscles work to oppose each other and thus cancel out any movement [5].

c)   Lengthening, eccentric or pliometric contractions, where the muscle lengthens during activation. It has been documented that lengthening contractions cause considerably more damage to skeletal muscle than shortening or isometric contractions [6]. The main difference between lengthening contractions and shortening and isometric contractions is that more tension per fiber is generated during lengthening contractions, since the cross-sectional area of the fiber is reduced [7]. However, the mechanisms underlying lengthening contraction-induced damage are not clearly understood.

Age-related changes in size and number of muscle fibers
Aging of mammals leads to a decrease in the total skeletal muscle mass. By the age of 70, the cross-sectional area of skeletal muscle is reduced by 30–40%. This reduction in muscle mass (also known as muscle atrophy) is the main cause of the age-related decrease in muscle strength and power [8]. In addition, it has been shown that the reduction of muscle mass is accompanied by a replacement with connective tissue and fat [9].

In order to understand the causes of muscle atrophy during aging, numerous attempts have been made to examine the morphology of muscles from aging individuals. Studies examining muscle atrophy in humans are difficult to undertake. In addition to the heterogeneity of the population, muscle biopsy studies are complicated by the small sample size provided and post mortem studies are extensive and tedious. The earliest studies on aging and human muscle reported alterations in the proportions of type I and type II muscle fibers, which results in an increase of the percentage of type I fibers with aging [10, 11].

However, more recent studies contradict these early findings. Grimby and Saltin (1983) examined the fiber distribution in the vastus lateralis (VL) muscles of 66- to 100-years-old individuals and found no age-related changes in type I fiber distribution [12]. Sato *et al.* (1984) reported that the total volume of type I fibers in the minor pectoral muscle of women did not change with age [13]. In addition, Lexell *et al.* (1986) measured the fiber type distribution of the VL muscle and found that the type I distribution was 49% for 24-year old men and 52% and 51% for men in their 50s and late 70s respectively [14]. These findings suggest that type I fibers may be little affected by aging.

The age-related change in muscle function may be explained, at least in part, by a reduction in the size of type II muscle fibers with age [15]. In addition to a significant decrease in the total number of fibers with age, Lexell *et al.* (1988) described a greater loss of contractile material of fast twitch type than of slow twitch type [15].

Studies on animals have provided more definitive results. For example, Caccia *et al.* (1979) have reported an age-related decrease in the relative proportion of type II fibers in rat soleus muscles [16]. Furthermore, Holloszy *et al.* (1991) has demonstrated a 37% decrease in the average cross-sectional area of the type IIB fibers in plantaris muscle of aged rats [17].

Thus, the reduction in muscle mass during aging has been postulated to be due to a decrease in fiber number, fiber size, or both [18]. As a result, muscle mass of the elderly is smaller and is weaker due, in part, to the reduction of size or number of type II fibers [19].

Age-related changes in mechanical characteristics of skeletal muscle
Loss of muscle mass results in a loss of total force production [19]. Since muscle mass declines with aging, there is a concomitant loss of total force production. Young *et al.* (1984) reported that the isometric forces of the quadriceps femoris muscles from women demonstrated a 35% reduction with aging [20]. McDonagh *et al.* (1984), have also demonstrated that the maximal isometric force of the triceps surae muscles from older men demonstrated a decrease of 40% compared with the maximal isometric force of the triceps surae muscles from younger men [21]. Studies on whole skeletal muscles of small rodents have shown an age-related decrease in strength and power due to the loss in muscle mass [22]. The maximum force developed by both slow and fast muscles of old mice and rats is 20–30% less than the force developed by muscles of adult animals [23, 24].

Another interesting phenomenon of aging is that the ability of muscle to resist fatigue is reduced with aging. Studies on extensor digitorum longus (EDL) muscles of young, adult and old mice have shown that the muscles of old mice more fatigable than those of young mice. This deficit appears to develop relatively early since muscles of adult mice were also more fatigable than muscles of young mice [6].

Susceptibility of muscles of older individuals to damage
During lengthening contractions the force per unit of active fiber area is greater, leading to the possibility of muscle fiber injury [19]. A study by Zebra *et al.* (1990) involving lengthening contractions of skeletal muscles from mice has shown that the muscles of old mice are more susceptible to damage than those of young and adult mice [25]. Brooks and Faulkner (1996) have also demonstrated that the magnitude of the injury induced by stretches of muscle fibers from rats and mice is increased with age and suggest that the increased susceptibility of muscles from old rodents to contraction-induced injury resides in part within the myofibrils [26].

Recovery of muscles of older individuals from damage
The ability of muscle to repair damage efficiently is critical to its survival. Repair occurs by activation and differentiation of stem cells from within the muscle bulk. However, this ability of muscle to repair declines considerably with age [27, 28].

Brooks and Faulkner (1990) have demonstrated that following contraction-induced injury, muscles from young and adult mice had recovered fully by 28 days following a severely damaging protocol, whereas muscles from old mice had not recover completely by up to 60 days following the same protocol [27]. The pattern of injury and recovery in mouse muscles is similar to that reported from human muscles [29]. Interestingly, Carlson and Faulkner (1988) have demonstrated that the ability of muscle to regenerate successfully is dependant on the "environment" of that muscle. When muscle was transplanted from an old rat into a young rat, then the muscle regenerated at an equivalent rate to young muscle in contrast to the impaired regeneration in muscles of old rats [27]. This suggested that there was no inherent difference between the ability of the stem cells from muscle of old rats to regenerate and that the environment of the muscle plays an important role.

In summary, several studies have shown that aging results in a reduction in muscle mass due to a decrease in muscle fiber number and fiber size and that skeletal muscles of older individuals are also weaker, are more susceptible to contraction-induced damage and take longer to recover from damage.

Free radical production by skeletal muscle

The cardiovascular benefits of regular muscular exercise have been well documented. However, physical exercise is associated with a dramatic increase in oxygen intake. Most of the oxygen is utilized in the mitochondrial electron transport chain [30], where about 2–3% of oxygen may be converted into highly reactive intermediates called free radicals. Free radicals are defined as atoms or compounds capable of independent existence and having at least one unpaired electron in the outer orbital [31]. Free radical species released from skeletal muscle include reactive oxygen species (ROS) such as superoxide O_2^- as well as reactive nitrogen species (RNS) such as nitric oxide NO˙ (Table 2).

Adaptive responses in skeletal muscle

During oxidative stress, ROS can oxidize virtually all macromolecules, including proteins, nucleic acids, carbohydrates and lipids. However, cells have developed mechanisms to respond to the increased ROS generation as an adaptation to protect themselves against potential subsequent damaging results. This complex machinery involves the transcription of antioxidant defense enzymes, such as catalase and superoxide dismutase, as well as the production of a family of proteins known as stress or heat shock proteins (HSPs). The production of these proteins is termed the "stress response."

Antioxidant defenses of skeletal muscle
Cellular antioxidant defenses are conventionally classified into two categories; enzymatic defenses and non-enzymatic antioxidants. The first includes antioxidant defense enzymes such as superoxide dismutase (SOD), glutathione peroxidase (GPX) and catalase (CAT) that are capable of neutralizing reactive oxygen metabolites [40, 41]. The second includes low molecular weight molecules such as glutathione, vitamin C, vitamin E and β-carotene. Table 3 summarizes the cellular location and function of each of the antioxidant defense enzymes. Table 4 summarizes the cellular location and properties of important non-enzymatic antioxidants.

Exercise and the antioxidant response in skeletal muscle
Evidence suggests that a period of exercise results in a rapid increase in the production of ROS. Davies *et al.*, (1982) and Jackson *et al.* (1985) found an increase

Table 2. Examples of reactive oxygen/nitrogen species in skeletal muscle and their major sources [Modified from ref. 30a]

R-O/N-S	Primary source	General Information
Hydroxyl radical ($\cdot$OH)	O_2^- H_2O_2	Hydroxyl radicals have an extremely short half-life and are not thought to migrate any significant distance within a cell. They damage any molecule present in their immediate vicinity, including proteins, carbohydrates, DNA and lipids [31, 32].
Superoxide anion (O_2^-)	Electron transport chain; xanthine oxidase; NAD(P)H oxidase	A major cellular source of O_2^- is the mitochondria where it is produced as a by-product of the mitochondrial electron transport chain [33, 34]. In skeletal muscle, superoxide radicals are found in the extracellular fluid and are produced at a relatively low rate with a significant increase during contractile activity [35, 36].
Nitric oxide (NO$\cdot$)	Nitric oxide synthase	Nitric oxide is generated continuously in skeletal muscle. Resting muscles produce low levels of NO and the production of NO$\cdot$ is increased during contractile activity [37].
Peroxynitrite (ONOO-)	O_2^- + NO$\cdot$	Nitric oxide can also react with O_2^- to form peroxynitrite (ONOO$^-$), a free radical species that is far more reactive than either of the parent radicals [38].
Hydrogen peroxide (H_2O_2)	O_2^-	Hydrogen peroxide is highly diffusible between and within cells and it is a weak oxidizing agent, less reactive than the superoxide radical [35]. H_2O_2 can also lead to $\cdot$OH generation, thus, despite its relatively weak reactivity, hydrogen peroxide is cytotoxic [39].

Table 3. Action and cellular location of antioxidant enzymes

Enzymes	Cellular locations	Antioxidant properties	General Information
Mn-SOD	Mitochondria	Dismutates superoxide into O_2 and H_2O_2	Mn-SOD accounts for 15–35% total SOD and Cu/Zn-SOD accounts for 65–85% total SOD in skeletal muscle Both Mn-SOD and Cu/Zn-SOD catalyse the dismutation of superoxide anion radicals with similar efficiency [42].
Cu/Zn-SOD	Cytosol	Dismutates superoxide into O_2 and H_2O_2	Skeletal muscles with high percentage of type I fibers have a greater SOD activity than muscles with high percentage of type IIB fibers [43].
Ec-SOD	Extracellular fluid	Dismutates superoxide into O_2 and H_2O_2	Removes O_2^- generated outside the cell membran
CAT	Mitochondria and cytosol	Dismutates H_2O_2 into O_2 and H_2O	Catalase is widely distributed in the cell, however the highest concentrations are found in both peroxisomes and mitochondria. In skeletal muscle, type I fibers display the highest catalase activity, followed by type IIA; type IIB muscle fibers have the lowest catalase activity [44].
GPX	Mitochondria and cytosol	Removes H_2O_2 and organic hydroperoxides	In skeletal muscles with high oxidative capacity (i.e., high percentage of type I fibers) have greater GPX activity.

Table 4. Properties and cellular location of non-enzymatic antioxidants

Non-enzymatic antioxidant	Cellular locations	Properties	General Information
GSH	Mitochondria and cytosol	Interacts with a variety of radicals by donating H	Glutathione plays a multifunctional role in protecting tissues from oxidative damage and keeping the intracellular environment in the reduced state. Muscles with high percentage of type I fibers contain six-fold higher glutathione content than muscles with high percentage of type IIB fibers. However, the GSH:GSSG ratio appears remarkably consistent across various fiber types [41].
Vitamin C	Cytosol	Quenches a wide variety of aqueous-phase ROS	Vitamin C can interact directly with O_2^- and $\cdot OH$ as well as lipid hydroperoxide radicals [40].
Vitamin E	Membranes	Major lipid peroxidation chain-breaking antioxidant	Vitamin E can scavenge O_2^- and $\cdot OH$ [40]. Interaction of vitamin E with ROS results in the formation of a vitamin E radical, which can be reduced by several other antioxidants including vitamin C.
Carotenoids	Primarily in membranes	Reduce lipid peroxidation	β-carotene, a major carotenoid precursor of vitamin A, is capable of combining with several forms of ROS, such as the singlet oxygen in order to form less active radicals [45]. β-carotene can also inhibit lipid peroxidation initiated by free radicals [44].

in the electron paramagnetic resonance (EPR) signals in contracting rat muscles compared with resting muscles [33, 46]. Reid *et al.* (1992) have reported that isolated strips of rat diaphragm released superoxide anion radicals into the external medium during isometric contraction [47]. In an *in situ* model, O'Neill *et al.* (1996) demonstrated that isometric contracting cat skeletal muscle generated ˙OH radicals that were detectable in the muscle microvasculature [48]. More recently, we have shown that fifteen minutes of isometric contractile activity induced a rapid release of superoxide anion radicals from mouse skeletal muscle *in vivo* and studies using contracting cultured primary skeletal muscle myotubes confirmed that this release was from muscle cells rather than other cell types present within the muscle [36].

Many authors have suggested that this increased production of free radicals is responsible for exercise-induced muscle damage. However, a comparison of the patterns of oxidant generation and muscle damage argues against this; the oxygen flux through the mitochondria is considerably greater during shortening and isometric contractions, whereas by far the most damaging form of contractile activity is by lengthening contractions [49] during which, oxygen flux through the mitochondria is relatively low.

The reason why these sudden changes in ROS production do not result in skeletal muscle damage may be due to the fact that skeletal muscle cells have developed a complex antioxidant defense mechanism to provide protection against fluctuations in the production of ROS at resting conditions and during increased activity.

Several studies have shown that an acute bout of exercise can increase the activities of different antioxidant defense enzymes. Khassaf *et al.* (2001) found that the SOD activity in human vastus lateralis muscles was increased following a single bout of non-damaging maximal isometric exercise [50]. Ji *et al.* (1992) found that an acute bout of exercise significantly increased the activities of catalase and GPX in the deep vastus lateralis (DVL) muscles from rats [51]. McArdle *et al.* (2001) have shown that a fifteen-minute period of aerobic contractile activity leads to an increase in the activities of catalase and SOD in muscles of mice [36].

Longer term exercise training also appears to result in an increase the activity of several antioxidant defense enzymes. Leeuwenburgh *et al.* (1994) found that a 10-week training program resulted in increased activities of GPX and SOD in DVL muscles of rats [52]. In another study Leeuwenburgh *et al.* (1997) have shown that GSH content was increased by 33% in DVL muscles from rats following training and that trained rats showed a 62% and 27% higher GPX and SOD activity respectively compared with muscles of non-trained rats [53]. In addition, Powers *et al.* (1994) showed an increase in GPX activity in red gastrocnemius muscles from rats following a 10-week training program. In the same study, the authors demonstrated an increase in SOD activity in rat soleus and red gastrocnemius muscles following training, whereas SOD activity in the white gastrocnemius muscles was not significantly altered [43]. This data suggests that training-induced changes in muscle antioxidant defense enzymes are fiber type specific, but also muscles composed of highly oxidative fibers such as soleus (primarily type I fibers) and red gastrocnemius (primarily type IIA fibers) appear to be more responsive to oxidative stress.

Thus, it has been proposed that the production of ROS during non-damaging exercise acts as a signal to muscle cells to adapt to provide protection against further and possibly damaging insults.

Reactive oxygen species and aging
It has been hypothesized that a main cause of the aging process and the development of chronic disease in older people may be the cumulative damage to lipids in cell membranes, DNA, and sub-cellular membranes and structures, by reactive forms of oxygen. This free radical theory of aging was first described by Harman in 1956 and is currently one of the most popular explanations for how aging occurs at the biological level [54].

It has been observed that free radical levels increase with age in species such as the housefly, rats and humans [55]. In skeletal muscle, studies on animals have provided evidence of an age related increased production of ROS. For example, Lass *et al.* (1998) have demonstrated that the rate of superoxide radical generation in skeletal muscle increases with age [56]. This increase in free radical production may lead to changes in the redox state of muscle cells with serious effects on muscle. Proteins with a high percentage of sulphydryl groups such as myosin and creating kinase can be oxidized and therefore they will not function efficiently. In addition, transcription factors containing redox-sensitive sites will be particularly susceptible to damage by the aberrant production of ROS.

Numerous studies have shown that aging cells accumulate increased levels of oxidant-damaged mitochondrial DNA. Increasing damage to mitochondrial DNA inevitably leads to abnormal mitochondrial function and integrity. Damaged mitochondria are thought to release more ROS that in turn leads to more DNA damage. Large mitochondrial deletions have been found to be increased as much as 10 000 fold with age in several human tissues including skeletal muscle [57]. In addition, oligonucleotide array analysis of resting skeletal muscle of aged mice has demonstrated an age-related decrease in the expression of genes involved in oxidative phosphorylation due to mitochondrial dysfunction in aged animals [58].

Thus, a pathological increase in production of free radicals would result in an accumulation of oxidation products and an accumulation of abnormal mitochondria. In return, dysfunctional mitochondria produce significant amounts of free radicals that may play a major role in the physiological and structural changes seen in skeletal muscle of aged mammals. Transgenic approaches with genes such as superoxide dismutase and catalase involved in detoxifying superoxide and hydrogen peroxide in the mitochondria and cytosol and pharmacological approaches using mimetics of these antioxidant enzymes have shown that the endogenous production of ROS by normal physiological processes is a major limit to lifespan [59].

Antioxidant defense enzymes and aging in resting skeletal muscle
This area of research is somewhat confusing. However, the consensus now seems to be that there is evidence of an attempt, particularly in skeletal muscle, to adapt to an age-related chronic increase in the production of ROS by adaptation of antioxidant protective enzymes. Ji *et al.* (1990) demonstrated an increase in the specific activities

of both mitochondrial and cytosolic SOD in rat skeletal muscle during aging [60]. CAT and GPX activities were also significantly higher in DVL muscle of aged versus young rats. Leeuwenburgh *et al.* (1994) found similar increases in the activities of all antioxidant protective enzymes with aging in DVL muscles in rats [52].

However, the response of antioxidants to aging appears to be highly tissue specific. Leeuwenburgh *et al.* (1994) have shown a 37% increase in GSH content in soleus muscles of aged rats, however DVL muscles from the same rats showed no significant alterations in GSH content with aging [52]. Lawler *et al.* (1993) reported a significant increase in GPX but not SOD activity in the soleus and gastrocnemius muscles of aged rats [61]. In addition, Oh-Ishi *et al.* (1995) reported an age-related increase in the activity and content of Cu/Zn SOD in rat soleus and extensor digitorum longus (EDL) muscles, whereas those of Mn SOD showed no difference between young and aged rats. Moreover, GPX activity indicated age-related increases only in soleus muscles with no significant differences in EDL muscles [62].

These observations indicate that age-related adaptation of cellular antioxidant defenses is evident but seems to be muscle fiber specific, with the most prominent increases found in type I muscles such as soleus, followed by type IIA muscles such as DVL, whereas type IIB muscle fibers show little effect [52, 62]. In addition, the fact that aged muscles already have higher antioxidant enzyme activities and GSH levels may also influence their response to exercise.

Influence of exercise on ROS production and antioxidant protective enzyme activity in tissues of aged mammals
Skeletal muscles of young/adult mammals rapidly adapt to acute bouts of exercise by increasing their antioxidant defenses. In contrast, muscles of aged mammals appear to be unable to adapt in this way. Ji *et al.* (1990), found no significant alterations in most antioxidant defense enzymes in the skeletal muscles from aged male rats following exercise [60]. Furthermore, Lawler *et al.* (1993) found no increase in the SOD and GPX activities in the gastrocnemius and soleus muscles from aged female rats after 40 minutes treadmill running [61].

Studies have shown that endurance training has some effect in the increase of some antioxidant defense enzyme activities in skeletal muscle of aged mammals. For example, Ji *et al.* (1991) reported an increase in the GPX activity in skeletal muscles from aged rats following a training regime, whereas CAT and SOD activities remained unchanged [63]. Leeuwenburgh *et al.* (1994) showed that training had little effect on the antioxidant defense enzyme activity in both DVL and soleus muscles from aged male rats [52].

This inability of muscle from aged mammals to adapt following exercise may play an important role in the development of age-related muscle weakness, increased susceptibility to damage, and poor recovery from damage.

The stress response in skeletal muscle

The second cytoprotective response in skeletal muscle and other tissues is the increased production of a family of proteins known as stress or heat shock proteins (HSPs).

Heat shock proteins (HSPs) are so called because their induction was first observed in response to hyperthermia. The cellular content of HSPs is increased following a variety of stresses including oxidative stress, viral infection, changes in pH or the incorporation of amino acid analogues into proteins. The induction of HSPs following stress has been termed as "the cellular stress response" [64].

Some HSPs are constitutively expressed in cells, which indicates that HSPs fulfil important tasks in cells under normal conditions. In an unstressed cell, HSPs act as molecular chaperones, necessary for facilitating protein folding, for blocking non-productive protein-protein interactions, for safely transporting newly synthesized proteins to their correct site of action and prevention of mislocation or aggregation [65] (Figure 1). As a result, HSPs play fundamental roles in maintaining cellular homeostasis.

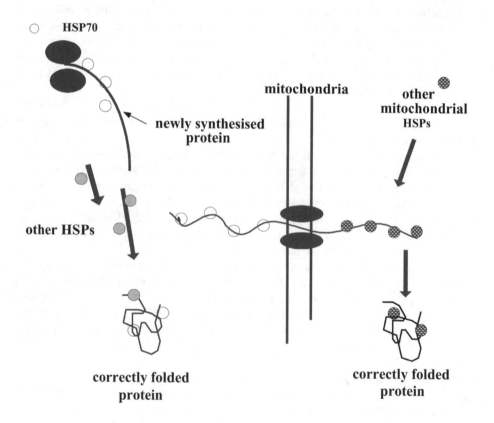

Figure 1. A schematic representation of the role of HSPs in the unstressed cell (reproduced from ref. 66).

The role of HSPs in the unstressed cell
HSPs are usually named according to their molecular weight and include the small HSPs such as HSP10 and HSP25, mitochondrial HSP60, the HSP70 family (which consists of the constitutively expressed HSC70 and highly inducible HSP70), and the larger HSPs such as HSP90 and HSP100 [67] (Table 5). Many other stress proteins have been identified in mammalian cells however, information regarding their role and expression in most tissues is limited. Since the greater number of exercise-related studies has focused on the expression of certain HSPs such as HSP25, HSP60, HSP70 and HSP90, these are the HSPs that will be discussed in this table.

The role of HSPs in the stressed cell
Several studies have shown that the cellular content of HSPs is increased following a variety of stresses such as incorporation of non-native amino acid analogues [72], alterations in pH [73], hypoxia [74], as well as oxidative stress [64, 71].

Early studies in the heart demonstrated that exposure of cardiac muscle to stress results in the increased production of HSPs. Salo *et al.* (1991) demonstrated that acute exhaustive exercise resulted in an increase in the HSP70 mRNA levels in cardiac muscle of rats [75]. In addition, Locke *et al.* (1995) reported an increase in the HSP70 mRNA levels after heat shock and exercise in cardiac muscle from rats [76]. Most importantly, recent studies have demonstrated that this increase in the HSP production is associated with myocardial protection. The development of transgenic mouse models overexpressing HSPs has demonstrated a direct protective effect by HSPs. For example, Marber *et al.* (1995) have reported that overexpression of HSP70 in transgenic mice increases the resistance of the heart to ischemic injury [77].

The ability of prior induction of the stress response to protect cells from damage has also been examined by several workers. Studies have shown that an initial mild stress, sufficient enough to induce HSPs, can provide protection against subsequent more severe stress (Figure 2).

Garramone *et al.* (1994) and Lepore *et al.* (2000) have demonstrated that a prior heat stress in rats, which resulted in increased muscle content of HSP70, provided protection to skeletal muscle against necrosis induced by ischemia-reperfusion [78, 79]. Furthermore, Suzuki *et al.* (2000) have shown that prior heat shock of muscle cells provides considerable protection against cell death following hypoxia and re-oxygenation *in vitro* [80]. It is thought that an increased cellular content of HSPs protects the cell by associating with misfolded cellular proteins during stress and facilitating the refolding of these proteins when conditions become more favorable. Thus, in this case, the induced HSPs act in a manner analogous to their molecular chaperone function (Figure 2).

Expression of HSPs in skeletal muscle following exercise
The possibility that HSPs provide cytoprotection to skeletal muscle following oxidative stress is receiving increasing attention. Salo *et al.* (1991) showed that acute exhaustive exercise results in an increased HSP expression in skeletal muscles of rats [75]. Neufer *et al.* (1996) demonstrated that chronic stimulation of rabbit anterior tibialis muscles results in the increase in HSP70 and HSP60 expression, demon-

Table 5. Size, name, function and cellular location of HSPs (Modified from ref. 68]

Size, kDa	Name	Cellular localization	Major function
Small HSPs (<35)	Hsp25/27 αβ crystallin	Cytosol and nucleus	Although the exact function of HSP25 remains unknown, it has been shown to be involved in signal transduction, differentiation and growth [64]. HSP25 has significant sequence homology with αβ crystallin (22 kDa), a member of the small HSPs [69]. αβ crystallin is a major structural protein in the lens of the eye, but is also expressed in other tissues, especially tissues with high oxidative capacity, including heart and type I skeletal muscle fibers [70]. It has a role in stabilization of microfilaments.
60	HSP60	Primarily mitochondria (with HSP10)	HSP60 is expressed constitutively under normal conditions and plays an important role in maintaining normal cell function by mediating the correct folding of polypeptides that enter the mitochondria and facilitating protein transport across intracellular membranes. HSP60 has been termed a molecular chaperone or chaperonin [64]. Under stress conditions, HSP60 is involved in the stabilization of pre-existing proteins.
70–73	HSP72 or HSP70i HSP73 or HSC70	Cytosol, nucleus, ER and mitochondria	The two most extensively studied proteins in the HSP70 family are the cognate isoform HSC70 and the highly inducible isoform HSP [70]. The two proteins have extremely high sequence homology (~095%) although they are encoded by different genes and they seem to have similar biological properties; they have been shown to act as molecular chaperones and are involved in transport, folding, protein synthesis, disassembly, prevention of protein aggregation and restoring the function of damaged proteins following stress [64, 71].
90	HSP90 HSP90α HSP90β GRP94	Cytosol, nucleus, ER	HSP90 is one of the most abundant cellular proteins found in both stressed and unstressed cells and appears to be involved in the general folding of various proteins. HSP90 is frequently found in complexes with other chaperones such as HSP70.

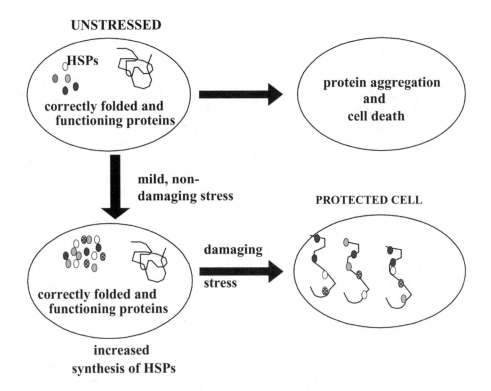

Figure 2. *Cytoprotective function of HSPs (reproduced from ref. 65).*

strated by changes in protein levels, mRNA and transcription rate [81]. McArdle *et al.* (2001) demonstrated that a 15-minute period of mild, non-damaging isometric contractions resulted in increased content of HSPs in soleus and EDL muscles from mice, whereby the increase in the soleus muscles appeared greater than that of the EDL muscles [36]. A study by Khassaf *et al.* (2001) in humans has shown that a single period of exhaustive, non-damaging aerobic exercise resulted in a significant increase in the HSP70 and HSP60 contents of vastus lateralis muscles at 3 to 6 days following the exercise protocol [50]. Interestingly, the authors observed a rapid increase in the HSP70 content in muscles with relatively low resting HSP70 content, whereas muscles with relatively high resting content of HSP70 did not respond to the same extent to the exercise protocol.

A number of workers have also examined the effect of endurance training on HSP content of muscle. Ornatsky *et al.* (1995) demonstrated that chronic contraction at a frequency of 10 Hz for 10 days resulted in an increase in the GRP75, HSP70 and HSP60 content of anterior tibialis muscles from rats [82]. Samelman (1999) showed an increase in the HSP70/HSC70 and HSP60 expression in soleus muscles from rats that were trained for 16–20 weeks on a motorised treadmill [83]. In addition, Gonzalez *et al.* (2000) demonstrated that treadmill running for 3 months increased the HSP70, GRP75 and GRP78 content in rat soleus muscles [84].

The HSP production in skeletal muscle appears to be fiber type specific since data from several studies suggest that, in general, the levels of HSPs in resting skeletal muscles composed predominantly of type I fibers are higher than that of muscles composed of type II fibers. Furthermore, the production of some HSPs appears to be greater following exercise in the soleus muscles (type I) compared with the production in the EDL muscles (type II) [36, 85].

Evidence suggests that this increased content of HSPs provides protection to skeletal muscle. McArdle *et al.* (1995) have demonstrated that a short period of non-damaging isometric exercise in mice leads to a rapid increase in HSP content and subsequent protection of skeletal muscle against a period of normally damaging contractile activity [86, 87]. However, the direct evidence for a role of HSP70 in providing protection against exercise-induced damage comes from studies using transgenic mice. McArdle *et al.* (2002b) have demonstrated that overexpression of HSP70 in transgenic mice provided protection against damage and resulted in an increase in the rate of recovery in EDL muscles following damaging lengthening contractions compared with muscles from wild-type mice [88].

HSPs, skeletal muscle and aging
A growing body of literature indicates that aging is associated with a reduced ability to express HSPs in several tissues following stress. In 1990, Fargnoli *et al.* reported an age-related decline in the expression of HSP70 in lung and skin fibroblasts of aged rats following thermal stress [89]. Wu *et al.* (1993) demonstrated that aging resulted in a decrease in the ability of hepatocytes from rats to synthesise HSP70 in response to hyperthermia [90]. In addition, Locke and Tanguay (1996a) showed that following heat stress, hearts from aged rats demonstrated a reduction in HSP70 mRNA and a reduction in HSP70 protein content, compared with hearts from adult rats [91].

Vasilaki *et al.* (2002) demonstrated that the production of HSP70 in response to a period of mild, non-damaging contractile activity was severely blunted in the gastrocnemius muscles from aged rats at 24 hours following the contraction protocol [92]. However, a study by Locke (2000) demonstrated no significant differences in the HSP70 content in both soleus and gastrocnemius muscles between adult and aged rats following whole-body hyperthermia [93]. In this case, it appears that the nature of the stress may be important and that the mechanisms responsible for activation of the stress response following hyperthermia or exercise may not be entirely comparable and may be differentially affected by age.

In response to training, the ability of muscles of aged rodents to adapt appears to be fiber type specific. Naito *et al.* (2001) have demonstrated that exercise training of rats on a treadmill for ten weeks resulted in a similar exercise-induced accumulation of HSP70 in highly oxidative skeletal muscles (such as soleus) of young and aged animals whereas, in fast skeletal muscles (such as white gastrocnemius), HSP70 expression remained blunted with aging [94].

These observations indicate that the blunted HSP expression in skeletal muscles of aged individuals seems to be muscle fiber specific as well as stress specific.

Transcription factors and activation of the adaptive response

Cells can respond to acute or chronic increases in RO/NS production by increased production of antioxidant defense enzymes and HSPs. In recent years, much attention has focused on the cellular mechanisms that initiate the adaptive response. In the vast majority of cases when a protein is produced in a tissue or in response to a signal this is achieved by control processes which ensure that the corresponding gene is transcribed only in that specific tissue or in response to the specific signal. In mammalian cells, the induction of antioxidant enzymes and HSPs is usually associated with the activation of transcription factors.

Transcription factors are proteins that control gene expression and are required to initiate or regulate transcription in eukaryotes. Generally, a stimulus such as an increase in the free radical production will trigger a signal transduction pathway and activate transcription factors that are then able to bind to the promoter region of target genes and initiate transcription. Furthermore, the rate of transcription initiation can be increased or decreased in response to this particular stimulus.

Cells contain several redox-sensitive transcription factors. Heat shock factor 1 (HSF1), nuclear factor kappa B (NF-κB) and activator protein-1 (AP-1) are three of the most well characterized transcription factors that play important roles in the response of cells to stress. These transcription factors will be discussed in more detail in the following sections.

NF-κB
Nuclear factor kappa B (NF-κB) is activated in cells by a great variety of stimuli that represent internal or external stress situations including pathological stimuli such as bacteria, viruses, hypoxia and inflammatory mediators as well as internal cellular stress such as endoplasmic reticulum overload and oxidative stress [95].

In higher eukaryotes, NF-κB is a multisubunit transcription factor. The currently known DNA-binding subunit members of the NF-κB family are p50, p52, p65 (RelA), c-Rel and Rel-B [96]. These subunits can homo- and heterodimerize in various combinations. The classical, and predominant form of NF-κB is a heterodimer consisting of the p50 and p65 (RelA) subunits [95]. The genes activated by NF-κB are diverse and include many used by the immune system. However NF-κB is also in part responsible for the induction of antioxidant defense enzymes such as Mn SOD [97].

A substantial body of experimental data links NF-κB activity to cellular oxidative status. It has been shown that NF-κB can be activated by H_2O_2 in some cell lines in the absence of any physiological stimulus and that NF-κB binding to DNA can be inhibited by a variety antioxidants [98]. In addition, agents that can activate NF-κB tend to trigger the formation of ROS or are oxidants themselves, such as the superoxide anion radical [98]. These observations have led to a general agreement that NF-κB activation is at least facilitated by some oxidative reactions.

AP-1

Like NF-κB, activator protein 1 (AP-1) is a well-characterized and ubiquitously expressed transcription factor capable of inducing expression of a large number of genes. AP-1 regulates the expression of genes associated with the control of proliferation, growth, differentiation of cells as well as stress-inducible genes such as the Mn SOD gene [95, 99].

AP-1 is a dimer composed of two DNA-binding subunits, c-Jun and c-Fos, which belong to the Jun and Fos proto-oncogene families. AP-1 can be either a Jun-Jun homodimer or a Fos-Jun heterodimer [99]. The classical form of AP-1 is the Fos-Jun heterodimer.

Oxidative stress induces AP-1 mediated transcription by increasing subunit synthesis and enhancing DNA binding activity [97]. Like NF-κB, the response of AP-1 to oxidative stress is cell type specific [97]. It has been shown that low concentrations of hydrophilic antioxidants can inhibit the redox-sensitive activation of AP-1 however, when the levels of intracellular antioxidants are increased above a certain level, they can act as a stimulus for AP-1 activation [66].

The HSF family of transcription factors

The inducible HSP expression in eukaryotic cells is regulated by the heat shock transcription factors (HSFs). In response to various inducers such as elevated temperatures, viral infections and oxidative stress, most HSFs acquire DNA binding activity thereby mediating transcription of the heat shock genes, which result in accumulation of HSPs [100].

Several different HSFs have been isolated from various species; a single HSF from yeasts and Drosophila, two (HSF1 and HSF2) from mouse, three (HSF1, HSF2, HSF3) from chicken and three (HSF1, HSF2 and HSF4) in human cells [101]. The role and the function of these HSFs are still incomplete; however, it appears that different HSFs respond to different stimuli in a different ways.

Both HSF1 and HSF2 act through a highly conserved upstream response element (heat shock element, HSE), which is located within the HSP gene promoters [102]. HSF1 undergoes rapid activation in response to a multitude of stress conditions, whereas HSF2 appears to be an important control mechanism during developmental and cellular differentiation processes [100]. In response to stress, HSF1 activation involves translocation to the nucleus and hyperphosphorylation on serine residues, whereas HSF2 appears not to undergo any changes in the phosphorylation upon activation [102].

HSF3 is expressed ubiquitously in avian cells. Like HSF1, HSF3 is a stress responsive transcription factor and has many characteristics similar to those of HSF1 [103]. It has been shown that the amount of HSF3 increases after severe heat shock, whereas HSF1 levels diminish suggesting that HSF3 has a role during severe and persistent stress in avian cells. The fact that no HSF3 homologue has been found in other than avian cells raises the possibility that the mechanisms responsible for activation of the stress response may be organism specific [100].

The most recently discovered mammalian HSF is HSF4. This is found in human cells and appears to be preferentially expressed in the human brain, heart, skeletal muscle and pancreas. HSF4 binds to DNA constitutively but lacks the properties of a transcriptional activator [104].

Activation of the stress response
The mechanism by which the stress results in an increased cellular content of HSPs is relatively well understood. In the unstressed cells HSF1 is present in the cytoplasm as a monomer, where it is thought to be associated with HSP70 or 90. Within seconds of the initiating stress, HSF1 is activated [105]. Activation of HSF1 is linked to the appearance of non-native proteins and the requirement for molecular chaperones (such as HSP70 and HSP90) to prevent the appearance of misfolded proteins. During stress, HSP70 and HSP90 have a greater binding affinity for destabilized proteins, releasing HSF1 and allowing it to migrate to the nucleus. HSF1 monomers oligomerise to a trimeric state, bind to the heat shock element (HSE) of the heat shock protein gene promoters and become inducibly hyperphosphorylated at serine residues. This results in increased transcription of the heat shock genes and subsequently, an increase in cellular content of HSPs. As the synthesis of HSPs increases, HSP70 and other molecular chaperones translocate to the nucleus where they are free to re-associate with HSF1. This negative feedback results in the inactivity of HSF1 [101, 104].

Transcription factors, oxidative stress and skeletal muscle
The activation of the heat shock genes in response to environmental stresses has been studied extensively in several tissues. A numbers of studies have reported that ROS play an important role in HSF1 activation as well as HSP gene expression. For example, Locke *et al.* (1995) have reported that a single bout of treadmill running is a sufficient stimulus to activate HSF1 and cause the subsequent accumulation of HSP70 mRNA in rat heart [106]. In addition, Nishizawa *et al.* (1999) demonstrated a significant activation of HSF1 and appearance of HSP70 and HSP90 mRNA with both ischemia-reperfusion and with H_2O_2 perfusion in isolated hearts [107]. It has been suggested that one pathway through which ROS activate HSF1 is via oxidation of protein thiols. Work by McDuffee *et al.* (1997) showed that the oxidative stress generated by menadione (a compound that redox-cycles to generate superoxide) results in the formation of non-native disulfides followed by protein destabilization, activation of HSF1, accumulation of HSP70 mRNA and increased synthesis of HSP70 in human hepatoma and chinese hamster lung cells [108].

In skeletal muscle, McArdle *et al.* (2001) have demonstrated that an increased muscle content of HSPs following a short protocol of isometric contractions is associated with the generation of superoxide anion radicals during the contractions and a subsequent, transient, non-damaging fall in the thiol content of cellular proteins [36]. Interestingly, there was no effect of the contraction protocol on muscle temperature, thus hyperthermia was not the main signal for activation of HSF1. Although the study did not investigate the activation of HSF1 following exercise, the data suggests an involvement of mild oxidative stress in the activation of HSF1, which results in the increased content of HSPs is the exercised muscles. In addition, activation of HSF1 in skeletal muscle has been shown following thermal stress [109]. However the effect of exercise on the HSF-1 activation remains uncharacterized.

The possibility that either NF-κB or AP-1 might be involved in the regulation of antioxidant defense enzyme induction in response to oxidative stress was based on the presence of NF-κB and AP-1 response elements in the promoter regions of genes encoding a number antioxidant defense enzymes [97]. Both NF-κB and AP-1 binding sites are present in the promoter of the mammalian Mn SOD gene and oxidative stress has been shown to activate their binding [110]. In addition, Zhou et al. (2001) has demonstrated that both NF-κB and AP-1 are important mediators of redox-responsible gene expression in skeletal muscle and that NF-κB is actively involved in the upregulation of glutathione peroxidase and catalase in response to oxidative stress [97]. These findings, together with the increasing evidence that NF-κB and AP-1 are redox responsive, suggest that combinations of these two transcription factors and possibly other redox-sensitive transcriptional regulators may determine which antioxidant defense enzymes are induced and to what extent, depending upon the tissue specific regulation.

Transcription factor activation and aging
The possibility that age-dependent changes in the cellular redox status due to the increased production of reactive oxygen intermediates and accumulation of oxidatively modified proteins can affect the function and regulation of the transcription factors has received increased attention.

There is now considerable evidence that aging affects transcription factor activation in several tissues. For example, Fawcett et al. (1994) demonstrated that while the levels of HSF1 remained constant in the adrenals of aged rats, the HSF1 exhibited a decreased ability to bind DNA following neurohormonal stress [111]. Liu et al. (1996) also reported a decrease in HSF1 DNA binding activity in aged human diploid lung fibroblasts following heat shock due to the reduced ability of heat shock to promote trimerization of HSF1 [112]. Furthermore, Locke and Tanguay (1996) demonstrated a 47% reduction in HSF1 activation in hearts from aged rats compared to hearts from young rats following thermal stress while myocardial HSF1 protein content was similar between the two experimental groups [109]. In addition, a study by Heydari et al. (2000) showed that the DNA binding activity of HSF1 from rat hepatocytes was decreased significantly following heat shock [113].

In skeletal muscle, a study by Locke (2000) demonstrated that heat shock activates HSF1 in soleus (fast) and white gastrocnemius (slow) muscles from aged rats [93]. However, the effect of exercise on HSF1 activation in skeletal muscle of aged mammals has not been studied in detail.

The influence of aging on the activation of NF-κB and AP-1 in skeletal muscle has been a subject of interest. Hollander et al. (2000) reported that DNA binding of both NF-κB and AP-1 was significantly decreased with age in the gastrocnemius, soleus and superficial vastus lateralis muscles of aged rats compared with the DNA binding of NF-κB and AP-1 in the gastrocnemius, soleus and superficial vastus lateralis muscles of young rats [110].

Summary

In summary, aging is associated with a substantial loss of skeletal muscle. In addition, the remaining muscle is weaker, more susceptible to damage and the ability to recover from damage is poor. Muscles of younger individuals adapts to exercise by structural remodelling. In general, this adaptation is accompanied by changes in gene expression and subsequent activation if the stress response. In contrast, muscles of older individuals do not adapt readily following exercise. This lack of ability to adapt may play a major role in the functional deficits seen in muscles of older individuals.

Acknowledgments

The authors would like to thank Research into Aging, the Wellcome Trust, European Commission, NIH and BBSRC for financial support and Professor Malcolm Jackson, Dr Frank McArdle and Dr David Pattwell for interesting discussions related to this work.

References

1. Kirkwood TBL, Austad SN (2000). Why do we age? *Nature* 408(9): 233–8.
2. Powers SK, Howley ET (1994). Skeletal muscle: structure and function. In: E Bartell, S Spoolman, eds. *Exercise Physiology: Theory and Application to Fitness and Performance.* Dubuque: Brown and Benchmark, pp. 145–70.
3. Voet D, Voet JG (1990). Motility: muscles, cilia and flagella. In: J Stiefel, ed. *Biochemistry.* Ontario, Canada: John Wiley & Sons, Inc., pp. 1118–28.
4. Ganong WF (1993). Excitable tissue: muscle. In: WF Ganong, ed. *Review of Medical Physiology,* 16th edn. Norwalk: Appleton and Lange, pp. 57–67.
5. Lodish H, Baltimore D, Berk A, Zipursky SL, Matsudaira P, Darnell J (1996). Muscle, a specialized contractile machine. In: J Darnell, ed. *Biology,* 3rd edn. New York: Scientific American Books, WH Freeman and Company, pp. 1021–32.
6. Faulkner JA, Brooks SV, Zebra E (1990b). Skeletal muscle weakness, fatigue and injury: inevitable concomitants of ageing? *Hermes (Leuven)* XXI: 269–80.
7. Armstrong RB, Warren GL, Warren JA (1991). Mechanisms of exercise-induced muscle fibre injury. *Sports Med.* 12: 184–207.
8. Porter MM, Vandervoort AA, Lexell J (1995). Ageing of human muscle: structure, function and adaptivity. *Scand J Med Sci Sports* 5: 129–42.
9. Lexell J (1995). Human ageing, muscle mass and fibre type composition. *J Gerontol.* 50A: 11–16.
10. Larsson L, Karlsson J (1978). Isometric and dynamic endurance as a function of age and skeletal muscle characteristics. *Acta Physiol Scand.* 104: 129–36.
11. Larsson L (1978). Morphological and functional characteristics of the ageing skeletal muscle in man: a cross-sectional study. *Acta Physiol Scand Suppl.* 457: 1–36.
12. Grimby G, Saltin B (1983). The ageing muscle. *Clin Physiol.* 3: 209–18.
13. Sato TH, Akatsuka H, Kito K, Tokoro Y, Tauchi H, Kato K (1984). Age changes in size and number of muscle fibres in human minor pectoral muscle. *Mech Ageing Dev.* 28: 99–109.

14. Lexell J, Downham D, Sjostrom M (1986). Distribution of different fibre types in human skeletal muscles: fibre type arrangement in muscle vastus lateralis from three groups of healthy men between 15 and 83 years. *J Neurol Sci.* 72: 211–22.

15. Lexell J, Taylor CC, Sjostrom M (1988). What is the cause of the ageing atrophy? Total number, size and proportion of different fibre types studied in whole vastus lateralis muscle from 15- to 83-year-old men. *J Neurol Sci.* 84: 275–94.

16. Caccia MR, Harris JB, Johnson MA (1979). Morphology and physiology of skeletal muscle in ageing rodents. *Muscle Nerve* 2: 202–12.

17. Holloszy JO, Chen M, Cartee GD, Young JC (1991). Skeletal muscle atrophy in old rats: differential changes in the three fibre types. *Mech Ageing Dev.* 60(2): 199–213.

18. Thompson LV (1994). Effects of age and training on skeletal muscle physiology and performance. *Phys Ther.* 74(1): 71–81.

19. Kirkendall, DT, Garrett WE (1998). The effects of ageing and training on skeletal muscle. *Am J Sports Sci.* 26(4): 598–602.

20. Young A, Stokes M, Crowe M (1984). Size and strength of the quadriceps muscles of old and young women. *Eur J Clin Invest.* 14: 282–7.

21. McDonagh MJ, White MJ, Davies CT (1984). Different effects of ageing on the mechanical properties of human arm and leg muscles. *Gerontology* 30: 49–54.

22. Faulkner JA, Brooks SV, Zebra E (1990a). Skeletal muscle weakness and fatigue in old mice: underlying mechanisms. In: VJ Cristofal, MP Lawton, eds. *Annual Review of Gerontology and Geriatrics.* New York: Springer Publishing Company, pp. 147–66.

23. Brooks SV, Faulkner JA (1988). Contractile properties of skeletal muscles from young, adult and aged mice. *J Physiol.* 404: 71–82.

24. Carlson BM, Faulkner JA (1988). Reinnervation of long-term denervated rat muscle freely grafted into an innervated limb. *Exp Neurol.* 102: 50–6.

25. Zebra E, Komorowski TE, Faulkner JA (1990). Free radical injury to skeletal muscles of young, adult, and old mice. *Am J Physiol.* 258: C429–35.

26. Brooks SV, Faulkner JA (1996). The magnitude of the initial injury induced by stretches of maximally activated muscle fibres of mice and rats is increases in old age. *J Physiol.* 497: 573–80.

27. Brooks SV, Faulkner JA (1990). Contraction-induced injury: recovery of skeletal muscles in young and old mice. *Am J Physiol.* 258: C436–42.

28. McBride TA, Gorin FA, Carlsen RC (1995). Prolonged recovery and reduced adaptation in aged rat muscle following eccentric exercise. *Mech Ageing Dev.* 83(3): 185–200.

29. Jones DA, Newham DJ, Round JM, Tolfree SE (1986). Experimental human muscle damage: morphological changes in relation to other indices of damage. *J Physiol.* 375: 435–48.

30. Ji LL (1995). Oxidative stress during exercise: implication of antioxidant nutrients. *Free Radic Biol Med.* 18(6): 1079–86.

30a. Lawler JM, Powers SK (1998). Oxidative stress, antioxidant status, and the contracting diaphragm. *Can J Appl Physiol.* 23: 23–55.

31. Halliwell B (1994). Free radicals and antioxidants: a personal review. *Nutr Rev.* 52(8): 253–65.

32. Michiels C, Raes M, Toussaint O, Remacle J (1994). Importance of Se-glutathione peroxidase, catalase and Cu/Zn-SOD for the cell survival against oxidative stress. *Free Rad Biol Med.* 17(3): 235–48.

33. Davies KJA, Quintanilha AT, Brooks GA, Parker L (1982). Free radicals and tissue damage produced by exercise. *Biochem Biophys Res Comm.* 107(4): 1198–205.

34. Halliwell B, Gutteridge JMC, Cross CE (1992). Free radicals, antioxidants, and human disease: where are we now? *J Lab Clin Med.* 119(6): 598–620.

35. Reid MB (2001). Plasticity in skeletal, cardiac and smooth muscle. Invited review: redox modulation of skeletal muscle contraction: what we know and what we don't. *J Appl Physiol.* 90: 724–31.

36. McArdle A, Pattwell D, Vasilaki A, Griffiths RD, Jackson MJ (2001). Contractile activity-induced oxidative stress: cellular origin and adaptive responses. *Am J Physiol Cell Physiol.* 280: C621–7.

37. Balon TW, Nadler JL (1994). Nitric oxide release is present from incubated skeletal muscle preparations. *J Appl Physiol.* 77: 2519–21.

38. Reid MB (1996). Reactive oxygen and nitric oxide in skeletal muscle. *News Physiol Sci.* 11: 114–19.

39. Halliwell B (1998). Free radicals and oxidative damage in biology and medicine: an introduction. In: AZ Reznick, L Parker, CK Sen, JO Holloszy, MJ Jackson, eds. *Oxidative Stress in Skeletal Muscle.* Basel, Switzerland: Birkhauser Verlag, pp. 1–27.

40. Powers SK, Lennon SL (1999). Analysis of cellular responses to free radicals: focus on exercise and skeletal muscle. *Proc Nutr Soc.* 58: 1025–33.

41. Powers SK, Ji LL, Leeuwenburgh C (1999). Exercise training-induced alterations in skeletal muscle antioxidant capacity: a brief review. *Med Sci Sports Exerc.* 31(7): 987–97.

42. Ohno H, Suzuki K, Fujii J, *et al.* (1994). Superoxide dismutases in exercise and disease. In: CK Sen, L Packer, O Hanninen, eds. *Exercise and Oxygen Toxicity.* New York: Elsevier Science, pp. 127–61.

43. Powers SK, Criswell D, Lawler JM, *et al.* (1994). Influence of exercise and fibre type on antioxidant enzyme activity in rat skeletal muscle. *Am J Physiol.* 266: R375–80.

44. Ji LL, Hollander J (2000). Antioxidant defence:: effects of ageing and exercise. In: Z Radak, ed. *Free Radicals in Exercise and Ageing.* Champaign Illinois, USA: Human Kinetics, pp. 35–72.

45. Goldfarb AH, Sen CK (1994). Antioxidant supplementation and the control of oxygen toxicity during exercise. In: CK Sen, L Packer, O Hanninen, eds. *Exercise and Oxygen Toxicity.* Amsterdam: Elsevier Science, pp. 163–89.

46. Jackson MJ, Edwards RHT, Symons MCR (1985). Electron spin resonance studies of intact mammalian skeletal muscle. *Biochim Biophys Acta* 847: 185–90.

47. Reid MB, Shoji T, Moody MR, Entman ML (1992). Reactive oxygen in skeletal muscle. II. Extracellular release of free radicals. *J Appl Physiol.* 73(5): 1805–9.

48. O'Neill CA, Stebbins CL, Bonigut S, Halliwell B, Longhurst JC (1996). Production of hydroxyl radicals in contracting skeletal muscle of cats. *J Appl Physiol.* 81: 1197–206.

49. Newham DL, Mills KR, Quigley BM, Edwards RHT (1983). Pain and fatigue after concentric and eccentric muscle contractions. *Muscle Nerve* 9: 59–63.

50. Khassaf M, Child RB, McArdle A, Brodie DA, Esanu C, Jackson MJ (2001). Time course of responses of human skeletal muscle to oxidative stress induced by nondamaging exercise. *J Appl Physiol.* 90: 1031–5.

51. Ji LL, Fu RG, Mitchell EW (1992). Glutathione and antioxidant enzymes in skeletal muscle: effects of fibre type and exercise intensity. *J Appl Physiol.* 73(5): 1854–9.

52. Leeuwenburgh C, Fiebig R, Chandwaney R, Ji LL (1994). Ageing and exercise training in skeletal muscle: responses of glutathione and antioxidant enzyme systems. *Am J Physiol.* 267: R439–45.

53. Leeuwenburgh C, Hollander J, Leichtweis S, Griffiths M, Gore M, Ji LL (1997). Adaptations of glutathione antioxidant system to endurance training are tissue and muscle fibre specific. *Am J Physiol.* 272: R363–9.

54. Harman D (1956). Ageing: theory based on free radical and radiation chemistry. *J Gerontol.* 11: 298–300.

55. Koward A, Kirkwood BL (1994). Towards a network theory of ageing: a model combining the free radical theory and the protein error theory. *J Theor Biol.* 168: 75–94.

56. Lass A, Sohal BH, Weindruch R, Forster MJ, Sohal RS (1998). Caloric restriction prevents age-associated accrual of oxidative damage to mouse skeletal muscle mitochondria. *Free Radic Biol Med.* 25: 1089–97.

57. De Flora S, Izzotti A, Randerath K, *et al.* (1996). DNA adducts and chronic degenerative disease. Pathogenic relevance and implications in preventive medicine. *Mutat Res.* 366: 197–238.

58. Lee CK, Klopp RG, Weindruch R, Prolla TA (1999). Gene expression profile of ageing and its retardation by caloric restriction. *Science* 285: 1390–3.

59. Melov S, Ravenscroft J, Malik S, *et al.* (2000). Extension of life-span with superoxide dismutase/catalase mimetics. *Science* 289: 1567–9.

60. Ji LL, Dillon D, Wu E (1990). Alteration of antioxidant enzymes with ageing in rat skeletal muscle and liver. *Am J Physiol.* 258: R918–23.

61. Lawler JM, Powers SK, Visser T, Kordus JM, Ji LL (1993). Acute exercise and skeletal muscle antioxidant and metabolic enzymes: effects of fibre type and age. *Am J Physiol.* 265: R1344–50.

62. Oh-Ishi S, Kizaki T, Yamashita H, *et al.* (1995). Alterations of superoxide dismutase isoenzyme activity, content and mRNA expression with ageing in rat skeletal muscle. *Mech Ageing Dev.* 84: 65–76.

63. Ji LL, Wu E, Thomas DP (1991). Effect of exercise training on antioxidant and metabolic functions in senescent rat skeletal muscle. *Gerontology* 37: 317–25.

64. Locke M (1997). The cellular stress response to exercise: role of stress proteins. *Exerc Sport Sci Rev.* 25: 105–36.

65. McArdle A, Vasilaki A, Jackson MJ (2002a). Exercise and skeletal muscle ageing: cellular and molecular mechanisms. *Ageing Res Rev.* 1: 79–93.

66. Jackson MJ, McArdle A, McArdle F (1998). Antioxidant micronutrients and gene expression. *Proc Nutr Soc.* 57: 301–5.

67. Gething MJ, ed. (1997). *Guidebook to Molecular Chaperones and Protein Folding Catalysts.* Oxford: Oxford University Press.

68. Moseley PL (1997). Heat shock proteins and heat adaptation of the whole organism. *J Appl Physiol.* 83(5): 1413–17.

69. Thomason DB, Menon V (2002). HSPs and protein synthesis in striated muscle. In: M Locke, EG Noble, eds. *Exercise and Stress Response: The Role of Stress Proteins.* Boca Raton, Florida: CRC Press, pp. 79–96.

70. Neufer PD, Benjamin IJ (1996). Differential expression of αB-crystallin and HSP27 in skeletal muscle during continuous contractile activity. *J Biol Chem.* 271: 24089–95.

71. Welch WJ (1992). Mammalian stress response: cell physiology, structure/function of stress proteins, and implications for medicine and disease. *Physiol Rev.* 72(4): 1063–77.

72. Welch WJ, Garrels JI, Thomas GP, Lin JJ-C, Feramisco JR (1983). Biochemical characterisation of the mammalian stress proteins and identification of two stress proteins as glucose- and Ca^{2+}-ionophore-regulated proteins. *J Biol Chem.* 258: 7102–11.

73. Whelan SA, Hightower LE (1985). Differential induction of glucose-regulated and heat shock proteins: effects of pH and sulfhydryl-reducing agents on chicken embryo cells. *J Cell Physiol.* 125: 251–8.

74. Heacock CS, Sutherland RM (1990). Enhanced synthesis of stress proteins caused by hypoxia and relation to altered cell growth and metabolism. *Br J Cancer* 62: 217–25.

75. Salo DC, Donovan CM, Davies KJA (1991). HSP70 and other possible heat shock or oxidative stress proteins are induced in skeletal muscle, heart and liver during exercise. *Free Radic Biol Med.* 11: 239–46.

76. Locke M, Noble EG (1995). Stress proteins: the exercise response. *Can J Appl Physiol.* 20(2): 155–67.

77. Marber MS, Mestril R, Chi SH, Sayen MR, Yellon DM, Dillmann WH (1995). Overexpression of the rat inducible 70-kD heat stress protein in a transgenic mouse increases the resistance of the heart to ischemic injury. *J Clin Invest.* 95(4): 1446–56.

78. Garramone RR Jr, Winters RM, Das DK, Deckers PJ (1994). Reduction of skeletal muscle injury through stress conditioning using the heat-shock response. *Plast Reconstr Surg.* 93(6): 1242–7.

79. Lepore DA, Hurley JV, Stewart AG, Morrison WA, Anderson RL (2000). Prior heat stress improves survival of ischemic-reperfused skeletal muscle *in vivo. Muscle Nerve* 23(12): 1847–55.

80. Suzuki K, Smolenski RT, Jayakumar J, Murtuza B, Brand NJ, Yacoub MH (2000). Heat shock treatment enhances graft cell survival in skeletal myoblast transplantation to the heart. *Circulation* 102(Suppl 3): III216–21.

81. Neufer PD, Ordway GA, Hand GA, *et al.* (1996). Continuous contractile activity induces fibre type specific expression of HSP70 in skeletal muscle. *Am J Physiol.* 271: C1828–37.

82. Ornatsky OI, Connor MK, Hood DA (1995). Expression of stress proteins and mitochondrial chaperonins in chronically stimulated skeletal muscle. *Biochem J.* 311: 119–23.

83. Samelman TR (1999). Heat shock protein expression is increased in cardiac and skeletal muscles of Fischer 344 rats after endurance training. *Exp Physiol.* 85: 97–102.

84. Gonzalez B, Hernando R, Manso R (2000). Stress proteins of 70 kDa in chronically exercised skeletal muscle. *Eur J Physiol.* 440: 42–9.

85. McArdle A, Jackson MJ (2002). Stress proteins and exercise-induced muscle damage. In: M Locke, EG Noble, eds. *Exercise and Stress Response: The Role of Stress Proteins.* Boca Raton, Florida: CRC Press, pp. 137–50.

86. McArdle A, Beaver A, Edwards RHT, Jackson MJ (1995). Prior contractile activity provides skeletal muscle with rapid protection against contraction-induced damage. *J Physiol.* 483P: 84.

87. McArdle A, McArdle F, Jackson MJ (1996). Stress proteins and protection of skeletal muscle against contraction-induced skeletal muscle damage in anaesthetised mice. *J Physiol.* 499P: 9.

88. McArdle A, Broome CS, Dillmann W, Mestril R, Faulkner JA, Jackson MJ (2002b). Lifelong overexpression of HSP70 in skeletal muscle of transgenic mice protects against age-related functional deficits. *J Physiol.* (in press).

89. Fargnoli J, Kunisada T, Fornace AJ, Schneider EL (1990). Decreased expression of heat shock protein 70 mRNA and protein after heat treatment in cells of aged rats. *Proc Natl Acad Sci USA* 87: 846–50.

90. Wu B, Gu MJ, Heydari AR, Richardson A (1993). The effect of age on the synthesis of two heat shock proteins in the HSP70 family. *J Gerontol.* 48: B50–6.

91. Locke M, Tanguay RM (1996a). Diminished heat shock response in the aged myocardium. *Cell Stress Chaperones* 1: 251–360.

92. Vasilaki A, Jackson MJ, McArdle A (2002). Attenuated HSP70 response in skeletal muscle of aged rats following contractile activity. *Muscle Nerve* 25(6): 902–5.

93. Locke M (2000). Heat shock transcription factor activation and HSP72 accumulation in aged skeletal muscle. *Cell Stress Chaperones* 5: 45–51.

94. Naito H, Powers SK, Demirel HA, Aoki J (2001). Exercise training increases heat shock protein in skeletal muscles of old rats. *Med Sci Sports Exerc.* 33: 729–34.

95. Mueller JM, Palh HL (2000). Assaying NF-κB and AP-1 DNA-binding and transcriptional activity. In: SM Keyse, ed. *Stress Response: Methods and Protocols*. Totowa, NJ: Humara Press Inc., pp. 205–215.

96. Bowie A, O'Neill LAJ (2000). Oxidative stress and nuclear factor-kB activation. *Biochem Pharmacol.* 59: 13–23.

97. Zhou LZ-H, Johnson AP, Rando TA (2001). NFκB and AP-1 mediate transcriptional responses to oxidative stress in skeletal muscle cells. *Free Radic Biol Med.* 31(11): 1405–16.

98. Flohe L, Brigelius-Flohe R, Saliou C, Traber MG, Packer L (1997). Redox regulation of NF-kappa B activation. *Free Radic Biol Med.* 22: 1115–26.

99. Latchman DS (1995). Cellular oncogenes and cancer. In: DS Latchman, ed. *Eukaryotic Transcription Factors*, 2nd edn. Cambridge: The University Press, pp. 167–72.

100. Pirkkala L, Nykanen P, Sistonen L (2001). Roles of the heat shock transcription factors in regulation of the heat shock response and beyond. *FASEB J.* 15: 1118–31.

101. Morimoto RI (1998). Regulation of the heat shock transcriptional response: cross talk between a family of heat shock factors, molecular chaperones and negative regulators. *Genes Dev.* 12: 3788–96.

102. Leppa S, Sistonen L (1997). Heat shock response-pathophysiological implications. *Ann Med.* 29: 73–8.

103. Cotto JJ, Kline M, Morimoto RI (1996). Activation of heat shock factor 1 DNA binding precedes stress-induced serine phosphorylation. *J Biol Chem.* 271: 3355–8.

104. Santoro MG (2000). Heat shock factors and the control of the stress response. *Biochem Pharmacol.* 59: 55–63.

105. Morimoto RI, Kroeger PE, Cotto JJ (1996). The transcriptional regulation of heat shock genes: a plethora of heat shock factors and regulatory conditions. In: U Feige, RI Morimoto, I Yahara, B Polla, eds. *Stress-Inducible Cellular Responses*. Berlin: Birkhauser Verlag, pp. 139–63.

106. Locke M, Noble EG, Tanguay RM, Field MR, Ianuzzo SE, Ianuzzo CD (1995). Activation of heat-shock transcription factor in rat heart after heat shock and exercise. *Am J Physiol.* 268(6): C1387–94.

107. Nishizawa J, Nakai A, Matsuda K, Komeda M, Ban T, Nagata K (1999). Reactive oxygen species play an important role in the activation of heat shock factor 1 in ischemic-reperfused heart. *Circulation* 99(7): 934–41.

108. McDuffee AT, Senisterra G, Huntley S, et al. (1997). Proteins containing non-native disulfide bonds generated by oxidative stress can act as signals for the induction of the heat shock response. *J Cell Physiol.* 171(2): 143–51.

109. Locke M, Tanguay RM (1996b). Increased HSF activation in muscles with a high constitutive HSP70 expression. *Cell Stress Chaperones* 1: 189–96.

110. Hollander J, Bejma J, Ookawara T, Ohno H, Ji LL (2000). Superoxide dismutase gene expression in skeletal muscle: fibre-specific effect of age. *Mech Ageing Dev.* 116: 33–45.

111. Fawcett TW, Sylvester SL, Sarge KD, Morimoto RI, Holbrook NJ (1994). Effects of neurohormonal stress and ageing on the activation of mammalian heat shock factor 1. *J Biol Chem.* 269: 32272–8.

112. Liu AYC, Lee YK, Manalo D, Huang LE (1996). Attenuated heat shock transcriptional response in ageing: molecular mechanism and implication in the biology of ageing. In: U Feige, RI Morimoto, I Yahara, B Polla, eds. *Stress-Inducible Cellular Responses.* Berlin: Birkhauser Verlag, pp. 393–408

113. Heydari AR, Shenghong Y, Takahashi R, Gutsmann-Conrad A, Sarge KD, Richardson A (2000). Age-related alterations in the activation of heat shock transcription factor 1 in rat hepatocytes. *Exp Cell Res.* 256: 83–93.

Aging of the Hematopoietic System

Peter M. Lansdorp

Terry Fox Laboratory, British Columbia Cancer Agency, 601 West 10th Avenue, Vancouver, BC, Canada, V5Z 1L3, and Department of Medicine, University of British Columbia, Vancouver, BC, Canada, V5Z 4E3

Introduction

The design of biological systems is not expected to be better than strictly needed for survival and reproduction under a particular set of circumstances. DNA itself is an unstable molecule that can accumulate a perplexing diversity of lesions arising from environmental agents (ranging from UV light and ionizing radiation to tobacco smoke), normal cellular metabolism (in particular, reactive oxygen species derived from oxidative respiration and lipid peroxidation) and spontaneous disintegration. [1]. The energy that an organism devotes to the prevention and repair of such DNA lesions is not spent on reproduction and other vital functions and is therefore subject to selective pressures. In multicellular organisms, the need to maintain genetic integrity varies between cell types and cell death is an additional option. Selective pressures on DNA damage responses and DNA repair pathways in relation to the need to maintain the genetic integrity in a particular cell type has resulted in cell-type specific differences in the sensitivity to DNA damage and the type of response that is triggered by a given level and type of DNA damage. This is illustrated in mammalian cells by the differences in response to lethal levels of DNA damage between e.g., human lymphocytes (apoptosis) and fibroblasts (senescence). HSCs are very sensitive to radiation [2] indicating a (very) low tolerance of DNA damage. Differences in DNA damage responses and DNA repair pathways between mammalian cell types greatly complicate studies on the role of such pathways in normal aging. Furthermore, because the number of cell divisions and the time before reproduction varies widely between mammalian species, it seems reasonable that DNA damage and repair pathways will show species-specific differences as well. This notion is supported by the observation that the specific activity of poly(ADP-ribose) polymerase-1 (PARP-1), an enzyme involved in DNA damage responses, varies with the

R. Aspinall (ed.), Aging of Organs and Systems, 101–114.

life-span of mammalian species [reviewed in ref. 3]. A link between DNA repair pathways, life-span and aging is supported by the premature aging in mice with specific DNA repair defects [4] and the recent demonstration that many traits linked to lifespan in that species map to genes implicated in cell cycle regulation and DNA repair [5, 6]. Taken together, these observations support the idea that the detrimental physical aspects of aging are a consequence of loss of cells or loss of cellular function as a result of DNA damage that escapes or exceeds the DNA repair capacity and cellular maintenance pathways available to a particular species. In this chapter, we review the role of telomeres as central genetic elements in DNA damage responses that control the proliferation of human hematopoietic cells. No attempt is made here to review the extensive literature dealing with descriptions of age-related changes in the number and function of various human hematopoietic cells or murine HSCs [6, 7]. Instead, the roles of telomeres, DNA damage responses and DNA repair processes in HSCs are highlighted as critical elements in maintaining the genetic integrity and function of HSCs over multiple cell divisions and time.

DNA damage responses, DNA repair and aging

While minimizing DNA damage (e.g., by caloric restriction) [see ref. 8] can be used to reduce the rate of aging in all species, interventions based on increasing the efficiency of DNA repair are more difficult to achieve. One of the dilemmas is that such pathways may differ between cells and between species. For example, programmed telomere shortening is a mechanism thought to act as a tumor suppressor pathway in cells from long-lived species that does not appear to exist in rodent cells [for review see ref. 9]. Thus, unlike primary human cells, rodent cells are easily transformed and immortalized [10]. The need for an additional tumor suppressor mechanism in long-lived species relative to rodents may originate from the increased numbers of cell divisions and increased time interval between cell divisions and reproductive cycles. This leaves us with a major dilemma: in order to study the role of tumor suppressor mechanisms that are specific for long-lived species in normal aging, we cannot use the model organisms that have been instrumental in dissecting common mechanisms of aging (such as yeast, flies, worms and mice). One solution to this dilemma (with many proponents) is to ignore the issue altogether and focus instead on general aspects of aging that are more easily studied in the laboratory. Another approach is to study aging in long-lived mammals with a similar need to suppress tumor growth. In such studies the distinction between pathways that evolved specifically to counter problems resulting from increases in life-span and cell mass and pathways involved in protection against DNA damage, DNA damage responses and DNA repair is problematic. Given the inherent complexity of the problem, no satisfactory solutions are expected in the near future. In the meantime, one can try to carefully delineate DNA damage responses and DNA repair pathways in primary human cells and study the age-related molecular changes in such pathways. In combination with *in vitro* studies, one can furthermore aim to understand the key events that limit normal cellular proliferation.

Organization of the hematopoietic system

The hematopoietic system has been subdivided into a hierarchy of three distinct populations of cells [11, 12]. In this model, the most mature cells are morphologically identifiable as belonging to a particular lineage and have very limited proliferative potential. The cells in this most mature compartment are derived from committed progenitor cells with a higher but finite proliferative potential. Committed progenitor cells in turn are produced by a population of HSCs that have typically been defined as pluripotent cells with "self-renewal" potential. A recurrent theme in experimental hematology over the last 40 years has been the (re-) definition of what HSCs are based on changes and improvements in assays. The trend has invariably been towards the identification of cells with a lower frequency than was previously reported or proposed [reviewed in ref. 13]. Initial observations showed that transplantation of a limited number of blood or bone marrow derived stem cells reproducibly results in complete regeneration of hematopoiesis in suitably conditioned recipients. This remarkable fact supported the original definition of HSCs as "pluripotent cells with self-renewal potential." Implicit in this definition is that the two daughter cells derived from the cell division of a single parental stem cell are essentially indistinguishable from that parental cell. Despite experimental evidence to the contrary, the central dogma and current paradigm in experimental hematology has remained that perhaps the frequency, but not HSCs themselves are subject to significant molecular changes.

An important question in stem cell biology is "how many times can stem cells divide"? Unfortunately, no techniques to address this important question are currently available, and possible answers range from <100 times [13] to >5000 times [14, 15]. Caution is needed in extrapolating findings with rodent cells [reviewed in ref. 16] in view of the issues that were mentioned above: murine cells are easily immortalized and mammalian species differ in their requirements and molecular mechanisms to limit cellular proliferation and prevent tumor growth. The total number of blood cells produced in the normal life of a human can be calculated to be in the order of 4.10^{16} cells ($\sim 10^{12}$ cells/day $\times$ 365 days $\times$ 100 years). In principle, only 55 divisions of a single cell would satisfy this need ($2^{55} = 4.10^{16}$). Because many hundreds of different HSCs are contributing to hematopoiesis at any given time, the required replicative potential in HSC could be less than fifty cell divisions. It seems that an absolute replicative potential of 50–100 cell divisions per HSC can easily accommodate both the known apoptosis of cytokine-deprived committed progenitor cells and the remarkable regenerative capacity of HSCs displayed in response to marrow injury and in transplantation experiments.

The notion that HSCs have a limited replicative potential is supported by several observations. Both the *in vivo* regenerative capacity [17–19] and the *in vitro* expansion potential [20, 21] of HSCs appear to be under developmental control. Most HSCs in adult tissue are quiescent cells in line with expectations if HSCs have only a limited replicative potential and are lacking self-renewal properties in an absolute sense. The loss of telomere repeats in adult hematopoietic cells (including purified "candidate" HSCs) relative to fetal hematopoietic cells [22], also fits a model that postulates a finite and limited replicative potential of HSCs [13, 23].

Telomere structure and function

Telomeres or the ends of linear chromosomes consist in all vertebrates of tandem repeats of $(TTAGGG/CCCTAA)_n$ and associated proteins [24, 25]. The length of the repeats varies between chromosomes and between species. In humans, the length of telomere repeats varies from 2–15 kilobase pairs depending on the tissue type, the age of the donor and the replicative history of the cells. Individual chromosome ends also vary in length and chromosome 17p typically has the shortest track of telomere repeats [26]. Telomeres prevent the ends of linear chromosomes from appearing as DNA breaks and protect chromosome ends from degradation and fusion. Telomeres also play a role in meiosis and the organization of chromosomes within the nucleus [27]. Telomeres contain DNA-binding proteins specific for duplex telomeric DNA, which include TRF1 and TRF2 [28] and a protein specific for the single strand overhang that is typically present at the 3′ ends of chromosomes [29]. In addition, many other proteins are known to indirectly bind to telomeres e.g., via TRF1 and TRF2 [reviewed in ref. 25]. The single-strand overhang at the 3′ end of telomeres folds back onto duplex telomeric DNA forming a protective "T-loop" [30]. The 3′ overhang associates with telomere repeats via TRF2 in a way that is incompletely understood but which appears important for telomere stability [31, 32].

Telomeres and replicative senescence

In contrast to embryonic stem cells or most tumor cell lines, HSCs are not immortal. Forty years ago, Hayflick suggested that most normal human cells are unable to divide indefinitely but are programmed for a given number of cell divisions [33]. In 1990, several papers described loss of telomeres with replication and with age and suggested that progressive telomere shortening could explain Hayflick's original observation [34–36]. This model was confirmed by subsequent studies showing that transfer of the telomerase reverse transcriptase gene could prevent telomere erosion and resulted in immortalizing of the cells that Hayflick studied in most detail: normal diploid human fibroblasts [37, 38]. Since then, many papers have appeared that are compatible with the notion that telomere shortening limits the number of times most normal human diploid cells can divide [for review see ref. 9]. An emerging consensus is that telomere shortening evolved as a checkpoint mechanism in long-lived mammals that controls unlimited and life-threatening proliferation of organ-specific stem cells and lymphocytes.

Checkpoints and genome integrity

DNA is under continuous assault by environmental agents as well as by-products of normal metabolism such as reactive oxygen species [for review see ref. 1]. To ensure that accurate copies of genetic information are copied to the next generation, the cell cycle machinery is overlaid with a series of surveillance pathways termed cell cycle checkpoints [39]. The overall function of these checkpoints is to detect damaged or abnormally structured DNA and to coordinate cell-cycle progression with DNA repair. The term "checkpoint" was originally defined as a "control mechanism enforcing dependency in the cell cycle" [40]. However, it has become clear that DNA damage checkpoints also control DNA repair and replication processes in addition to imposing cell cycle delay [41]. Cells from different organisms have furthermore adopted different strategies to respond to stress and DNA damage. Relative to unicellular organisms, multicellular organisms not only require a higher fidelity of DNA replication and repair but also more diversity in the response to DNA damage. Whereas, for example, the response in yeast can be described as binary between cell cycling and arrest, the response in mammals is more complex and includes outcomes such as apoptosis and senescence. The more complex response in multicellular organisms is enabled by a new regulatory control module (that does not exist in yeast) involving the tumor suppressor gene p53 [41]. In addition, it now appears that long-lived mammals such as humans have acquired an extra layer of genome integrity surveillance that is not operating in short lived-mammals such as rodents. This additional checkpoint is related to the length and function of telomeres. The term "telomere checkpoint" can be used to describe this tumor suppressor function.

Loss of telomeric DNA

Telomeric DNA is lost in human cells via several mechanisms that are related to DNA replication, remodeling and repair. Causes of telomere loss include the "end replication problem" [42, 43], nucleolytic processing of 5′ template strands following DNA replication to create a 3′ single strand overhang [44, 45] and failed repair of oxidative DNA damage to telomeric DNA [46, 47]. The relative contribution of these different causes of telomere shortening to the overall decline in telomere length with age is not known and most likely varies between cell types and with age. That telomeres shorten as a result of oxidative damage has only recently been realized. It has been shown that telomeric DNA, with its G-rich repeats, is 5–10 fold more vulnerable to oxidative damage than non-telomeric, genomic DNA [48, 49]. Repair of oxidative damage to nucleotides is typically achieved using nucleotide excision repair pathways which may involve a DNA polymerase template switch [1]. This essential mechanism may fail for lesions near chromosome ends: once the replication fork reaches the end of a chromosome, the physical linkage between template and newly synthesized DNA strands is presumably lost. In general, the contribution of oxidative damage to telomere shortening and the importance of the redox state in cells to prevent such damage remains to be precisely defined.

Telomere signaling pathways

The mechanism by which short telomeres signal a DNA damage response is poorly understood [47, 50]. Recent studies have highlighted the dynamic structure of telomeres [reviewed in ref. 51]. It now appears that individual telomeres can be either "on" or "off" in terms of signaling downstream DNA damage pathways. It seems likely that the terminal 3′ end is involved in generating such signals but details other than that ATM [39, 52] and p53 are likely to be involved are lacking.

An important question is when telomeres signal during the cell cycle. DNA replication inevitably involves remodeling of the telomere structure. Because telomere loss is known to occur during DNA replication, the inability to form functional telomeres following replication is expected to generate a DNA damage signal during the S or G2 phase of the cell cycle. However, when diploid human fibroblasts reach replicative senescence, they typically enter an irreversible growth arrest in G1. It has furthermore been shown that disruption of TFR2 binding to telomeres generates a DNA damage signal that is independent of DNA replication [31]. Perhaps telomeres cycle continuously between "on" and "off" states even when the cells are in G0/G1, with the likelihood of the "on" state inversely and indirectly correlated with the length of telomere repeats. According to this idea, the strength of DNA damage signals generated by telomeres gradually increases with overall telomere shortening. Anti-apoptotic effects of long telomeres or telomerase expression as well as increased levels of p53 (and the increased sensitivity to apoptosis) in "older" cells (with shorter telomeres) are in agreement with this model. Differences between individuals and cells in the telomere length required to activate downstream signaling pathways complicate studies attempting to use telomere length as an absolute predictor of cellular responses [53]. No doubt such differences explain, in part, the marked variation in telomere length between similar cells form normal individuals of the same age [54, 55].

Role of telomerase and ALT

To compensate for the loss of telomere repeats, cells require expression of functional telomerase. Telomerase is a ribonucleoprotein containing the reverse transcriptase telomerase protein (hTERT) and the telomerase RNA template (hTERC) as essential components. In addition, a number of proteins have been described that are important for telomerase assembly, nuclear localization and stability [reviewed in ref. 25]. Telomerase is capable of extending the 3′ ends of telomeres. Telomerase levels are typically high in immortal cells that maintain a constant telomere length such as the stem cells of the germline in the testis and embryonic stem cells. Interestingly, certain rare stem cells such as mesenchymal stem cells also express sufficient telomerase to maintain telomeres at a constant length [56]. The resulting unlimited proliferative potential provides considerable therapeutic potential because it allows extensive genetic manipulation and selection of karyotypically normal cells without the restrictions that are typically imposed by a limited replicative lifespan. For reasons that remain to be precisely defined, telomerase levels are insufficient to maintain the telomere length in hematopoietic stem cells (HSC). Nevertheless, existing telomerase levels in HSC are functionally important as is highlighted in patients with the disorder dyskeratosis congenita. Patients with the autosomal dominant form of this disease have one normal and one mutated copy of the telomerase RNA template gene. As expected, such patients show a modest reduction in telomerase levels, yet they typically suffer from progressive aplastic anemia, immune deficiencies or cancer and rarely live past the age of 50 [25, 57–59]. These findings are in stark contrast to those in the mouse where complete lack of telomerase activity is tolerated for up to six generations [60]. Together with the age-related decline in telomere length in leukocytes, these observations have provided strong support for the idea that telomerase levels are extremely tightly regulated and limiting in human HSCs.

The role of so-called alternative ("ALT") pathways [61] in the elongation and/or maintenance of telomeres in hematopoietic cells is not clear. Most likely, such pathways are not very efficient in elongating telomeres in cells that express telomerase. Because telomere lengthening via telomerase and/or ALT appears to be limiting in hematopoietic cells, molecular defects that result in accelerated telomere shortening may result in aplastic anemia. In many cases, it is not possible or straightforward to distinguish direct from indirect causes of telomere shortening. In dyskeratosis, the cause appears to be directly related to telomerase deficiency, whereas, for example in Fanconi's anemia [62], telomere shortening could be caused directly by defective repair of telomeric DNA or indirectly because loss of stem cells (resulting from defective DNA repair) will result in increased (compensatory) proliferation of remaining stem cells.

The telomere checkpoint

Following loss of telomere repeats, a DNA damage signal is generated that signals cell cycle arrest most likely via increases in levels of p53. When short telomeres are subsequently elongated by "telomere repair" pathways involving telomerase and/or recombination, the cell cycle arrest is expected to be transient. However, the continued loss of telomeric DNA eventually generates too many short telomeres for the limited capacity of telomere repair pathways (Figure 1). At this point, otherwise normal cells, as well as pre-malignant cells, are destined to die by apoptosis or convert to an unresponsive state ("replicative senescence") depending on the response to high and sustained levels of p53 in a particular cell type. Possibly, the replicative lifespan of cells in long-lived species has been under selective pressure to 1) permit sufficient divisions for the maintenance of cellular function during a normal lifespan while 2) act as a brake to prevent excessive cellular proliferation and tumor development. It appears that in most hematopoietic cells, the required balance is achieved by regulating telomerase activity at levels that are sufficient to maintain a minimal length in only a proportion of the 92 telomeres in a human cell. Limiting levels of telomerase and progressive telomere shortening could contribute to organismal aging in at least two ways. First, an increasing proportion of cells could reach the end of their programmed proliferative lifespan in old age. For example, T cell responses could be compromised in the elderly as a result of such limits in proliferative potential. Second, gene expression in cells near or at senescence may be abnormal, resulting in aberrant secretion of molecules including enzymes and cytokines [see also ref. 63]. Both factors could contribute to impaired immune responses in old individuals. In general, the study of telomere biology in relation to human aging is in its infancy. Major challenges are difficulties related to longitudinal studies in humans and the limitations of rodent models that were mentioned.

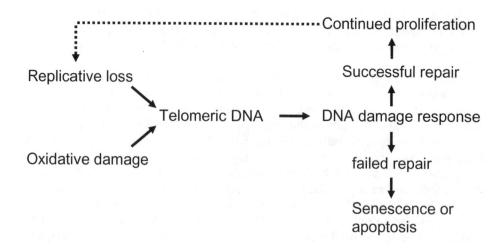

Figure 1. The telomere checkpoint controls the replicative lifespan of human somatic cells. Every time a human somatic cell is triggered to divide, all 92 chromosome ends need to be duplicated and all resulting 184 ends need to properly "capped" before cells can complete cell division. Most likely, the efficiency of this telomere "capping process" decreases as the telomere length decreases. Telomeres may switch between capped and uncapped states with the likelihood of the capped state linked to the length of telomere repeats. In such a dynamic model of telomeres, the DNA damage signal generated by uncapped telomeres could gradually increase, perhaps explaining the increased sensitivity to apoptosis in "older" cells. While telomerase in most HSCs and lymphocytes is capable of extending a limited number of short telomeres, this telomere repair function of telomerase appears to be limited. As a result, telomeres execute an important tumor suppressor function by imposing an absolute a limit on the total number of times cells are allowed to divide.

Telomere shortening in human nucleated blood cells

Since the important original observation that telomeres in adult blood leukocytes are significantly shorter as compared to germ-line material (sperm) from the same donor [64], the decline of somatic telomeres has been documented in three ways. The original observation was confirmed [36, 65], it was shown that telomeres in various tissues were shorter in older donors [34, 35] and telomere shortening was documented during *in vitro* culture of human cells [35, 66]. In the decade that followed these initial reports, a large number of papers have appeared that have greatly refined our understanding of telomere shortening in human nucleated blood cells [reviewed in ref. 67]. Studies in this general area have been facilitated by the development of quantitative fluorescence *in situ* hybridization techniques to measure the telomere length in suspension cells using flow cytometry (flow FISH [68]). With this technique, it was shown that the age-related decline in telomere length in lymphocytes is much more pronounced than in granulocytes and that rapid telomere shortening early in life is followed by a much more gradual decline thereafter [55]. The telomere length in granulocytes can be used as a surrogate marker for the telomere length in HSCs if one assumes that the number of cell divisions between HSCs and a granulocyte is relatively constant throughout life and that telomere shortening in HSCs is: (1) primarily resulting from replication, and (2) relatively constant with each cell division. This approach has been used to model the turnover of human hematopoietic stem cells on the basis of existing telomere length data [69].

A striking observation is that telomere length at any given age in humans is very heterogeneous [54, 68]. This variation appears to be primarily genetic [55, 70]. For example, monozygous twins of over 70 years of age were shown to have very similar telomere length in both granulocytes and lymphocytes, whereas dizygotic twins differed more but not as much as unrelated individuals [55]. Using further refinements in the flow FISH method (Baerlocher and Lansdorp, unpublished observations), it was recently shown that the rapid decline early in life is followed by a slow decline until the age of 50–60 years after which the decline again accelerates. The decline in both granulocytes and lymphocytes was found to be non-linear and fits a cubic curve. The pronounced decline in telomere length in T and NK cells could activate the telomere checkpoint in such cells during a normal lifetime. Immune responses in the elderly could be compromised as a result.

In view of the modest age-related decline in telomere length in granulocytes, it seems unlikely that most HSCs encounter their Hayflick limit as a result of telomere shortening during normal hematopoiesis. Most likely, HSCs are lost from the stem cell pool primarily by differentiation into progenitor cells and not replicative senescence. Furthermore, the occasional loss of individual HSCs via telomere shortening is not expected to impact on overall hematopoiesis (or overall telomere length in granulocytes) in the presence of a large reserve of additional HSCs. That normal HSCs and tissues have extensive replicative potential is also in agreement with extensive experience using allogeneic and autologous stem cell transplantation. Nevertheless, the telomere checkpoint does appear to operate in HSCs as is indicated by the age-related loss of telomeres in granulocytes, the (modest) loss of telomeres following allogeneic transplantation [67, 71] and the aplastic anemia that follows partial telomerase deficiency [25]. Recent studies have shown that the number of HSCs can be altered by manipulating decisions that control self-renewal and differentiation [72]. Indeed, the number of mature "end" cells such as granulocytes produced by individual stem cells is most likely highly variable and primarily determined by the processes that regulate self-renewal vs. differentiation at the level of individual stem cells. Even a limited number of additional self-renewal divisions in a stem cell will greatly increase cell output. As a result, individual stem cells can produce staggering numbers of cells. This is illustrated in clonal proliferative disorders such PNH and chronic myeloid leukemia (CML). However, even in CML, clonally expanded Philadelphia positive stem cells eventually appear to encounter the telomere checkpoint [73]. Unfortunately, with a large number of cells to select from, the genetic instability triggered by the loss of functional telomeres appears to favor the selection of a subclone with additional genetic abnormalities and more malignant properties.

Not all cells in the hematopoietic system are programmed to encounter the telomere checkpoint. B cells appear to be a particularly interesting exception as the telomere length in B cells is increasingly heterogenous with age. Apparently, some B cells express sufficient telomerase (and other factors) to effectively elongate telomeres. Perhaps the many cell divisions required for effective selection and "affinity maturation" of antibody responses favored inactivation of the telomere checkpoint. It is tempting to speculate that B cells are, as a result, at a higher risk for tumor development which could explain the much higher incidence of B versus T cell lymphomas in the human population.

Conclusions

Based on observations from several areas, telomeres have emerged as important regulatory elements that control the number of times most normal human somatic cells can divide. Activation of the telomere checkpoint results from loss of telomeric DNA with replication and from oxidative damage to telomeric DNA (Figure 1). The DNA damage response that is triggered by activation of the telomere checkpoint can be resolved by telomere elongation pathways that involve either telomerase or recombination. However, in most somatic cells, including HSCs, the capacity of such telomere repair pathways appears to be limiting and telomere shortening effectively limits the proliferative potential of such cells. Most likely, the telomere checkpoint evolved as a tumor suppressor mechanism in long-lived species. The function of the telomere checkpoint may help explain poorly understood aspects of stem cell biology including stem cell "exhaustion" in aplastic anemia and other proliferative disorders. Cells may bypass the telomere checkpoint by expressing high levels of telomerase or by inactivating downstream signaling events, e.g., by loss of p53 function. The resulting genetic instability accelerates malignant progression as is illustrated in chronic myeloid leukemia where the onset of blast crisis and additional genetic changes is inversely correlated with the length of telomeres in Philadelphia-positive chronic phase cells [73]. The realization that telomeres occupy a central role in the replicative lifespan of HSCs opens up new avenues to research and therapy. Telomere length measurements in subsets of leukocytes are expected to further elucidate the role of telomere shortening in normal again and modulation of telomere length has the potential to extend the replicative lifespan beyond the limits that were imposed by natural selection.

Acknowledgments

Work in the laboratory of the author is supported by grants from the National Institutes of Health (AI29524), the Canadian Institute of Health Research (MOP38075) and the National Cancer Institute of Canada (with support from the Terry Fox Run).

References

1. Hoeijmakers JHJ (2001). Genome maintenance mechanisms for preventing cancer. *Nature* 411: 366–74.
2. van Bekkum DW (1991). Radiation sensitivity of the hemopoietic stem cell. *Radiat Res.* 128: S4–8.
3. Burkle A (2000). Poly(ADP-ribosyl)ation: a posttranslational protein modification linked with genome protection and mammalian longevity. *Biogerontology* 1: 41–6.
4. de Boer J, Andressoo JO, de Wit J, *et al.* (2002). Premature aging in mice deficient in DNA repair and transcription. *Science* 296: 1276–9.
5. Geiger H, True JM, de Haan G, Van Zant G (2001). Age- and stage-specific regulation patterns in the hematopoietic stem cell hierarchy. *Blood* 98: 2966–72.
6. Geiger H, Van Zant G (2002). The aging of lympho-hematopoietic stem cells. *Nat Immunol.* 3: 329–33.
7. Sudo K, Ema H, Morita Y, Nakauchi H (2000). Age-associated characteristics of murine hematopoietic stem cells. *J Exp Med.* 192: 1273–80.
8. Lin SJ, Kaeberlein M, Andalis AA, *et al.* (2002). Calorie restriction extends Saccharomyces cerevisiae lifespan by increasing respiration. *Nature* 418: 344–8.
9. Mathon NF, Lloyd AC (2001). Cell senescence and cancer. *Nature Rev Cancer* 1: 203–13.
10. Trott DA, Cuthbert AP, Overell RW, Russo I, Newbold RF (1995). Mechanisms involved in the immortalization of mammalian cells by ionizing radiation and chemical carcinogens. *Carcinogenesis* 16: 193–204.
11. Till JE, McCulloch EA, Siminovitch L (1964). A stochastic model of stem cell proliferation, based on the growth of spleen colony-forming cells. *Proc Natl Acad Sci USA* 51: 29–36.
12. Metcalf D (1984) *The Hemopoietic Colony Stimulating Factors.* Amsterdam: Elsevier.
13. Lansdorp PM (1997). Self-renewal of stem cells. *Biol Blood Marrow Transplant.* 3: 171–8.
14. Potten CS, Loeffler M (1990). Stem cells: attributes, cycles, spirals, pitfalls and uncertainties. Lessons for and from the crypt. *Development* 10: 1001–20.
15. Rubin H (2002). The disparity between human cell senescence *in vitro* and lifelong replication *in vivo*. *Nat Biotechnol.* 20: 675–81.
16. Iscove NN, Nawa K (1997). Hematopoietic stem cells expand during serial transplantation *in vivo* without apparent exhaustion. *Curr Biol.* 7: 805–8.
17. Moore MAS, Metcalf D (1970). Ontogeny of the haemopoietic system; yolk sac origin of *in vivo* and *in vitro* colony forming cell in the developing mouse embryo. *Br J Haematol.* 18: 279–86.
18. Rebel VI, Miller CL, Eaves CJ, Lansdorp PM (1996). The repopulation potential of fetal liver hematopoietic stem cells in mice exceeds that of their adult bone marrow counterparts. *Blood* 87: 3500–7.
19. Pawliuk R, Eaves C, Humphries RK (1996). Evidence of both ontogeny and transplant dose-regulated expansion of hematopoietic stem cells *in vivo*. *Blood* 88: 2852–8.
20. Lansdorp PM, Dragowska W, Mayani H (1993). Ontogeny-related changes in proliferative potential of human hematopoietic cells. *J Exp Med.* 178: 787–91.
21. Lansdorp PM (1995). Developmental changes in the function of hematopoietic stem cells. *Exp Hematol.* 23: 187–91.
22. Vaziri H, Dragowska W, Allsopp RC, Thomas TE, Harley CB, Lansdorp PM (1994). Evidence for a mitotic clock in human hematopoietic stem cells: loss of telomeric DNA with age. *Proc Natl Acad Sci USA* 91: 9857–60.

23. Lansdorp PM (1995). Telomere length and proliferation potential of hematopoietic stem cells. *J Cell Sci.* 108: 1–6.

24. Moyzis RK, Buckingham JM, Cram LS, *et al.* (1988). A highly conserved repetitive DNA sequence, $(TTAGGG)_n$, present at the telomeres of human chromosomes. *Proc Natl Acad Sci USA* 85: 6622–6.

25. Collins K, Mitchell JR (2002). Telomerase in the human organism. *Oncogene* 21: 564–79.

26. Martens UM, Zijlmans JMJM, Poon SSS, *et al.* (1998). Short telomeres on human chromosome 17p. *Nat Genet.* 18: 76–80.

27. Blackburn EH (2000). Telomere states and cell fates. *Nature* 408: 53–6.

28. Smogorzewska A, van Steensel B, Bianchi A, *et al.* (2000). Control of human telomere length by TRF1 and TRF2. *Mol Cell Biol.* 20: 1659–68.

29. Baumann P, Cech TR (2001). Pot1, the putative telomere end-binding protein in fission yeast and humans. *Science* 292: 1171–5.

30. Griffith JD, Comeau L, Rosenfield S, *et al.* (1999). Mammalian telomeres end in a large duplex loop. *Cell* 97: 503–14.

31. Karlseder J, Broccoli D, Dai Y, Hardy S, de Lange T (1999). p53- and ATM-dependent apoptosis induced by telomeres lacking TRF2. *Science* 283: 1321–5.

32. Zhu XD, Kuster B, Mann M, Petrini JH, Lange T (2000). Cell-cycle-regulated association of RAD50/MRE11/NBS1 with TRF2 and human telomeres. *Nat Genet.* 25: 347–52.

33. Hayflick L, Moorhead PS (1961). The serial cultivation of human diploid strains. *Exp Cell Res.* 25: 585–621.

34. Hastie ND, Dempster M, Dunlop MG, Thompson AM, Green DK, Allshire RC (1990). Telomere reduction in human colorectal carcinoma and with ageing. *Nature* 346: 866–8.

35. Harley CB, Futcher AB, Greider CW (1990). Telomeres shorten during ageing of human fibroblasts. *Nature* 345: 458–60.

36. de Lange T, Shiue L, Myers R, *et al.* (1990). Structure and variability of human chromosome ends. *Mol Cell Biol.* 10: 518–27.

37. Bodnar AG, Ouellette M, Frolkis M, *et al.* (1998). Extension of life-span by introduction of telomerase into normal human cells. *Science* 279: 349–53.

38. Vaziri H, Benchimol S (1998). Reconstitution of telomerase activity in normal human cells leads to elongation of telomeres and extended replicative life span. *Curr Biol.* 8: 279–82.

39. Abraham RT (2001). Cell cycle checkpoint signaling through the ATM and ATR kinases. *Genes Dev.* 15: 2177–96.

40. Hartwell LH, Weinert TA (1989). Checkpoints: controls that ensure the order of cell cycle events. *Science* 246: 629–34.

41. Wahl GM, Carr AM (2001). The evolution of diverse biological responses to DNA damage: insights from yeast and p53. *Nat Cell Biol.* 3: E277–86

42. Watson JD (1972). Origin of concatameric T4 DNA. *Nat New Biol.* 239: 197–201.

43. Olovnikov AM (1973). A theory of marginotomy. The incomplete copying of template margin in enzymic synthesis of polynucleotides and biological significance of the phenomenon. *J Theor Biol.* 41: 181–90.

44. Wellinger RJ, Ethier K, Labrecque P, Zakian VA (1996). Evidence for a new step in telomere maintenance. *Cell* 85: 423–33.

45. Makarov VL, Hirose Y, Langmore JP (1997). Long G tails at both ends of human chromosomes suggest a C strand degradation mechanism for telomere shortening. *Cell* 88: 657–66.

46. Petersen S, Saretzki G, von Zglinicki T (1998). Preferential accumulation of single-stranded regions in telomeres of human fibroblasts. *Exp Cell Res.* 239: 152–60.

47. Lansdorp PM (2000). Repair of telomeric DNA prior to replicative senescence. *Mech Ageing Dev.* 118: 23–34.
48. Henle ES, Han Z, Tang N, Rai P, Luo Y, Linn S (1999). Sequence-specific DNA cleavage by Fe^{2+}-mediated fenton reactions has possible biological implications. *J Biol Chem.* 274: 962–71.
49. Oikawa S, Kawanishi S (1999). Site-specific DNA damage at GGG sequence by oxidative stress may accelerate telomere shortening. *FEBS Lett.* 453: 365–8.
50. Hemann MT, Strong MA, Hao LY, Greider CW (2001). The shortest telomere, not average telomere length, is critical for cell viability and chromosome stability. *Cell* 107: 67–77.
51. Blackburn EH (2001). Switching and signaling at the telomere. *Cell* 106: 661–73.
52. Hande MP, Balajee AS, Tchirkov A, Wynshaw-Boris A, Lansdorp PM (2001). Extra-chromosomal telomeric DNA in cells from $Atm^{-/-}$ mice and patients with ataxia-telangiectasia. *Hum Mol Genet.* 10: 519–28.
53. Serra V, von Zglinicki T (2002). Human fibroblasts *in vitro* senescence with a donor-specific telomere length. *FEBS Lett.* 516: 71–4.
54. Frenck RW Jr, Blackburn EH, Shannon KM (1998). The rate of telomere sequence loss in human leukocytes varies with age. *Proc Natl Acad Sci USA* 95: 5607–10.
55. Rufer N, Brummendorf TH, Kolvraa S, *et al.* (1999). Telomere fluorescence measurements in granulocytes and T lymphocyte subsets point to a high turnover of hematopoietic stem cells and memory T cells in early childhood. *J Exp Med.* 190: 157–67.
56. Reyes M, Lund T, Lenvik T, Aguiar D, Koodie L, Verfaillie CM (2001). Purification and *ex vivo* expansion of postnatal human marrow mesodermal progenitor cells. *Blood* 98: 2615–25.
57. Mitchell JR, Wood E, Collins K (1999). A telomerase component is defective in the human disease dyskeratosis congenita. *Nature* 402: 551–5.
58. Vulliamy T, Marrone A, Goldman F, *et al.* (2001). The RNA component of telomerase is mutated in autosomal dominant dyskeratosis congenita. *Nature* 413: 432–5.
59. Dokal I (2001). Dyskeratosis congenita. A disease of premature ageing. *Lancet* 358(Suppl): S27.
60. Blasco MA, Lee H-W, Hande MP, *et al.* (1997). Telomere shortening and tumor formation by mouse cells lacking telomerase RNA. *Cell* 91: 25–34.
61. Henson JD, Neumann AA, Yeager TR, Reddel RR (2002). Alternative lengthening of telomeres in mammalian cells. *Oncogene* 21: 598–610.
62. Grompe M, D'Andrea A (2001). Fanconi anemia and DNA repair. *Hum Mol Genet.* 10: 2253–9.
63. Kirkwood TB, Holliday R (1979). The evolution of ageing and longevity. *Proc R Soc Lond B Biol Sci.* 205: 531–46.
64. Cooke HJ, Smith BA (1986). Variability at the telomeres of the human X/Y pseudoauto-somal region. *Cold Spring Harb Symp Quant Biol.* 51: 213–19.
65. Allshire RC, Gosden JR, Cross SH, *et al.* (1988). Telomeric repeat from T. thermophila cross-hybridizes with human telomeres. *Nature* 332: 656–9.
66. Counter CM, Avilion AA, LeFeuvre CE, *et al.* (1992). Telomere shortening associated with chromosome instability is arrested in immortal cells which express telomerase activity. *EMBO J.* 11: 1921–9.
67. Ohyashiki JH, Sashida G, Tauchi T, Ohyashiki K (2002). Telomeres and telomerase in hematologic neoplasia. *Oncogene* 21: 680–7.

68. Rufer N, Dragowska W, Thornbury G, Roosnek E, Lansdorp PM (1998). Telomere length dynamics in human lymphocyte subpopulations measured by flow cytometry. *Nat Biotechnol.* 16: 743–7.

69. Edelstein-Keshet L, Israel A, Lansdorp P (2001). Modelling perspectives on aging: can mathematics help us stay young? *J Theor Biol.* 213: 509–25.

70. Slagboom PE, Droog S, Boomsma DI (1994). Genetic determination of telomere size in humans: a twin study of three age groups. *Am J Hum Genet.* 55: 876–82.

71. Rufer N, Brummendorf TH, Chapuis B, Helg C, Lansdorp PM, Roosnek E (2001). Accelerated telomere shortening in hematological lineages is limited to the first year following stem cell transplantation. *Blood* 97: 575–7.

72. Antonchuk J, Sauvageau G, Humphries RK (2002) *HoxB4*-induced expansion of adult hematopoietic stem cells *ex vivo. Cell* 109: 39–45.

73. Brummendorf TH, Mak J, Baerlocher GM, Sabo K, Abkowitz JL, Lansdorp PM (2000). Longitudinal studies of telomere length in feline blood cells point to a rapid turnover of stem cells in the first year of life. *Blood,* 96: 455a (Abstract).

Aging of the Human Skeleton and its Contribution to Osteoporotic Fractures

Moustapha Kassem, Kim Brixen and Leif Mosekilde

*Department of Endocrinology, University Hospital of Odense, Odense, and Department of
Endocrinology, University Hospital of Aarhus, Aarhus, Denmark*

Introduction

One of the cardinal manifestations of old age in humans is bone loss leading to
fragility of the skeleton and increased risk of fractures, a disease known as
osteoporosis. Osteoporosis is a systemic bone disease characterized by low bone
mass and structural deterioration of the skeleton leading to bone fragility and
increased fracture risk. Osteoporosis is a major public health problem. Its socio-
economic burden is enormous because of the costs associated with treatment of
complications of fractures including prolonged hospital stay and loss of independence. The problem of osteoporosis is aggravated by increasing life expectancy of the
world population and thus increasing the number of people at risk for the disease.
Furthermore, a real increase in the age-adjusted incidence of osteoporosis has been
observed [1].

Bone is a complex organ that serves triple functions. First it has a biomechanical
function facilitating locomotion, protecting inner organs. Second, it supports
hematopoiesis. Third, it is a major reservoir for divalent ions and an active
participant in mineral homeostasis. Thus, the skeleton is exposed to a multitude of
biomechanical and hormonal signals. In order to understand the mechanisms of age-
related bone loss and biomechanical deterioration, this complex context should be
appreciated. The aim of this chapter is to give a description of age-related changes of
the skeleton and to review the current understanding of the pathophysiological
mechanisms mediating the senescent skeletal phenotype.

R. Aspinall (ed.), Aging of Organs and Systems, 115–136.
© *2003 Kluwer Academic Publishers. Printed in Great Britain.*

Patterns of age-related changes in bone mass, structure and strength

Bone mass of the whole skeleton or of a particular region of interest can be measured by a number of different technologies e.g., single photon absorptiometry (SPA), dual photon absorptiometry (DPA), dual energy X-ray absorptiometry (DEXA), and single or dual energy quantitative computer tomography (QCT). Bone mass is usually expressed as area bone mineral density (BMD) and bone mineral content (BMC). Measurement of BMD and BMC employing DEXA machines has become widely used in clinical assessment of fracture risk and the diagnosis of osteoporosis [2]. QCT has the advantage of providing information of the relative distribution of cortical and trabecular bone but it requires a high radiation dose and is therefore not used routinely in clinical practice.

Studies measuring bone mass by DEXA in large cohorts of normal populations have revealed that BMD increases during childhood and adolescence with higher values in males compared with females. However, this is an artifact resulting from differences in bone size between the two genders [3]. BMD measured by DEXA is usually calculated as BMC divided by projected bone area in cm^2. Since the third dimension is not taken into account, BMD is dependent on bone size. If volumetric BMD is calculated (BMC/bone volume cm^3), differences between males and females disappear [4–6]. The highest BMD or bone mass known as peak bone mass is reached during the 3rd–4th decade of life. Bone loss starts shortly thereafter at some skeletal sites (lumbar spine and proximal femur) and a decade later at other skeletal sites [7]. As shown in Figure 1, two patterns of bone loss are recognized. A continuous, slow, age-related bone loss is observed in both men and women and results in an overall bone loss of 20–25% of both cortical and trabecular bone. In the perimenopausal period in women, a rapid phase of bone loss is observed during a period of 5–10 years around menopause. This phase leads to bone loss up to 14% but it varies in different studies. A decade after the menopause, the rapid phase of bone loss terminates and merges with the slow but progressive aged-related bone loss. The rate of bone loss varies between skeletal sites and is generally most pronounced in the spine being rich in the metabolically active trabecular bone. It is least pronounced in the hip and other sites rich in cortical bone [8].

Age related changes of structure and strength
The skeleton is composed of cortical bone (80%) making the outer dense envelop of most bones and trabecular bone (20%) located internal to the cortical bone at the end of long bones and in the vertebrae and other short or irregular bones. Trabecular bone forms a 3-dimensional network of trabeculae separated by marrow space. Trabecular bone has a higher surface to volume ratio and is more metabolically active than cortical bone. Structurally, bone tissue is sheathed by several envelopes: an outer periosteal envelope, the cortical Haversian envelope, the inner endosteal envelope (separated into the endocortical envelope and the trabecular envelope) [9, 10]. During growth and aging, bone tissue is formed or resorbed differently at these envelopes leading to the characteristic shape of each bone (a process termed bone modeling). Also the renewal of existing bone (a process termed bone remodeling) is

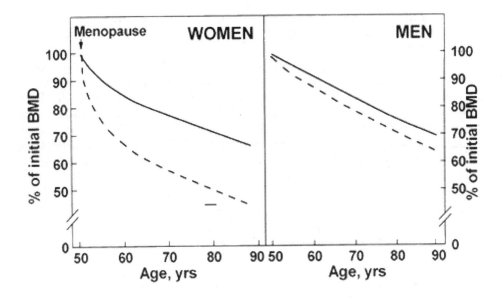

Figure 1. Schematic representation of changes in bone mass over life in cancellous (broken line) and cortical (solid line) bone in women (left panel) and men (right panel) from age 50 onward. In men only one phase of continuous bone loss is observed but in women two phase are recognized: a perimenopausal accelerated phase of bone loss and a late slow phase. Note also that the accelerated phase, but not the slow phase, involves disproportionate loss of cancellous bone [8].

initiated from these envelopes. This concept suggests that bone cells located at different skeletal envelopes react differently to stimuli, i.e., mechanical loading, hormones and cytokines [9, 10].

Trabecular bone mass and ash weight decrease with increasing age in both males and females (Table 1) [11, 12] and there is no difference between males and females with respect to trabecular bone volume or compressive strength. The 25% difference in peak bone mass observed between males and females appears to be caused mainly by differences in skeletal size [13]. Interestingly, cross-sectional studies on post mortem biopsies were not able to reveal differences in the rates of age-related trabecular bone loss between genders which may be explained by the lack of information on menopausal age and the effect of studying cohorts with different ages.

Most bones grow continuously during adulthood because of periosteal apposition [14]. The cross sectional area of vertebral bodies increases by 25% from age 30 to age 70 [11, 12]. However, specific skeletal sites lacking a periosteal coverage (like the femoral neck inside the joint capsule) is not affected by this process. The increase in bone size with age is thought to be more pronounced in men compared with women [15]. In contrast to the periosteal growth, there is a continuous endocortical bone resorption leading to an expansion of the marrow space and a decrease in cortical thickness with increasing age. These changes may lead to decreased bone strength. However, the increase in bone dimensions counteracts the decrease in structural thickness (Table 1).

The trabecular bone structure may be isotropic (equally structured and distributed in the 3-dimensional space) or anisotropic. In loaded areas (i.e., spine, femoral neck) the structure is usually anisotropic with thicker trabecular elements along the load trajectories and thinner supporting struts [16] which gives bones a considerable strength with smallest possible material use. During bone remodeling the thinner supporting struts are preferentially removed by perforations leaving the larger load bearing elements unsupported and subjected to enhanced buckling forces (Figure 2).

Table 1. Absolute and relative changes in trabecular bone mass, structure and biomechanical competence. The relative changes are expressed in percentage of value at peak bone mass

	Absolute Unit/year	Relative %/year	References
Iliac cresr (isotropic)			
Trabecular bone			
Bone volume (%)	−0.16	−0.73	11
Compressive strength (MPa)	−0.05	−1.06	13
Spine (anisotropic)			
Whole vertebral body			
L_2 Cross sectional area cm^2	0.06	0.95	13
Compressive strength (MPa)	−0.04	−0.34	13
Trabecular bone			15
Volume (%)	−0.12	−0.82	11
Ash weight (mg/cm^3)	−1.74	−0.89	15
Vertical direction			
Compressive strength (MPa)	−0.06	−1.43	15
Energy absorption (mJ/mm^2)	−0.01	−1.24	15
Strain (%)	+0.03	+0.45	15
Trabecular thickness (μm)	+0.13	+0.06	16
Trabecular distance (μm)	+6.74	+1.23	16
Horizontal direction			
Compressive strength (MPa)	−0.03	−1.60	15
Energy absorption (mJ/mm^2)	−0.01	−1.44	15
Strain (%)	+0.02	+0.32	15
Trabecular thickness (μm)	−1.03	−0.59	16
Trabecular distance (μm)	+13.74	+2.50	16

Strain indicates compressibility: (deformation at failure point divided by originally length) × 100%

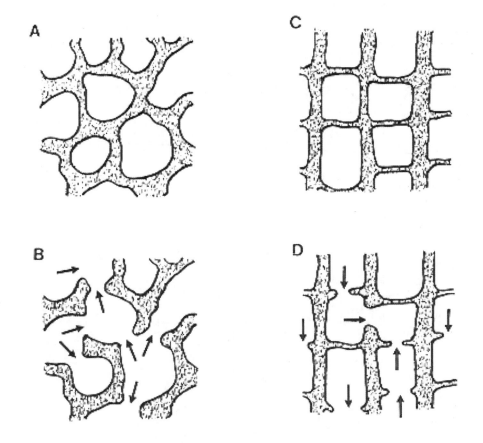

Figure 2. *Effect of repeated trabecular perforations on trabecular structure and integrity in isotropic (A–B) and anisotropic (C–D) trabecular bone.*

Gender-related structural differences are observed during aging. Several studies have demonstrated that trabecular thinning occurs with age in men, but not in women [17, 18]. In women, the main structural change is the loss of trabecular connectivity due to trabecular perforations [17, 18]. These structural changes combined with the thin cortical shell covering the epiphyseal regions may explain the occurrence of osteoporotic fractures at these sites and the sex differences in fracture risk.

Age-related changes in fracture risk

The risk of fractures increases with aging in both sexes. Osteoporotic fractures are defined as low energy fractures occurring spontaneously or during the forces of daily life including falls on the same level. They include spine, hip and forearm fractures and most fractures of the proximal humerus, pelvis, ribs, and ankle. Figure 3 shows

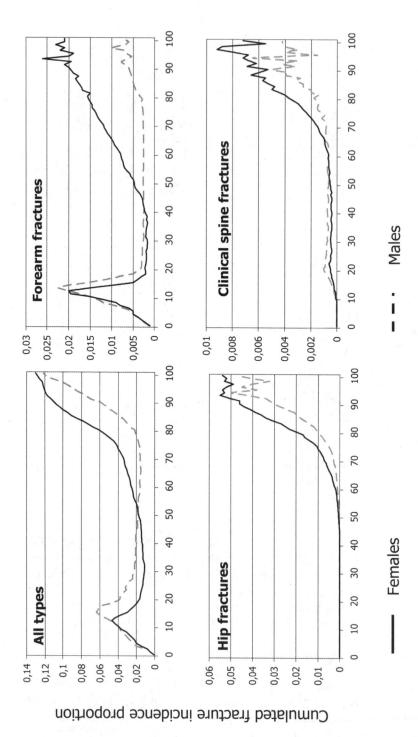

Figure 3. *Fracture risk in Denmark (population 5 million) 1995–99 according age and sex. Based on patients admitted to Danish Hospitals (Danish Hospital Central Register). (P. Vestergaard and Leif Mosekilde, unpublished data).*

the age- and gender-specific incidence of all fractures and selected osteoporotic fractures in Denmark. After 50 years of age the fracture risk increases exponentially in both sexes. However, the increase in fracture risk takes place approximately10 years later in males compared with females. The first fracture type to increase after the menopause is the forearm fracture (Figure 3) which often is related to falls during forward movement, where the energy of the fall is conveyed to the stretched forearm. Hip fractures often occur in elderly people during falls on the side when standing or walking slowly [19].

Cellular basis of bone loss and structural deterioration with aging

Our current understanding of the cellular mechanisms responsible for the above described age-related changes in bone mass and structure are based on quantitative studies of bone cell activities in bone biopsies obtained from iliac crest or vertebral bodies of aging human population and by employing histomorphometric techniques [9, 20, 21]. Bone as a tissue, is composed of bone matrix and bone cells. Bone matrix is built up of type I collagen (90%) and the remaining 10% is composed of a large number of non-collagenous proteins (e.g., osteocalcin, osteonectin, bone sialopro-teins and various proteoglycans). Non-collagenous proteins participate in the process of matrix maturation, mineralization and may regulate the functional activity of bone cells. Two main types of bone cells have been identified. Osteoblasts (bone forming cells) and osteoclasts (bone resorbing cells). These cells together with their precursor cells and associated cells (e.g., endothelial cells, nerve cells) are organized in specialized units called bone multicellular units (BMU) [20] that perform bone modeling and remodeling activities.

As mentioned above, the skeleton is continuously "remodeled" during adult life. Bone remodeling is a bone replacement mechanism maintaining the integrity of the skeleton by removing old bone of high mineral density and high prevalence of fatigue microfractures and replacing it with young bone of low mineral density and better mechanical properties [20]. This process is important for the biomechanical function of the skeleton and it also supports the role of the skeleton as an active participant in the divalent ion homeostasis. Bone remodeling consists of a specific sequence of cellular events with a defined temporal sequence occurring at the same anatomical location (Figure 4). It is the same sequence in both trabecular and cortical bone. The remodeling sequence is termed ARF sequence. "A" refers to the attraction of osteoclast precursors to specific bone sites where remodeling will take place. These sites are determined by specific mechanical needs or mechanical signals, the nature of which is not known. This is followed by activation to the osteoclast precursor cells to fuse and form functional multinucleated osteoclasts. "R" indicates the resorptive phase, where osteoclasts remove a certain thickness of mineralized bone tissue which can be measured histomorphometrically and known as erosion depth. This phase usually lasts 4–6 weeks (Figure 4). "F" refers to the formative phase where osteoblasts are recruited from stem cells and precursor cells in the bone marrow. They recreate the amount of bone matrix removed by the osteoclasts and secure a proper mineralization of the newly formed osteoid tissue. The amount of new bone formed

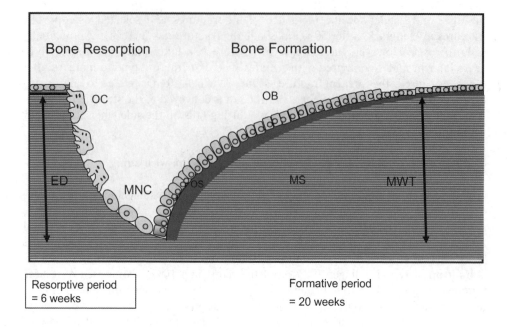

Figure 4. Trabecular bone remodeling following the A-R-F sequence (activation of osteoclasts, resorption by osteoclasts (OC) and mononuclear cells (MNC) and formation by osteoblasts (OB). ED, erosion depth; MWT, mean wall thickness; OS, osteoid (unmineralized bone) surface; MS, mineralized surface.

can also be measured histomorphometrically and known as mean wall thickness. The duration of the formative phase is usually 3–4 months. The end product of the process is a Bone Structural Unit (BSU). In cortical bone, the BSU is a cylindrical osteon formed by concentric lamella of bone surrounding a Haversian canal and enclosed by a cement line. In trabecular bone the BSU is termed a "packet" or a trabecular osteon which looks like an irregular pancake-like structure with a certain thickness (the mean wall thickness) enveloping the irregular trabecular structure.

Based on understanding of bone remodeling dynamics maintenance of stable bone mass depends on: (i) the balance between the osteoclastic activity indicated by the erosion depth and osteoblastic activity indicated by the mean wall thickness, and (ii) the number of remodeling cycles initiated in unit time per unit bone volume (termed the activation frequency). In the young adult, there is a balance between the amount of bone removed by osteoclasts and the amount of bone formed by osteoblast and bone mass is unchanged. Both the erosion depth [22] and the mean wall thickness [23] decrease with increasing age. However, in perimenopausal women estrogen deficiency is associated with hyperactive osteoclasts and increased bone resorption compared to bone formation [24]. On the other hand, age-related decreased mean wall thickness and impaired osteoblast functions have been observed in several histomorphometric studies in the elderly [25, 26].

In addition to age-related decrease in bone mass caused by imbalance of bone resorption and bone formation, aging is associated with architectural deterioration of the skeleton as outlined above. These changes are also caused by age-related changes in bone remodeling dynamics. An age-related increase in the activation frequency (turnover) or in resorption depth will by itself threaten the integrity of the 3-dimensional trabecular network [16]. During bone resorption, deep osteoclastic lacunae may hit thin trabecular structures leading to trabecular perforations. Concomitant remodeling processes on the opposite sides of thicker trabeculae may have the same consequence. The thinning of trabecular structures with age due to the imbalance between bone resorption and bone formation may also increase the risk of perforations. After perforation the unloaded remaining trabecular struts are rapidly removed by osteoclastic resorption [27]. The consequence of this process is a progressive loss of trabecular elements, deterioration of bones three-dimensional structure and a loss of mechanical strength with age. Complex calculations from trabecular density and intertrabecular distances suggest that age-related trabecular perforations and structural changes contribute more to the age-related decrease in bone strength compared with age-related decrease in bone mass [28]. Another mechanism leading to mechanical failure of the aged skeleton is the increased incidence of "stress raisers." Age-related increase in activation frequency and in the number of bone resorption sites lead to decreased trabecular thickness and create "stress raisers" between neighboring resorption sites resulting in decreased strength of the trabecular structure. The effect of antiresorptive treatment of osteoporosis (e.g., bisphosphonate) on decreasing fracture risk may be caused by reduction in activation frequency and bone turnover that lead to decreased trabecular perforations and the occurrence of stress raisers.

Mechanisms of age-related changes of the skeleton

As mentioned above aging is associated with several changes of behavior of bone cells. There is an increase in osteoclastic activity around menopause and decrease in osteoblastic activities in aged males and females. These changes are caused by a large number of factors. Intrinsic age-related changes in bone cells and age-related changes in the endocrine system are the universal factors determining the activity of bone cells. Also, genetic, environmental and behavioral factors may determine the variations observed between individuals in their peak bone mass, rates of bone loss or bone structural changes.

Age-related changes in bone cells

Similar to other cellular compartments in the aging body, bone cells undergo a multitude of age-related changes that lead to bone loss. *In vitro* models of bone cells provide a powerful tool for examining the effect of age *per se* on the biological functions of cells. Several investigators have examined the effect of age on the number and biological activities of bone cells obtained from donors at different ages. The results of these extensive studies are reviewed by Kassem *et al.* in "Aging of Cells In and Outside the Body (Volume 2, of this series).

Age-related changes in the endocrine system

Aging is associated with several changes in the endocrine system which in turn affects different organs in the body including the skeleton. Some of the best studied endocrine systems with respect to their impact on bone are: sex steroids, growth hormone (GH)/insulin-like growth factor (IGF) system and vitamin D (Figure 5).

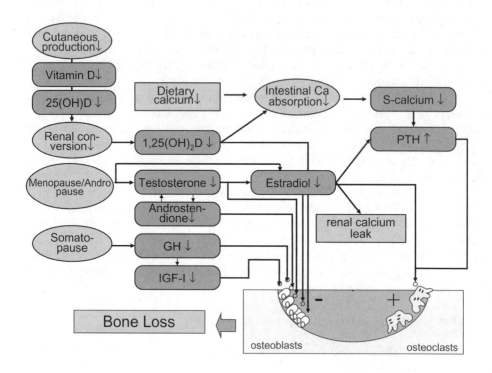

Figure 5. *Age-changes in the endocrine system and its contribution to the observed age-related bone loss. 25(OH)D, 25-hydroxyvitamin D; 1,25(OH)₂D, 1,25-dihydroxyvitamin D; PTH, parathyroid hormone; GH, growth hormone; IGF, insulin-like growth factor; Ca, calcium. All the changes in the endocrine system lead finally to increase (+) in osteoclastic bone resorption and inhibition (−) of osteoblastic bone formation leading to remodelling imbalance and bone loss.*

Sex steroids

Aging is associated with dramatic changes in the production of sex steroids. However, the pattern of changes is not similar between men and women. Women experience significant changes in sex steroid production around the menopause. Men on the other hand experience a gradual decline in sex steroid production. These changes exert profound effects on the skeleton.

In premenopausal women, more than 95% of serum estradiol (E_2) and estrone (E_1) are derived from ovarian secretion and the remainder is derived from peripheral conversion of adrenal androgens. In postmenopausal women, the ovarian estrogen production decreases and nearly all of the circulating estrogen (E) level is derived from peripheral conversion of steroid precursors. For serum testosterone in pre-menopausal women, 25% is derived from ovarian secretion, 25% from adrenal secretion and 50% from peripheral conversion. This is similar to postmenopausal women except that ovarian secretion of testosterone (T) decreases. The net result is a decrease in total E_1 from 221 pmol/L in young women to 133 pmol/L in elderly women and in E_2 from 338 pmol/L in young to 78 pmol/L in elderly women. There is a slight drop in T levels from 1.4 in young to 1.1 nmol/L in elderly women [29, 30].

In men the major potent circulating androgen, T is derived from testicular secretion. The main source of androgenic activity is 5α-dihydrotesterone (DHT) derived from T through the action of the enzyme 5α-reductase. Almost all circulating DHT arises from this extragonadal conversion and not a direct gonadal secretion. An age-related decrease in serum levels of total T from 20 nmol/L in young males to 16 nmol/L in elderly males is observed. However, DHT levels are maintained between young and old males (2.8 vs 3.2 nmol/L respectively) [29, 30]. Because of the binding characteristics of E and T to circulating binding proteins, bio-available hormone levels are better indicators for the biologically available hormone. Bioavailable E_2 decreases in women from 108 to 20 pmol/L and in men from 70 to 43 pmol/L between young and old age. Similarly, bio-available T decreases from 0.3 to 0.2 nmol/L in women and from 6.6 to 3.3 nmol/L in men between young and old age [29, 30]. The mechanisms of the observed decrease in free steroid hormone differs between women and men. In women, it is caused by an age-related ovarian failure and in men by an age-related increase in levels of sex hormone binding globulin (SHBG) [30].

Estrogen deficiency and bone loss in women
The rapid decrease of estrogen metabolites in the postmenopausal period leads to increased bone turnover, and osteoclast activity [24]. This leads to bone remodeling imbalance and consequently bone loss. The molecular basis of increased osteoclastic activity has recently been a topic of intensive investigation. Both osteoclasts and osteoblasts contain functional E receptors [31, 32]. E regulates several aspects of osteoclast recruitment, differentiation and function through changes in cytokine production in bone microenvironment and also through direct effects [33, 34]. E deficiency has been shown to increase the production of osteoclast-activating cytokines (IL-1, TNF-α, IL-6) and E treatment led to the inhibition of their production [33, 35]. *In vitro*, E treatment of osteoblastic cells increases gene expression of osteoprotegerin (OPG) which is the a principal regulator of osteoclast functions [36]. Also, E is able to induce apoptosis in osteoclasts and shortens their life span [37]. The direct effects of E on osteoblastic cell functions are less clear. Histomorphometric studies on bone biopsies obtained from perimenopausal women showed that E deficiency and E therapy affect bone resorption but not bone formation [24]. In some studies, however, E increases bone formation [38]. *In vitro*,

E can regulate osteoblast proliferation and the production of the growth factors known to affect osteoblast functions e.g., insulin-like growth factor (IGF)-1 and IGF binding proteins [39].

The slow age-related bone loss has usually been attributed to age-related secondary hyperparathyroidism and age-related impairment of osteoblast functions. Age-related secondary hyperparathyroidism is caused by age-related impaired mechanisms of calcium conservation. With increasing age, intestinal calcium absorption is impaired because of decreased production of 1,25-dihydroxyvitamine D [34]. Also, an age-related increased urinary calcium excretion (urinary calcium leak) has been reported [34]. Recently, Riggs et al. [34] have suggested that the age-related impaired mechanisms of calcium conservation and homeostasis are caused by the effects of E deficiency on intestine and kidneys [41, 42].

Testosterone and estrogen deficiency and bone loss in males

Men experience age-related decrease in bioavailable E and T. It has been generally accepted that age-related bone loss is caused by E deficiency in women and T deficiency in men. For examples, hypogonadal males exhibit severe bone loss and osteoporosis [43, 44] and serum T correlates with bone mass in some studies of elderly males [43, 45, 46]. Also, in animal models, pharmacologic blockade of androgen receptors led to bone loss [44, 45] and the non-aromatizable androgen dihydrotesterone (DHT) prevents orchidectomy-associated bone loss [45]. Also, testosterone and DHT exerts direct effects on bone cells with inhibition of osteoclast and stimulation of osteoblast functions [47, 48]. These data demonstrate the fundamental importance of androgen in skeletal homeostasis. However, several new studies suggest that E is more important than T for the maintenance of integrity enance of male skeleton and that age-related E deficiency in male is the main cause of bone loss. Three patients with either ER-α mutation or mutation in P-450 aromatase gene presented with un-fused epiphysis and osteopenia in spite of high levels of T have been reported [49, 50]. E treatment of a patient with aromatase gene mutation led to increased BMD suggesting that E and not T is the determinant for bone loss [51]. Furthermore, several population-based studies have demonstrated a stronger correlation between the bioavailable levels of E compared to T as a predictor of bone mineral density in men [30, 52, 53]. In a recent study by Falahti-Nini et al. [54] the authors performed a pharmacological blockade of endogenous production of sex steroids as well as inhibiting P-450 aromatase enzyme in a group of elderly men. Administration of E and not T was more effective in abolishing the sex-steroid deficiency-induced increased bone turnover [54]. Male skeleton, similar to that described above for female skeleton, responds to sex-hormone deficiency with increased bone-resorption and decreased bone formation leading to bone loss.

Growth hormone and insulin-like growth factors

Growth hormone (GH) promotes linear growth in children but also has a number of important physiological effects on the intermediary fuel metabolism, fluid homeostasis, body composition and bone metabolism in adults. GH is produced in the anterior pituitary gland in a pulsetile fashion and has both direct and indirect effects on target organs. Direct effects are mediated by a specific GH receptor [55] and the indirect effects are mediated by insulin-like growth factor-I (IGF-I). In circulation, most GH is bound to a specific GH binding protein (GHBP) which modifies the half-life of GH and its interactions with the GH-receptor. GH secretion is controlled by GH releasing hormone (GHRH) and somatostatin (SRIF) and possibly Ghrelin. GHRH and SRIF are in turn controlled by hypothalamus and through a negative feedback from serum IGF-I levels [56]. Serum levels of GH exhibit a pronounced circadian rhythm and GH secretion occurs in spikes observed mainly during night [56].

Serum levels of GH reach its peak in late puberty, and show a pronounced decline with aging. The age-related decline in serum GH can be explained by decreased secretion rate [57, 58] evidenced by a decreased number and amplitude of GH secretion spikes [57–59]. There is also an age-related increased in GH clearance rate [60]. Decreased GH secretion may be due to age-related failure of pituitary cells and impaired responsiveness to GHRH [61]. Other factors may also contribute to the age-related decline in GH production. Aging is associated with increased total and visceral fat and serum GH decreases with increasing body mass index independent of age [58, 62]. Moreover, the age-related decline in physical fitness contributes to the decline in serum GH with age [63]. In women, GH levels decrease following the menopause [64]. In men, the relationship between serum levels of testosterone and GH is unclear [56].

Both IGF-I and IGF-II act as hormones and as auto- and paracrine growth factors. Effects of IGF-I and II are mediated by specific receptors. Circulating IGF-I and II are bound to binding proteins, six of which (IGFBP-1 through 6) have been described [56]. The IGFBPs modify the half-life, distribution space of the IGFs and also modulate the effects of IGFs on target tissues and serve as carriers transporting IGFs selectively to their target tissues including bone [65]. Most tissues produce IGF-I, IGF-II, or both.

Serum concentrations of IGF-I largely parallel serum GH with a peak at puberty and a decrease with aging. Serum IGF-I, but not IGF-II correlates closely to 24-h integrated GH secretion [66]. Similarly, serum levels of IGF-I and not IGF-II decrease with age in both men and women [56, 58, 59, 67, 68]. Serum concentrations of IGF-I vary inversely with body mass index [59, 67] and directly with aerobic capacity [69, 70]. Levels are also decreased in postmenopausal compared with pre-menopausal women [71, 72].

The age-related decline in GH and IGF-I parallels the decline in BMD suggesting that changes in serum GH and IGF-I are responsible for the age-related bone-loss. Several *in vitro* studies have demonstrated that GH can regulate the proliferation and differentiation of osteoblasts [73, 74]. Moreover, osteoblasts produce both IGF-I [74] and IGF-II [73, 74] and some studies have shown that GH increases osteoblastic IGF production. IGF-I is also an important controller of osteoblast cell proliferation and differentiation [75–79]. Also, GH and IGF can stimulate osteoclast differentiation and activity through changes in the osteblast production of cytokines e.g., osteoprotegerin (OPG) and RANK-L [80–83]. The effects of GH on osteoblasts and osteoclasts functions have been demonstrated *in vivo* by showing increased levels of biochemical markers of bone formation (e.g., osteocalcine) and bone resorption (collagen degradation products) after administration of GH to human subjects [84, 85].

It seems unlikely that GH and IGF are major factors contributing to the skeletal phenotype of senescence. Administration of GH to healthy elderly persons was unable to restore and only increased bone mass slightly [86, 87]. Moreover, the age-related decline in serum GH and IGF-I cannot explain the increase in bone remodeling observed with aging since *in vivo* administration of GH or IGF-I increases bone turnover [84, 85]. On the other hand open-labeled long-term studies in patients with GH-deficiency suggest that GH substitution in this group of patients have some positive effects on bone mass [88]. It is therefore plausible that GH and IGF-I responsible for bone loss in subgroup of osteoporotic patients with abnormally low levels of the hormones.

Vitamin D system

Depending on the intensity of solar radiation about 80% of body vitamin D is produced in the skin and the rest is absorbed from the intestine. Hence, for most populations in the world the cutaneous production is the main source of vitamin D with the exception of people living on high northern or southern latitudes [89, 90]. Vitamin D is produced in the skin from 7-dehydrocholesterol (provitamin D_3) by irradiation with UVB (wavelength 290–315 nm). The photochemical process leads to initial formation of previtamin D_3, which by a temperature sensitive process is later rearranged to cholecalciferol or vitamin D_3. The cutaneous production of vitamin D decreases with increasing age. Following a standardized dermal solar exposure, young adults can produce 2–3 times more previtamin D_3 in the skin than older people [91]. Consequently, the epidermal content of provitamin D_3 decreases with age [91].

The dietary vitamin D content is generally low and dispersed in few foods, mainly fatty fish and codfish liver oil. Smaller amounts are found in liver, eggs, and milk. After ingestion of a single dose of 50 000 IU of vitamin D, the plasma concentration increases after 4 hours, reaches is maximum after 12 hours and returns to baseline after 72 hours [92]. Old age does not by itself influence the intestinal absorption [93]. Vitamin D is hydroxylated in the liver to 25-hydroxyvitamin D (25-OHD) and in the kidney to 1,25-dihydroxyvitamin D (1,25-$(OH)_2$D). The hepatic conversion is usually not limited and plasma concentrations of 25-OHD are the best estimate of individual vitamin D status. Based on these measurements a growing amount of data demonstrates a high prevalence of vitamin D deficiency in the elderly, particularly in those not living independently [94, 95]. The prevalence of vitamin D deficiency depends on sunlight intensity (latitude), sunlight exposure and dietary practices and appears to be more common in Europe than in North America [96].

The renal 25(OH)D-1α-hydroxylase is a mitochondrial cytochrome P-450 oxidase localized in the epithelial cells in the proximal convolutes tubules. Its synthesis is stimulated by PTH and hypophosphatemia and inhibited by phosphate and calcium loading. Therefore, under normal circumstances only small measured amounts of this active metabolite are produced. However, the renal production of 1,25-$(OH)_2$D decreases with increasing age because of the reduction in renal mass [97]. Further-more, the increase in plasma 1,25-$(OH)_2D$ following PTH infusion decreases with age in normal individuals [97] and in osteoporotic patients [98, 99]. The production of 1,25-$(OH)_2D$ may be stimulated by estrogen, GH or IGF-1 which decreases with age and postmenopausal state. Thus, it appears that age-related reduction in 1,25-$(OH)_2D$ may be caused by several partly independent mechanisms.

Vitamin D deficiency exerts deleterious effects on the skeleton. It leads to secondary hyperparathyroidism especially during winter periods and increased bone turnover (activation frequency) leading to bone loss and reduced bone mass [100]. Patients sustaining hip fractures have lower plasma 25-OHD concentrations than age matched controls [95] and may exhibit frank osteomalacia [101]. The importance of vitamin D (and calcium) for the age related changes in the skeleton is further emphasized by the effect of vitamin D and calcium supplementation on decreasing fracture incidence in the elderly [21, 102]. We have recently shown that a dose of 400 units of vitamin D and 800 mg calcium as carbonate was more efficient in preventing osteoporotic fractures among independently living elderly Danes compared to a fall and fracture preventing program or no intervention (Figure 6) (Larsen and Mosekilde, unpublished results).

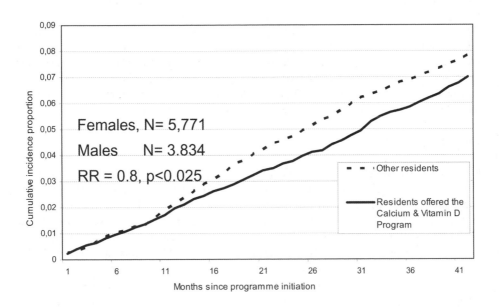

Figure 6. The Randers City Study. The effect of calcium and vitamin D supplementation on the risk of osteoporotic fractures among elderly residents (>65 yrs.). Proportional hazard model and intention to prevent analysis were employed.

Genetic, environmental and individual risk factors
Bone mass in the elderly is the result of an interaction between peak bone mass and the rate of bone loss. Peak bone mass and the rate of bone loss are affected by a multitude of factors including genetic, behavioral and dietary. They are also affected by diseases and medications received by the persons throughout their life history.

Genetic factors
Based on a number of twin studies, the heritable component of peak bone mass has been estimated to be 50–70% but the heritable component of age-related bone loss is thought to be much less [103, 104]. Several studies have shown that part of the varaitions of bone mass of adult skeleton can be explained by polymorphic traits in a number of key extracellular matrix components (collagen type I), hormones receptors (vitamin D receptors, ER, AR, PTH/PTHrp receptors), cytokines (OPG, RANKL, TGF-β). However, the relative contributions of each of these polymorphic traits to age-related bone loss need to be determined [103,104].

Environmental, life style and individual risk factors
Smoking, large alcohol intake, exercise levels, decreased in muscle strength due to aging or specific neuromuscular disorders, diet and diseases affecting the skeleton (e.g., hyperthyroidism, anorexia nervosa, chronic exposure to glucocorticoids) are some of a potentially long list of factors that are capable of affecting bone mass and skeletal integrity. These factors can interact with the universal mechanisms of age-related bone loss described above. For example, smoking accelerates bone loss in both sexes after middle age and increases the risk of hip fractures [105]. Furthermore, in epidemiological studies the risk of hip fracture depends on milk intake and present and past sun exposure [106]. Finally, supplementation with calcium and vitamin D has protective effects on the incidence of osteoporotic fractures as mentioned above.

Concluding remarks

Understanding the pathophysiological mechanisms responsible for skeletal aging is important to develop effective prevention approaches to avoid the devastating consequences of bone fractures. Genetic, environmental, nutritional, biomechanical and hormonal factors determine the integrity of the skeleton and age-related bone loss. Gaps still exist in our knowledge about the relative contributions of all these factors to the observed fragility of the aged skeleton. Hopefully studies using the modern cell biology, molecular biology techniques and gene modified animal models together with insights obtained from epidemiological studies of large cohorts of normal aging population will provide us with the necessary information required for developing effective strategies for maintaining a healthy skeleton in old age.

References

1. Riggs BL, Melton LJ III (1986). Involutional osteoporosis. *N Engl J Med.* 314: 1676–86.
2. Marshall D, Johnell O, Wedel H (1996). Meta-analysis of how well measures of bone mineral density predict occurrence of osteoporotic fractures. *Br Med J.* 312: 1254–9.
3. Seeman E, Delmas PD (2001). Reconstructing the skeleton with intermittent parathyroid hormone. *Trends Endocrinol Metab.* 12: 281–3.
4. Gilsanz V, Kovanlikaya A, Costin G, Roe TF, Sayre J, Kaufman F (1997). Differential effect of gender on the sizes of the bones in the axial and appendicular skeletons. *J Clin Endocrinol Metab.* 82: 1603–7.
5. Gilsanz V, Skaggs DL, Kovanlikaya A, *et al.* (1998). Differential effect of race on the axial and appendicular skeletons of children. *J Clin Endocrinol Metab.* 83: 1420–7.
6. Lu PW, Cowell CT, LLoyd-Jones SA, Briody JN, Howman-Giles R (1996). Volumetric bone mineral density in normal subjects, aged 5–27 years. *J Clin Endocrinol Metab.* 81: 1586–90.
7. Matkovic V, Jelic T, Wardlaw GM, *et al.* (1994). Timing of peak bone mass in Caucasian females and its implication for the prevention of osteoporosis. Inference from a cross-sectional model. *J Clin Invest.* 93: 799–808.
8. Riggs BL, Khosla S, Melton LJ III (1998). A unitary model for involutional osteoporosis: estrogen deficiency causes both type I and type II osteoporosis in postmenopausal women and contributes to bone loss in aging men. *J Bone Miner Res.* 13: 763–73.
9. Frost HM (2001). The Utah paradigm of skeletal physiology: what is it? *Vet Comp Orthop Traumatol.* 14: 179–84.
10. Frost HM, Burr DB (1997). On our age-related bone loss: insights from a new paradigm. *J Bone Miner Res.* 12: 1539–51.
11. Mosekilde L, Mosekilde L (1988). Iliac crest trabecular bone volume as predictor for vertebral compressive strength, ash density and trabecular bone volume in normal individuals. *Bone* 9: 195–9.
12. Mosekilde L, Mosekilde L (1986). Normal vertebral body size and compressive strength: relations to age and to vertebral and iliac trabecular bone compressive strength. *Bone* 7: 207–12.
13. Mosekilde L, Mosekilde L, Danielsen CC (1987). Biomechanical competence of vertebral trabecular bone in relation to ash density and age in normal individuals. *Bone* 8: 79–85.
14. Garn SM, Rohmann CG, Wagner B, Ascoli W (1967). Continuing bone growth throughout life: a general phenomenon. *Am J Phys Anthropol.* 26: 313–17.
15. Ebbesen EN, Thomsen JS, Beck-Nielsen H, Nepper-Rasmussen HJ, Mosekilde L (1999). Age- and gender-related differences in vertebral bone mass, density, and strength. *J Bone Miner Res.* 14: 1394–403.
16. Mosekilde L (1988). Age-related changes in vertebral trabecular bone architecture–assessed by a new method. *Bone* 9: 247–50.
17. Aaron JE, Makins NB, Sagreiya K (1987). The microanatomy of trabecular bone loss in normal aging men and women. *Clin Orthop.* 260–71.
18. Mellish RW, Garrahan NJ, Compston JE (1989). Age-related changes in trabecular width and spacing in human iliac crest biopsies. *Bone Miner.* 6: 331–8.
19. Cummings SR, Nevitt MC (1989). A hypothesis: the causes of hip fractures. *J Gerontol.* 44: M107–11.

20.   Parfitt AM (1991). Bone forming cells in clinical conditions. In: BK Hall, ed. *Bone: The Osteoblast and Osteocyte*. London: The Telford Press, pp. 351–426.

21.   Frost HM, Vilanueva AR, Jett S, Eyring E (1969). Tetracycline-based analysis of bone remodelling in osteopetrosis. *Clin Orthop*. 65: 203–17.

22.   Eriksen EF, Melsen F, Mosekilde L (1984). Reconstruction of the resorptive site in iliac trabecular bone: a kinetic model for bone resorption in 20 normal individuals. *Metab Bone Dis Related Res*. 5: 235–42.

23.   Eriksen EF, Gundersen HJ, Melsen F, Mosekilde L (1984). Reconstruction of the formative site in iliac trabecular bone in 20 normal individuals employing a kinetic model for matrix and mineral apposition. *Metab Bone Dis Related Res*. 5: 243–52.

24.   Eriksen EF, Langdahl B, Vesterby A, Rungby J, Kassem M (1999). Hormone replacement therapy prevents osteoclastic hyperactivity: a histomorphometric study in early postmenopausal women. *J Bone Miner Res*. 14: 1217–21.

25.   Cohen-Solal ME, Shih MS, Lundy MW, Parfitt AM (1991). A new method for measuring cancellous bone erosion depth: application to the cellular mechanisms of bone loss in postmenopausal osteoporosis. *J Bone Miner Res*. 6: 1331–8.

26.   Eriksen EF, Hodgson SF, Eastell R, Cedel SL, O'Fallon WM, Riggs BL (1990). Cancellous bone remodeling in type I (postmenopausal) osteoporosis: quantitative assessment of rates of formation, resorption, and bone loss at tissue and cellular levels. *J Bone Miner Res*. 5: 311–19.

27.   Mosekilde L (1990). Consequences of the remodelling process for vertebral trabecular bone structure: a scanning electron microscopy study (uncoupling of unloaded structures). *Bone Miner*. 10: 13–35.

28.   Parfitt AM, Mathews CH, Villanueva AR, Kleerekoper M, Frame B, Rao DS (1983). Relationships between surface, volume, and thickness of iliac trabecular bone in aging and in osteoporosis. Implications for the microanatomic and cellular mechanisms of bone loss. *J Clin Invest*. 72: 1396–409.

29.   Labrie F, Belanger A, Cusan L, Gomez JL, Candas B (1997). Marked decline in serum concentrations of adrenal C19 sex steroid precursors and conjugated androgen metabolites during aging. *J Clin Endocrinol Metab*. 82: 2396–402.

30.   Khosla S, Melton LJ III, Atkinson EJ, O'Fallon WM, Klee GG, Riggs BL (1998). Relationship of serum sex steroid levels and bone turnover markers with bone mineral density in men and women: a key role for bioavailable estrogen. *J Clin Endocrinol Metab*. 83: 2266–74.

31.   Eriksen EF, Colvard DS, Berg NJ, *et al*. (1988). Evidence of estrogen receptors in normal human osteoblast-like cells. *Science* 241: 84–6.

32.   Oursler MJ, Osdoby P, Pyfferoen J, Riggs BL, Spelsberg TC (1991). Avian osteoclasts as estrogen target cells. *Proc Natl Acad Sci USA* 88: 6613–17.

33.   Manolagas SC (2000). Birth and death of bone cells: basic regulatory mechanisms and implications for the pathogenesis and treatment of osteoporosis. *Endocr Rev*. 21: 115–37.

34.   Riggs BL, Khosla S, Melton LJ (2002). Sex steroids and the construction and conservation of the adult skeleton. *Endocr Rev*. 23: 279–302.

35.   Pacifici R (1996). Estrogen, cytokines, and pathogenesis of postmenopausal osteoporosis. *J Bone Miner Res*. 11: 1043–51.

36.   Hofbauer LC, Khosla S, Dunstan CR, Lacey DL, Spelsberg TC, Riggs BL (1999). Estrogen stimulates gene expression and protein production of osteoprotegerin in human osteoblastic cells. *Endocrinology* 140: 4367–70.

37.   Hughes DE, Dai A, Tiffee JC, Li HH, Mundy GR, Boyce BF (1996). Estrogen promotes apoptosis of murine osteoclasts mediated by TGF-beta. *Nat Med.* 2: 1132–6.

38.   Tobias JH, Compston JE (1999). Does estrogen stimulate osteoblast function in postmenopausal women? *Bone* 24: 121–4.

39.   Kassem M, Okazaki R, De Leon D, *et al.* (1996). Potential mechanism of estrogen-mediated decrease in bone formation: estrogen increases production of inhibitory insulin-like growth factor- binding protein-4. *Proc Assoc Am Physicians* 108: 155–64.

40.   Oursler MJ, Cortese C, Keeting P, *et al.* (1991). Modulation of transforming growth factor-beta production in normal human osteoblast-like cells by 17 beta-estradiol and parathyroid hormone. *Endocrinology* 129: 3313-3320.

41.   Heshmati HM, Khosla S, Robins SP, Geller N, McAlister CA, Riggs BL (1997). Endogenous residual estrogen levels determine bone resorption even in late postmeno-pausal women. *J Bone Miner Res.* 12: 76.

42.   McKane WR, Khosla S, Risteli J, Robins SP, Muhs JM, Riggs BL (1997). Role of estrogen deficiency in pathogenesis of secondary hyperparathyroidism and increased bone resorption in elderly women. *Proc Assoc Am Physicians* 109: 174–80.

43.   Murphy S, Khaw KT, Cassidy A, Compston JE (1993). Sex hormones and bone mineral density in elderly men. *Bone Miner.* 20: 133–40.

44.   Orwoll E (1999). Androgens. In: Rosen CJ, Glowacki J, Bilezikian JP, eds. *The Aging Skeleton.* New York: Academic Press, pp. 521–39.

45.   Vanderschueren D, Boonen S, Bouillon R (1998). Action of androgens versus estrogens in male skeletal homeostasis. *Bone* 23: 391–4.

46.   Rudman D, Drinka PJ, Wilson CR, *et al.* (1994). Relations of endogenous anabolic hormones and physical activity to bone mineral density and lean body mass in elderly men. *Clin Endocrinol.* 40: 653-61.

47.   Oursler MJ, Kassem M, Turner R, Riggs BL, Spelsberg TC (1996). *Regulation of Bone Cell Function by Gonadal Steroids: Osteoporosis.* New York: Academic Press, pp. 237–60.

48.   Kasperk C, Fitzsimmons R, Strong D, *et al.* (1990). Studies of the mechanism by which androgens enhance mitogenesis and differentiation in bone cells. *J Clin Endocrinol Metab.* 71: 1322–9.

49.   Morishima A, Grumbach MM, Simpson ER, Fisher C, Qin K (1995). Aromatase deficiency in male and female siblings caused by a novel mutation and the physiological role of estrogens. *J Clin Endocrinol Metab.* 80: 3689–98.

50.   Smith EP, Boyd J, Frank GR, *et al.* (1994). Estrogen resistance caused by a mutation in the estrogen-receptor gene in a man. *N Engl J Med.* 331: 1056–61.

51.   Bilezikian JP, Morishima A, Bell J, Grumbach MM (1998). Increased bone mass as a result of estrogen therapy in a man with aromatase deficiency. *N Engl J Med.* 339: 599–603.

52.   Slemenda CW, Longcope C, Zhou L, Hui SL, Peacock M, Johnston CC (1997). Sex steroids and bone mass in older men. Positive associations with serum estrogens and negative associations with androgens. *J Clin Invest.* 100: 1755–9.

53.   Greendale GA, Edelstein S, Barrett-Connor E (1997). Endogenous sex steroids and bone mineral density in older women and men: the Rancho Bernardo Study. *J Bone Miner Res.* 12: 1833–43.

54.   Falahati-Nini A, Riggs BL, Atkinson EJ, O'Fallon WM, Eastell R, Khosla S (2000). Relative contributions of testosterone and estrogen in regulating bone resorption and formation in normal elderly men. *J Clin Invest.* 106: 1553–60.

55.   Leung DW, Spencer SA, Cachianes G, *et al.* (1987). Growth hormone receptor and serum binding protein: purification, cloning and expression. *Nature* 330: 537–43.

56. Corpas E, Harman SM, Blackman MR (1993). Human growth hormone and human aging. *Endocr Rev.* 14: 20–39.

57. Finkelstein JW, Roffwarg HP, Boyar RM, Kream J, Hellman L (1972). Age-related change in the twenty-four-hour spontaneous secretion of growth hormone. *J Clin Endocrinol Metab.* 35: 665–70.

58. Ho KY, Evans WS, Blizzard RM, *et al.* (1987). Effects of sex and age on the 24-hour profile of growth hormone secretion in man: importance of endogenous estradiol concentrations. *J Clin Endocrinol Metab.* 64: 51–8.

59. Rudman D, Kutner MH, Rogers CM, Lubin MF, Fleming GA, Bain RP (1981). Impaired growth hormone secretion in the adult population: relation to age and adiposity. *J Clin Invest.* 67: 1361–9.

60. Iranmanesh A, Lizarralde G, Veldhuis JD (1991). Age and relative adiposity are specific negative determinants of the frequency and amplitude of growth hormone (GH) secretory bursts and the half-life of endogenous GH in healthy men. *J Clin Endocrinol Metab.* 73: 1081–8.

61. Bellantoni MF, Harman SM, Cho DE, Blackman MR (1991). Effects of progestin-opposed transdermal estrogen administration on growth hormone and insulin-like growth factor-I in postmenopausal women of different ages. *J Clin Endocrinol Metab.* 72: 172–8.

62. Clasey JL, Weltman A, Patrie J, *et al.* (2001). Abdominal visceral fat and fasting insulin are important predictors of 24-hour GH release independent of age, gender, and other physiological factors. *J Clin Endocrinol Metab.* 86: 3845–52.

63. Holt RI, Webb E, Pentecost C, Sonksen PH (2001). Aging and physical fitness are more important than obesity in determining exercise-induced generation of GH. *J Clin Endocrinol Metab.* 86: 5715–20.

64. Weissberger AJ, Ho KK, Lazarus L (1991). Contrasting effects of oral and transdermal routes of estrogen replacement therapy on 24-hour growth hormone (GH) secretion, insulin-like growth factor I, and GH-binding protein in postmenopausal women. *J Clin Endocrinol Metab.* 72: 374–81.

65. Schmid C, Ernst M (1992). Insulin-like growth factors. In: Green H, ed. *Cytokines and Bone Metabolism.* London: CRC Press, pp. 229–65.

66. Florini JR, Prinz PN, Vitiello MV, Hintz RL (1985). Somatomedin-C levels in healthy young and old men: relationship to peak and 24-hour integrated levels of growth hormone. *J Gerontol.* 40: 2–7.

67. Copeland KC, Colletti RB, Devlin JT, McAuliffe TL (1990). The relationship between insulin-like growth factor-I, adiposity, and aging. *Metabolism* 39: 584–7.

68. Bennett AE, Wahner HW, Riggs BL, Hintz RL (1984). Insulin-like growth factors I and II: aging and bone density in women. *J Clin Endocrinol Metab.* 59: 701–4.

69. Poehlman ET, Copeland KC (1990). Influence of physical activity on insulin-like growth factor-I in healthy younger and older men. *J Clin Endocrinol Metab.* 71: 1468–73.

70. Kelly PJ, Eisman JA, Stuart MC, Pocock NA, Sambrook PN, Gwinn TH (1990). Somatomedin-C, physical fitness, and bone density. *J Clin Endocrinol Metab.* 70: 718–23.

71. Franchimont P, Urbain-Choffray D, Lambelin P, Fontaine MA, Frangin G, Reginster JY (1989). Effects of repetitive administration of growth hormone-releasing hormone on growth hormone secretion, insulin-like growth factor I, and bone metabolism in postmenopausal women. *Acta Endocrinol (Copenh).* 120: 121–8.

AGING OF THE HUMAN SKELETON 135

72. Romagnoli E, Minisola S, Carnevale V, *et al.* (1993). Effect of estrogen deficiency on IGF-I plasma levels: relationship with bone mineral density in perimenopausal women. *Calcif Tissue Int.* 53: 1–6.

73. Kassem M, Blum W, Ristelli J, Mosekilde L, Eriksen EF (1993). Growth hormone stimulates proliferation and differentiation of normal human osteoblast-like cells *in vitro. Calcif Tissue Int.* 52: 222–6.

74. Slootweg MC, Most WW, van Beek E, Schot LP, Papapoulos SE, Lowik CW (1992). Osteoclast formation together with interleukin-6 production in mouse long bones is increased by insulin-like growth factor-I. *J Endocrinol.* 132: 433–8.

75. Mohan S, Strong DD, Lempert UG, Tremollieres F, Wergedal JE, Baylink DJ (1992). Studies on regulation of insulin-like growth factor binding protein (IGFBP)-3 and IGFBP-4 production in human bone cells. *Acta Endocrinol (Copenh).* 127: 555–64.

76. Wergedal JE, Mohan S, Lundy M, Baylink DJ (1990). Skeletal growth factor and other growth factors known to be present in bone matrix stimulate proliferation and protein synthesis in human bone cells. *J Bone Miner Res.* 5: 179–86.

77. Chenu C, Valentin-Opran A, Chavassieux P, Saez S, Meunier PJ, Delmas PD (1990). Insulin like growth factor I hormonal regulation by growth hormone and by 1,25(OH)2D3 and activity on human osteoblast-like cells in short-term cultures. *Bone* 11: 81–6.

78. Ernst M, Froesch ER (1987). Osteoblastlike cells in a serum-free methylcellulose medium form colonies: effects of insulin and insulinlike growth factor I. *Calcif Tissue Int.* 40: 27–34.

79. Schmid C, Guler HP, Rowe D, Froesch ER (1989). Insulin-like growth factor I regulates type I procollagen messenger ribonucleic acid steady state levels in bone of rats. *Endocrinology* 125: 1575–80.

80. Rubin J, Ackert-Bicknell CL, Zhu L, *et al.* (2002). IGF-I regulates osteoprotegerin (OPG) and receptor activator of nuclear factor-kappaB ligand *in vitro* and OPG *in vivo. J Clin Endocrinol Metab.* 87: 4273–9.

81. Zhang CZ, Young WG, Li H, Clayden AM, Garcia-Aragon J, Waters MJ (1992). Expression of growth hormone receptor by immunocytochemistry in rat molar root formation and alveolar bone remodeling. *Calcif Tissue Int.* 50: 541–6.

82. Lewinson D, Shenzer P, Hochberg Z (1993). Growth hormone involvement in the regulation of tartrate-resistant acid phosphatase-positive cells that are active in cartilage and bone resorption. *Calcif Tissue Int.* 52: 216–21.

83. Andrew JG, Hoyland J, Freemont AJ, Marsh D (1993). Insulinlike growth factor gene expression in human fracture callus. *Calcif Tissue Int.* 53: 97–102.

84. Brixen K, Kassem M, Nielsen HK, Loft AG, Flyvbjerg A, Mosekilde L (1995). Short-term treatment with growth hormone stimulates osteoblastic and osteoclastic activity in osteopenic postmenopausal women: a dose response study. *J Bone Miner Res.* 10: 1865–74.

85. Kassem M, Brixen K, Blum WF, Mosekilde L, Eriksen EF (1994). Normal osteoclastic and osteoblastic responses to exogenous growth hormone in patients with postmenopausal spinal osteoporosis. *J Bone Miner Res.* 9: 1365–70.

86. Christmas C, O'Connor KG, Harman SM, *et al.* (2002). Growth hormone and sex steroid effects on bone metabolism and bone mineral density in healthy aged women and men. *J Gerontol A Biol Sci Med Sci.* 57: M12–18.

87. Rudman D, Feller AG, Nagraj HS, *et al.* (1990). Effects of human growth hormone in men over 60 years old. *N Engl J Med.* 323: 1–6.

88.   Kann P, Piepkorn B, Schehler B, *et al.* (1998). Effect of long-term treatment with GH on bone metabolism, bone mineral density and bone elasticity in GH-deficient adults. *Clin Endocrinol (Oxf)*. 48: 561–8.

89.   Ladizesky M, Lu Z, Oliveri B, *et al.* (1995). Solar ultraviolet B radiation and photoproduction of vitamin D3 in central and southern areas of Argentina. *J Bone Miner Res*. 10: 545–9.

90.   Holick MF (1994). Sunlight, vitamin D and human health. In: Holick MF, Jung EG, eds. *Proceedings, Symposium on the Biological Effects of Light*. Berlin: Walter De Gruyter & Co, pp. 3–15.

91.   MacLaughlin J, Holick MF (1985). Aging decreases the capacity of human skin to produce vitamin D3. *J Clin Invest*. 76: 1536–8.

92.   Lo CW, Paris PW, Clemens TL, *et al.* (1985). Vitamin D adsorption in healthy subjects and in patients with intestinal malabsorption syndromes. *Am J Clin Nutr*. 42: 644–9.

93.   Holick MF (1986). Vitamin D requirements for the elderly. *Clin Nutr*. 5: 121–9.

94.   Webb AR, Pilbeam C, Hanafin N, Holick MF (1990). An evaluation of the relative contributions of exposure to sunlight and of diet to the circulating concentrations of 25-hydroxyvitamin D in an elderly nursing home population in Boston. *Am J Clin Nutr*. 51: 1075–81.

95.   Lips P, van Ginkel FC, Jongen MJ, Rubertus F, van der Vijgh WJ, Netelenbos JC (1987). Determinants of vitamin D status in patients with hip fracture and in elderly control subjects. *Am J Clin Nutr*. 46: 1005–10.

96.   McKenna MJ (1992). Differences in vitamin D status between countries in young adults and the elderly. *Am J Med*. 93: 69–77.

97.   Tsai KS, Heath H, III, Kumar R, Riggs BL (1984). Impaired vitamin D metabolism with aging in women. Possible role in pathogenesis of senile osteoporosis. *J Clin Invest*. 73: 1668–72.

98.   Riggs BL, Hamstra A, DeLuca HF (1981). Assessment of 25-hydroxyvitamin D 1 alpha-hydroxylase reserve in postmenopausal osteoporosis by administration of parathyroid extract. *J Clin Endocrinol Metab*. 53: 833–5.

99.   Slovik DM, Adams JS, Neer RM, Holick MF, Potts JT Jr (1981). Deficient production of 1,25-dihydroxyvitamin D in elderly osteoporotic patients. *N Engl J Med*. 305: 372–4.

100. Khaw KT, Sneyd MJ, Compston J (1992). Bone density parathyroid hormone and 25-hydroxyvitamin D concentrations in middle aged women. *Br Med J*. 305: 273–7.

101. Meunier P, Aaron J, Edouard C, Vignon G (1971). Osteoporosis and the replacement of cell populations of the marrow by adipose tissue. A quantitative study of 84 iliac bone biopsies. *Clin Orthop*. 80: 147–54.

102. Baeksgaard L, Andersen KP, Hyldstrup L (1998). Calcium and vitamin D supplementation increases spinal BMD in healthy, postmenopausal women. *Osteoporosis Int*. 8: 255–60.

103. Nguyen TV, Blangero J, Eisman JA (2000). Genetic epidemiological approaches to the search for osteoporosis genes. *J Bone Miner Res*. 15: 392–401.

104. Ralston SH (2002). Genetic control of susceptibility to osteoporosis. *J Clin Endocrinol Metab*. 87: 2460–6.

105. Law MR, Hackshaw AK (1997). A meta-analysis of cigarette smoking, bone mineral density and risk of hip fracture: recognition of a major effect. *Br Med J*. 315: 841–6.

106. Johnell O, Gullberg B, Kanis JA, *et al.* (1995). Risk factors for hip fracture in European women: the MEDOS Study. Mediterranean Osteoporosis Study. *J Bone Miner Res*. 10: 1802–15.

Aging of the Cardiovascular System

James D. Cameron and Christopher J. Bulpitt

Department of Vascular Sciences, Dandenong Hospital Southern Health Network and Monash University Melbourne Australia and Section of Care of the Elderly, Hammersmith Campus, Imperial College London, UK

Aging can be considered as the time-related loss of optimum function of a system and as being associated with increased risk of failure. In biological systems a number of theories of aging have been proposed ranging from the molecular to cellular to the whole organism [1]. In practice the underlying cause of the effects we associate with age and the inevitable death of the organism are clearly multifactorial and contain environmental elements as well as random shocks and genetic programing. Although life expectancy is increasing for both men and women, the eventual cause of death of between 40 and 50 percent of individuals in westernized societies is classified as cardiovascular disease (ICD-9 codes 390-459 [2]).

A number of characteristics of the cardiovascular system are known to diminish with age, including conduit arterial function and vascular endothelial function. Superimposed on normal aging is the process of atherosclerosis, a predominant cause of functional and structural arterial deterioration and the primary precipitant of heart attack and stroke. These functions and processes increase with age and are not independent as it is considered that decreased endothelial function, whether accelerated by exogenous factors (e.g., smoking, hyper LDL-cholesterolemia) or not, is a ubiquitous precursor to clinical atheromatous disease [3].

Blood pressure is the best known and most commonly utilized index of cardiovascular function and also changes with increasing age. Brachial blood pressure, related to increased cardiovascular risk [4], is now recognized as predominantly consequential on underlying changes in systemic vascular properties that affect systolic and diastolic blood pressure in specific ways. An understanding of the relationship of age to cardiovascular function is therefore critical to the implementation of primary and secondary health interventions aimed at prolonging quantity and quality of life.

The cardiovascular system can be subdivided into the heart and the vascular system with further division into the systemic and pulmonary systems. Aging changes

137

R. Aspinall (ed.), Aging of Organs and Systems, 137–152.

affecting the heart or the central or peripheral vasculature should not be considered in isolation; both sides of the ventricular:vascular interface influence the other. Since the left heart and systemic vascular system operate under a pressure load some 12 times that of the right heart and pulmonary system it is not surprising that the predominant age related influences and associated adverse consequences are more manifest in the left/systemic side of the circulation. Effects of normal aging, free of concurrent disease, can be summarized as shown in Table 1.

Sequelae of cardiovascular aging

Changes with aging occur in both the heart and vascular system. Mortality associated with cardiovascular aging is predominantly associated with catastrophic failure of either the heart or large arteries leading to, for example, stroke, acute myocardial infarction or fulminant heart failure. Usually these acute events represent a final stage in progressive dysfunction of the cardiovascular system, the degree of which is often age-related.

Cardiac function and aging

Maintenance of cardiac function requires adequate inotropic and chronotropic responses as well as a functioning Frank-Starling mechanism. To maintain appropriate cardiac output ventricular mechanisms must be adequate to cope with a given vascular afterload. Change on the cardiac side of this equilibrium may occur due to age or disease-related change in ventricular parameters. With aging, the major determinant of left ventricular load is progressive age-related loss of compliance (increased stiffness) of the left-ventricular outflow tract, aortic arch and large conduit arteries; with these large artery changes also having a marked effect on blood pressure (see below).

Reports over the last few decades have indicated that in so-called "normal" aging in otherwise disease free individuals there is probably relatively little change in cardiac structure or function [5, 6]. Age associated changes that do occur in cardiac function affect particular aspects of the myocardium instead of being a generalized deterioration. With aging there is an increase in extracellular matrix and myocyte size while the overall number of myocytes decreases with a compensatory increase in contraction and relaxation time [7]. This mild hypertrophy associated with normal disease free aging is probably secondary to the presence of increased aortic stiffness associated with aging.

The inotropic response of aged myocytes to direct stimulation with Ca^{2+} is maintained with the predominant cellular mechanistic change being a decrease in contractile response of myocytes to post-synaptic β-adrenoreceptor stimulation [8, 9]. Age does not seem to affect the ability of isolated cardiac muscle to develop tension (either maximum or rate [8]), however non-invasive studies in humans demonstrate an increased contraction period with slower relaxation phase in older compared to younger individuals [10]. This decreased β-adrenergic response is also associated with a decreased chronotropic response in older individuals.

Table 1. Effects of normal aging on the cardiovascular system

---

Cardiovascular response to normal ageing (from various sources [1,12,31,33, 36,70]):

A. Structural/functional level
Systolic function
1. No change in maximum capacity of the coronary flow bed [41]
2. Moderate left ventricular hypertrophy
3. Maintenance of ability to generate wall-tension
4. Decreased velocity of myocardial shortening
5. Increased myocardial stiffness
6. Prolonged duration of (systolic) contraction
7. Increased left ventricular cavity diameter
8. No change in stroke volume, heart rate, cardiac output or ejection fraction at rest
9. Decline in maximum heart rate and maximum oxygen uptake with exercise
10. Greater use of the Frank-Starling mechanism
11. Increased ventricular stiffness
12. Decreased ventricular relaxation

Diastolic function
13. Delayed relaxation
14. Diastolic peak filling rate (normalized to left ventricular end-diastolic volume) decreases with age, approximately halves between age 30 years to age 70 years.
15. Decreased peak velocity of early filling while atrial filling fraction (atrial contribution to total left ventricular filling) increases with age.
16. Ratio of early peak to atrial peak (E/A ratio) flow velocity decreases with age (approximately halving between 20 years and 60 years of age).

Arterial function
17. Increased arterial stiffness.
18. Decreased endothelial function.
19. Increased blood pressure.

B. Molecular/cellular level
1. Increased catecholamine levels.
2. Decrease in beta-adrenoreceptor mediated responses (no decrease in number/density but decreased sensitivity).
3. Decreased response to the renin-angiotensin-aldosterone system.
4. Maintenance of peak amplitude of force generation (in the presence of adequate Ca^{2+}).
5. (in rats) Increased duration of the myoplasmic Ca^{2+} transient during excitation-contraction coupling
6. (in rats) Prolongation of the ventricular transmembrane action potential.
7. Cell drop-out and compensatory cellular hypertrophy.

In healthy elderly people, increased cardiac output at a given workload is un-changed compared to the young. Differences are pronounced however in the mechanisms employed to achieve this response. The Frank-Starling mechanism seems to remain basically unaffected by aging [1, 11, 12] with older people seeming to compensate for a lesser increase in heart rate with exercise by an increase in stroke volume and greater end-diastolic volume [12].

Cardiovascular mortality and age

In spite of the relatively small effect of normal aging on cardiac structure and function it is well known that cardiovascular mortality and morbidity increase with age. Figure 1 shows the Australian experience which mirrors that of other developed westernized countries.

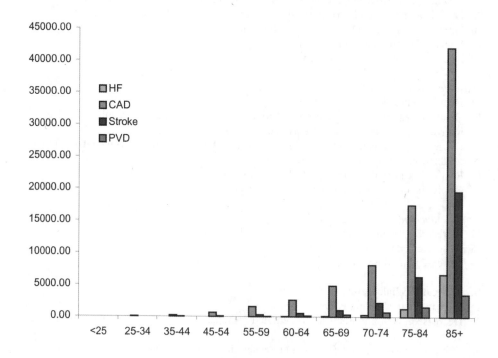

Figure 1. Death rate per 1 000 000 men in Australia in 1999 due to cardiovascular disease (ICD-9). (Adapted from National Cardiovascular Disease Database. Cardiovascular Deaths Report. Australian Institute of Health and Welfare.)

The major cause of the increased incidence of cardiac related morbidity and mortality with increasing age can be assigned to the effects of other underlying age-related processes, the most ubiquitous being atherosclerosis. Other age-related increases in predisposing disease state include hypertension, diabetes, obesity and other less common conditions such as thyroid disease. Environmental factors are also important with increasing age associated with increased exposure to potential myocardial toxins and associated with other factors such as duration of cigarette smoking. Autopsy studies suggest that significant cardiovascular disease is present in approximately 50% of apparently normal individuals over the age of 60 years [13].

Heart failure

Cardiac dysfunction or failure is not a normal accompaniment of aging. While coronary artery disease is the predominant cause of cardiovascular death; heart failure is associated with the greatest number of cardiovascular related doctor–patient interactions and represents a greater cost of diagnosis and treatment than any other condition [14]. Clinical heart failure is predominantly a disease of the elderly with 6–10% of people over 65 exhibiting heart failure [6, 15, 16].

Heart failure is best considered a clinical syndrome related to ventricular abnormalities that prevent the adequate pump action of the heart. Functionally the failing heart is characterized by an inability of the ventricle to adequately receive and/or eject blood. In disease free individuals at rest there is very little apparent age-related change in cardiac volumes, ejection fraction or left-ventricular function [17]. Those at high risk of progressing to heart failure are those with concurrent hypertension, coronary artery disease, diabetes mellitus or with exposure to cardiotoxins. Approximately two-thirds of heart failure patients have coronary artery disease [18] and one third have diabetes mellitus [19]. The prevalence of these underlying conditions increases with age, as does duration of exposure, thus accounting for the age associated increase in symptomatic heart failure.

Recent descriptions of progressive heart failure have described a continuum of cardiac function passing from normal, to ventricular dysfunction to symptomatic heart failure. Ventricular dysfunction can be defined as an abnormality of mechanical function but without the clinical syndrome associated with heart failure [20]. Structural changes predisposing to eventual left ventricular failure include the mild hypertrophy and change in inter-cellular matrix associated with age and the strongly age associated increase in systemic arterial input impedance (LV afterload) associated with decreased compliance/increased stiffness of the aorta.

Age, diastolic function, dysfunction and failure
Diastolic heart failure, ie heart failure with normal LV function, primarily affects elderly women [21]. The major abnormality in diastolic ventricular dysfunction is an increase in diastolic stiffness and the inability of the ventricle to adequately relax (a reduced rate of LV relaxation) and thus to fill normally during the diastolic phase of the cardiac cycle. The ability of the myocardium to relax in diastole can be affected by changes in myocyte size and number and changes with age in extracellular matrix

volume and composition. Aging may have a greater affect on diastolic than systolic function [22], possibly associated with the concomitant deterioration in compliance of the large conduit vessels as well as the raised systolic blood pressure this induces.

The prevalence of individuals with diastolic dysfunction but no clinical symptoms of heart failure is unknown. Table 2 shows estimated prevalence of diastolic heart failure by age in those presenting with clinical heart failure [23]. It is clear that both prevalence and prognosis of diastolic heart failure depend strongly on age, with prevalence in hospitalized or outpatient attending patients over 70 years of age being approximately 50% [24].

Table 2. Association of age and diastolic heart failure. Adapted from ref. 23

Age (years)	< 50	50–70	> 70
Prevalence (%)	15	33	50
Mortality (%) (5-year mortality rate)	15	33	50
Morbidity (%) (1-year rate of hospital admission for heart failure)	25	50	50

The velocity of left ventricular filling during the early filling phase (E-wave) decreases with age [25]. Other diastolic filling parameters that have been shown to change with age include:

1. A doubling of the atrial contribution to total ventricular filling from the third to 8th decade of age [26].

2. A decrease in the E/A (early peak to atrial peak) flow velocity ratio from age 20 to age 70 years [27].

3. Peak diastolic filling rate normalized to either end-diastolic or mitral stroke volume, decreases with age [28, 29].

In the Framingham study [30] age was the major determinant of Doppler indices of diastolic function. In this group the Pearson correlation coefficient between age and E/A ratio was –0. 80 [31].

Age, systolic function, dysfunction and failure
Estimated decreases of approximately 8.4 ml/min/m^2 [32] or 1% [33] per year in stroke index have been reported. In the absence of concurrent precipitating disease, however, left ventricular function is usually not considered to deteriorate significantly with age [5, 12, 34]. Resting left ventricular ejection fraction (stroke volume/

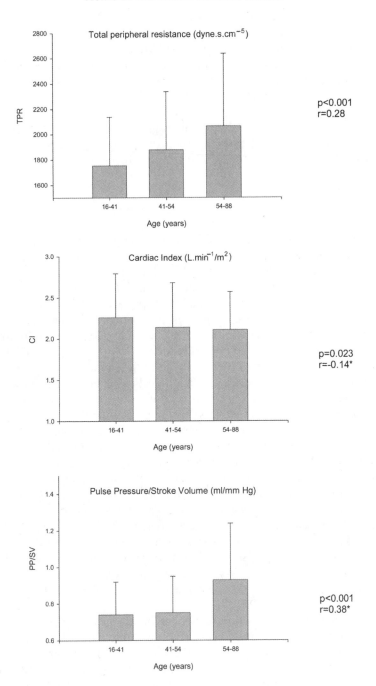

Figure 2*. Changes in systemic hemodynamics grouped by age (adapted from ref. 5). No change in heart rate, stroke volume or index or cardiac output. r = correlation coefficient (* = controlled for gender).*

end-diastolic volume) also does not change with age however exercise induced increase in ejection fraction is less in the elderly by virtue of a lesser decrease in end-systolic volume. This is probably associated with decreased β-adrenergic sensitivity and results in decreased ability in the elderly to increase cardiac output in response to exercise or other stress [35].

In patients with left ventricular dysfunction the underlying cause of heart failure is coronary artery disease (a manifestation of progressive age related atherosclerosis) in approximately 66% of cases. Other underlying causes with prevalence related to age include hypertension, thyroid disease and valvular disease [17].

Blood pressure and aging

Normal blood pressure response with aging
Not all individuals become hypertensive, therefore hypertension in not an inevitable consequence of the aging process. However in view of its well-established tracking processes [36], it appears that those who will achieve a blood pressure exceeding that defined as hypertension are largely predetermined. At least in industrialized societies alterations in the cardiovascular system with aging are clearly such that increased systolic BP occurs. Since modern definitions of hypertension are not age related this corresponds to a steady increase in the prevalence of hypertension in the community with increasing age. Systolic blood pressure increases with age from birth to puberty and onwards [37, 38]. On the other hand diastolic blood pressure increases until approximately the sixth decade, after which it declines in both men and women. The result of these divergent trends is a progressive widening of pulse pressure with age in both men and women. In the subjects from the Framingham cohort studied by Franklin *et al.* [36], of those men and women with initial systolic BP less than 120 mmHg in their fourth decade, the men in this limited cohort had a 3 to 4 mmHg greater systolic BP than did women, a difference lost by the seventh decade and reversed in the eighth. For those with initial systolic BP greater than 160 mmHg in their fourth decade, men had a 13 to 14 mmHg greater systolic BP than women. However, for this cohort in the eighth decade, women had on average a 4 to 5 mmHg greater systolic BP than their male contemparies. In the fourth decade diastolic BP was 2–3 mmHg and 8–9 mmHg lower in women than men in the less than 120 mmHg and more than 160 mmHg systolic BP groups respectively. This sex difference in diastolic BP disappeared by the seventh decade of age but, because of their higher systolic pressures, by the ninth decade women had a greater pulse pressure than men at all ranges of initial systolic BP.

Other gender differences in progression of blood pressure with age have been noted. Although difficult to assess because of the confounding influences of factors such as socio-economic class, differences in smoking activity and body mass index, it is generally considered that blood pressure increases at both menarche and menopause. In pre-menopausal women over the age of 20 years blood pressure is usually lower than for their male counterparts, with this tendency being reversed after middle age [39]. The effect of menopause on systolic blood pressure appears to be an increase in the rate of change of SBP with increasing post-menopausal age, such that the lower relative initial systolic BP in pre-menopausal women compared with men of the same age is overcome [40]. A recent study by Smulyan and colleagues [41] confirmed that women have on average, a higher heart rate than men (2 beats/minute), but that this difference disappears over the age of 55. After age 50 pulse pressure is also higher in women although pulse wave velocity in the aorta increases markedly in both sexes with age by a similar extent, so that there is no difference in this measure of aortic compliance between the sexes. These authors therefore concluded that the increased pulse pressure in older women compared to men was due to a smaller body size and arterial dimensions causing an earlier return of reflected waves. Men and women also had similar common carotid distensibilities and there was no evidence for any hormonal effects on vascular dynamics.

Age-related differences also seem to exist for risk of coronary artery events based on blood pressure parameters. Franklin et al. [42] have reported an age dependence of the blood pressure parameter most predictive for coronary heart disease with an apparent change between the ages of 50 to 59. In this model the hazard ratio for a 10 mmHg increase in BP is greater for a diastolic than systolic pressure in those less than 50 years of age with the reverse evident in those over 60 years of age. Hazard ratios decreased with increasing age for diastolic BP but increased for pulse pressure. When both systolic and diastolic BP were included in predictive models diastolic but not systolic BP was predictive in those less than 50 years old while in those older than 60 a combination of both systolic and diastolic pressures was slightly more predictive than systolic BP alone. Age associated differences in prediction are likely to be due to changes in underlying mechanisms of blood pressure elevation occurring in the vasculature at different stages of the aging process.

Since there is no appreciable change in stroke volume or heart rate (and therefore cardiac output) with aging the predominant cause of an age-related increase in blood pressure is likely to relate to changes in the systemic vasculature. The most pronounced changes are age-related and are seen to occur in the large, proximal conduit arteries and include [43]:

1. Increase in diameter and tortuosity.

2. Increase in intimal smooth muscle proliferation.

3. Progressive fragmentation of elastin fibers.

4. Increase in collagen load bearing (and increased collagen:smooth muscle ratio) together with accumulation of glycosaminoglycans.

5. Increase in operating blood pressure.

Changes as above tend towards a vicious cycle as increased blood pressure causes increased stiffness through passive changes in the non-linear pressure–volume operating characteristics of the artery and hence provokes a further increase in blood pressure [44].

Aging changes also affect the vascular endothelium, the most important of which is a relative decrease in nitric oxide mediated vasodilatation [45]. The net effect of these functional and progressively structural changes is a decrease in compliance of the conduit arteries. Impaired compliance of the proximal aorta increases the input impedance of the systemic vasculature and therefore left ventricular afterload. Decreased arterial compliance (increased stiffness) is associated with increased pressure wave velocity and hence the earlier return from distal reflecting sites of the forward traveling pressure pulse wave generated by left ventricular contraction. As arteries lose their compliance this reflected pressure wave is superimposed on the forward traveling wave in systole rather than diastole, thus augmenting left ventricular generated pressure and increasing systolic blood pressure. At the same time reduced compliance increases the rate (decreases the time constant) of diastolic pressure run-off with the effect of decreasing diastolic blood pressure and decreasing myocardial blood flow in diastole.

Age-associated increases in afterload (increased energy requirement) and decrease in myocardial blood flow (decreased energy supply) deleteriously change the energy supply-demand balance of myocardial energetics as well as increasing measured pulse pressure [46, 47]. Increased left ventricular afterload is also associated with ventricular hypertrophy, further adding to the energy supply-demand imbalance.

In view of the differences in pattern of age-related changes in SBP, mean BP and DBP described above, it has been suggested that normal aging changes can be usefully sub-divided into three relatively distinct hemodynamic mechanisms at three different age ranges [48]:

A. Less than 50 years of age	– Progressive rise in DBP, mean BP and SBP	– Increased vascular resistance predominates
B. Between 50 and 60 years of age	– Flat DBP – Decreasing rate of rise in mean BP – Increasing SBP	– Increased pulse pressure – Increasing vascular resistance – Increasing large artery stiffness
C. Greater than 60 years of age	– Decreasing DBP – Flat mean BP – Increasing SBP	– Arterial stiffening predominates, greatly increasing pulse pressure

Isolated systolic hypertension

Isolated systolic hypertension (ISH) can be defined as an isolated elevation of SBP to above 140 mmHg with a DBP less than 90 mmHg [49] and is the most common form of hypertension in the elderly. Approximately 25% of people aged over 80 years have ISH [38]. Data from the National Health and Nutrition Examination Survey [50] suggests that in hypertensive individuals less than 50 years of age diastolic hypertension predominates whereas in hypertensives between 50 and 59 years of age 54% have ISH increasing to 87% in hypertensives over 60 years of age. ISH is thought to be predominantly a consequence of increased central arterial stiffness with increased systolic blood pressure occurring through a mechanism of pressure augmentation due to wave reflection as described above, hence the strong positive age-dependence of prevalence of ISH.

Arterial function and aging

Broadly speaking the function of the arteries is to distribute the cardiac output to working tissue. Ideally the vascular system provides a low driving point impedance to blood flow and the larger conduit arteries accommodate the pulsatile stroke volume and buffer this pulsitility to provide a steady forward flow at the level of cellular respiration. As discussed above cardiac afterload increases when the aorta and other large conduit vessels lose their compliance and hence their ability to accept a bolus volume of blood with an acceptable pressure increase. Arterial stiffness affects compliance through geometric factors but in general decreased compliance can be equated with increased stiffness. Age-related decrease in compliance and increase in stiffness are terms often used reciprocally [51].

It is generally perceived that arteries progressively stiffen with age [52–54]. This is broadly correct and even in the absence of concurrent disease the aging process is associated with decreased arterial compliance. Only a few studies have looked at arterial compliance in young and adolescent individuals. Laogin and Gosling [55] have measured compliance in a pre-adolescent group and shown that aortic compliance initially increases from birth, reaching a peak around 10 years of age and then declines. Waddell et al. [56] reported that while aortic compliance and distensibility was less in middle and older aged men and women than in younger groups, the observed age-related decline seemed to be greater in women than in men. Cheung et al. measured pulse wave velocity (inversely related to arterial stiffness) over the brachioradial arterial segment in individuals between the ages of 6 and 20 years and showed a linear increase from approximately 6 meters/second to 8 meters/second over this age range [57].

Superimposed processes, often secondarily related to age, also influence the rate of change of arterial properties with stiffening also influenced by lifestyle, underlying disease processes and undoubtedly by genetic factors [58].

On top of these normal aging processes is superimposed other environmental and disease related factors such as high salt intake [59], smoking [44], diabetes [60], hyper LDL-cholesterolemia [3, 61], and lack of aerobic exercise [62]. Prevalence of these factors increases with age and contributes to observed aging changes in arterial properties. Coronary artery disease increases with age and increased arterial stiffness has been demonstrated in association with coronary disease [54, 63]. Whether this is best considered a marker of generalized disease or a causative factor is unclear. Secondary to mechanical effects, aging also results in changes in the shape and relative proportions of the pressure pulse [64]. The increased sharpness of the systolic peak, increased pulse pressure and alteration in the systolic inflection point are consequences of age-related conduit artery stiffening and secondary changes in pulse wave velocity and the effect of reflected pressure waves.

Endothelial dysfunction

Normal endothelial function is considered an important protective mechanism against atherosclerotic disease. Endothelially mediated relaxation decreases with age as does acetylcholine induced coronary blood flow [65]. Brachial artery response to increased shear stress, an endothelial dependent mechanism, has been shown to decrease with age in both men and women, with the effect occurring in men from approximately 40 years of age but not until their fifties in women [45].

Chronological versus biological age

Indices of arterial mechanics associated with elastic arteries have shown a close correlation with age (Figure 3). The strength of association of mechanical indices is in fact greater than other indices associated with aging e.g., skin elasticity and greying of hair [66] and arterial mechanics can be considered a candidate marker of biological, as opposed to chronological, age. Not all arterial segments, however, show similar age-related changes, with muscular arteries changing little as regards their properties with age [68, 69].

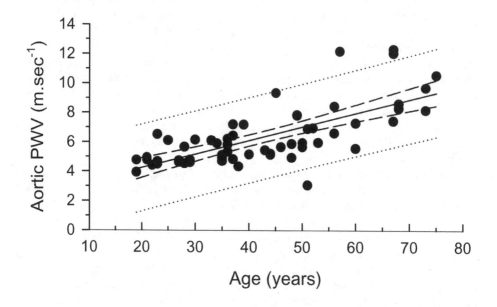

Figure 3. *Relationship of aortic-femoral PWV to age in 26 male and 26 female individuals with no known cardiovascular disease [ref 67, with permission].*

The future

Therapy targeted at normal age-related changes in the cardiovascular system are theoretically appealing. If deterioration of the structure and/or function of the cardiovascular system were preventable or able to be delayed the substantial financial cost burden of cardiovascular morbidity would be decreased, as would the individual cost in terms of quantity and quality of life lost. In clear contrast to evidence supporting treatment of primary risk factors and secondary preventive therapy to delay or prevent development of clinical disease, there are as yet no convincing data that any particular therapeutic agent, except perhaps well-known lifestyle interventions, modulates the affect of normal cardiovascular aging.

Conclusion

It is difficult to separate "normal" cardiovascular aging from the effects of superimposed pathology. The most obvious change with age is an increase in systolic blood pressure associated with increased stiffness and deterioration of central artery function. Independent as well as secondarily associated changes also occur in the myocardium, also related to increasing ventricular stiffness and particularly affecting the ability to respond to hemodynamic stress.

References

1. Braunwald E (1992). *Heart Disease – A Textbook of Cardiovascular Medicine*, 4th edn. Philadelphia: WB Saunders, chapter 52.
2. Manual of the International Statistical Classification of Diseases, Injuries, and Causes of Death (1977). Geneva: WHO, Geneva.
3. Ross R (1999). Atherosclerosis an inflammatory disease. *N Eng J Med.* 340(2): 115–26.
4. McMahon S, Peto R, Cutler J, *et al.* (1990). Blood pressure, stroke, and coronary heart disease: part 1. Prolonged differences in blood pressure: prospective observational studies corrected for regression dilution bias. *Lancet* 335: 765–74.
5. Slotwiner DJ, Devereux RB, Schwartz JE, *et al.* (1998). Relation of age to left ventricular function in clinically normal adults. *Am J Cardiol.* 82: 621–6.
6. Kannel WB, Belanger AJ (1999). Epidemiology of heart failure. *Am Heart J.* 137: 352–60.
7. Lakatta EG, Gerstenblith G, Angell CS, Shock NW, Weisfeldt ML (1975). Prolonged contraction duration in aged myocardium. *J Clin Invest.* 55: 61.
8. Weisfeldt ML, Loeven WA, Shock NW (1977). Resting and active mechanical properties of carnae from aged male rats. *Am J Physiol.* 220: 1921.
9. Lakatta EG, Yin FCP (1975). Myocardial ageing: functional alterations and related cellular mechanisms. *Am J Physiol.* 242: H927.
10. Harrison TR, *et al.* (1964). The relation of age to the duration of contraction, ejection and relaxation of the normal human heart. *Am Heart J.* 67: 189.
11. Weisfeldt M (1998). Aging, changes in the cardiovascular system, and responses to stress. *Am J Hypertens.* 11: 41S–5S.
12. Rodeheffer RJ, Gerstenblith G, Becker LC, Fleg JL, Weisfeldt ML, Lakatta EG (1984). Exercise cardiac output is maintained with advancing age in healthy human subjects:

cardiac dilatation and increased stroke volume compensate for a diminished heart rate. *Circulation* 69: 203–13.

13. Elveback L, Lie JT (1984). Continued high incidence of coronary artery disease at autopsy in Olmsted County, Minnesota, 1950–1979. *Circulation* 70: 345–9.

14. Hunt SA, Baker DW, Chin MH, *et al.* (2001). ACC/AHA guidelines for the evaluation and management of chronic heart failure in the adult: executive summary. *Circulation* 104: 2996–3007.

15. Kannel WB, Belanger AJ (1991). Epidemiology of heart failure. *Am Heart J.* 121: 951–7.

16. Kannel WB (1987). Epidemiology and prevention of cardiac failure: Framingham study insights. *Eur Heart J.* 8(Suppl F): 23–6.

17. Schulmann SP, Weisfeldt ML (1994). Cardiovascular ageing and adaptation to disease. In: Schlant RC, Alexander RW, eds. *The Heart*, 8th edn. New York: McGraw-Hill, Inc.

18. Gheorghiade M, Bonow RO (1998). Chronic heart failure in the United States: a manifestation of coronary artery disease. *Circulation* 97: 282–9.

19. Solong L, Malmberg K, Ryden L (1999). Diabetes mellitus and congestive heart failure: further knowledge needed. *Eur Heart J.* 20: 789–95.

20. Zile MR, Brutsaert DL (2002). New concepts in diastolic dysfunction and diastolic heart failure: part 1. *Circulation* 105: 1387–93.

21. Davie AP, Francis CM, Caruana L, Sutherland GR, McMurray JJ (1997). The prevalence of left ventricular diastolic filling abnormalities in patients with suspected heart failure. *Eur Heart J.* 18: 981–4.

22. Brutsaert DL, Sys SU, Gillebert TC (1993). Diastolic failure: pathophysiology and therapeutic implications [published errata *J Am Coll Cardiol.* 1993; 22(4): 1272] *J Am Coll Cardiol.* 22: 318–25.

23. Zile MR, Brutsaert DL (2002). Clinical cardiology: new frontiers. New concepts in diastolic dysfunction and diastolic heart failure: Part I. Diagnosis, prognosis, and measurements of diastolic function. *Circulation* 105: 1387.

24. O'Connor CM, Gattis WA, Shaw L, Cuffe MS, Califf RM (2000). Clinical characteristics and long-term outcome of patients with heart failure and preserved systolic function. *Am J Cardiol.* 86: 863–7.

25. Gerstenblith G, Frederiksen J, Yin FCP, Fortium NJ, Lakatta EG, Weisfeldt ML (1977). Echocardiographic assessment of a nomal adult aging population. *Circulation* 56: 273–8.

26. Kuo LC, Quinones MA, Rokey R, Sartori M, Abinader EG, Zoghbi WA (1987). Quantification of atrial contribution to left ventricular filling by pulsed Doppler echocardiography and the effect of age in normal and diseased hearts. *Am J Cardiol.* 59: 1174–8

27. Bryg RJ, Williams GA, Labovitz AJ (1987). Effect of aging on left ventricular diastolic filling in normal subjects. *Am J Cardiol.* 59: 971–4.

28. Miller TR, Grossman SJ, Schectman KB, Biello DR, Ludbrook PA, Ehsani AA (1986). Left ventricular diastolic filling and its association with age. *Am J Cardiol.* 58: 531–5.

29. Bowman LK, Lee FA, Jaffe C, Mattera J, Wackers FJT, Zaret BL (1988). Peak filling rate normalized to mitral stroke volume: a new Doppler echocardiographic filling index validated by radionuclide angiographic techniques. *Am Coll Cardiol.* 12: 937–43.

30. Benjamin EJ, Levy D, Anderson KM, *et al.* (1992) Determinants of Doppler indices of left ventricular diastolic function in normal subjects [the Framingham study]. *Am J Cardiol.* 70: 508–15.

31. Oparil S, Weber MA (2000). *Hypertension: A Companion to Brenner and Rector's* The Kidney. Philadelphia: WB Saunders, pp. 255.

32. Starr I, Donald JS, Margolies A, *et al.* (1934). Studies on the heart and circulation in disease: estimation of basal cardiac output, metabolism, heart size and blood pressure in 235 subjects. *J Clin Invest.* 13: 561 (Quoted in ref. 40, p. 552).

33. Brandfonbreuer M, Landowne M, Shock NW (1955). Changes in cardiac output with age. *Circulation* 12: 557.

34. Yin FC, Raizes GS, Guarnieri T, *et al.* (1978). Age-associated decrease in ventricular response to haemodynamic stress during beta-adrenergic blockade. *Br Heart J.* 40: 1349–55.

35. Lakatta EG (1993). Cardiovascular regulatory mechanisms in advanced age. *Physiol Rev.* 73: 413–67.

36. Franklin SS, Gustin W, Wong ND, *et al.* (1997). Hemodynamic patterns of age-related changes in blood pressure. The Framingham Heart Study. *Circulation* 96: 308–15.

37. Kaplan NM (1994). *Clinical Hypertension*, 6th edn. Baltimore: Williams and Wilkins, pp. 4–13.

38. Staesson J, Amery A, Fagard R (1990). Isolated systolic hypertension in the elderly. *J Hypertens.* 8: 393–405.

39. Krakoff LR (1995). *Management of the Hypertensive Patient.* New York: Churchill Livingston.

40. Staessen J, Bulpitt CJ, Fagard R, *et al.* (1989). The influence of menopause on blood pressure. *J Hum Hypertens.* 3: 427–33.

41. Smulyan H, Asmer RG, Rudnicks A, London GH, Safar ME (2001). Comparative effects of ageing in men and women on the properties of the arterial tree. *JACC* 37: 1374–80.

42. Franklin SS, Larson MG, Khan SA, *et al.* (2001). Does the relation of blood pressure to coronary heart disease risk change with ageing? The Framingham Heart Study. *Circulation* 103: 1245–9.

43. Loscalzo J, Creager MA, Dzau VJ, eds. (1992). *Vascular Medicine.* Boston: Little, Brown and Company.

44. Liang Y-L, Shiel LM, Teede H, *et al.* (2001). Effects of blood pressure, smoking and their interaction on carotid artery structure and function. *Hypertension* 37: 6–11.

45. Celermajer DS, Sorensen KE, Spiegelhalter DJ, Georgakopoulos D, Robinson J, Deanfield JE (1994). Ageing is associated with endothelial dysfunction in healthy men years before age-related decline in women. *J Am Coll Cardiol.* 24(2): 471–6.

46. Kass DA, Saeki AS, Tunin RS, Recchia FA (1996). Adverse influence of systemic vascular stiffening on cardiac dysfunction and adaptation to acute coronary occlusion. *Circulation* 93: 1533–41.

47. Ohtsuka S, Kakihana M, Watanabe H, Sugishita Y (1994). Chronically decreased aortic distensibility causes deterioration of coronary perfusion during increased left ventricular contraction. *J Am Coll Cardiol.* 24: 1406–14.

48. Franklin SS (2000). New interpretations of blood pressure: the importance of pulse pressure. In: Oparil S, Weber MA, eds. *Hypertension: A Companion to Brenner and Rector's The Kidney.* Philadelphia: WB Saunders, pp. 228–9.

49. The Sixth Report of the Joint National Committee on Prevention, Detection, Evaluation, and Treatment of High Blood Pressure. National Institutes of Health. National Heart, Lung, and Blood Institute (1997). NIH Publication No 98-4080.

50. Franklin SS, Jacobs MJ, Wong ND, L'Italien GJ, Lapuerta P (2001). Predominance of isolated systolic hypertension among middle-aged and elderly US hypertensives (NHANES) III. *Hypertension* 37: 869–74.

51. Cameron JD (1999). Estimation of arterial mechanics in clinical practice and as a research technique. *Clin Exp Pharmacol Physiol.* 26: 285–94.

52. Avolio AP, Chen SG, Wang RP, Zhang CL, Li MF, O'Rourke MF (1983). Effects of aging on changing arterial compliance and left ventricular load in a northern Chinese urban community. *Circulation* 68: 50–8.

53. Avolio AP, Den FQ, Li WQ, *et al.* (1985). Effects of aging on arterial distensibility in populations with high and low prevalence of hypertension: comparison between urban and rural communities in China. *Circulation* 71: 202–10.

54. Dart AM, Lacombe F, Yeoh JK, *et al.* (1991). Aortic distensibility in patients with isolated hypercholesterolaemia, coronary artery disease, or cardiac transplant. *Lancet* 338: 270–3.

55. Laogin AA, Gosling RG (1982). *In vivo* arterial compliance in man. *Clin Phys Physiol Meas.* 3: 201–12.

56. Waddell TK, Dart AM, Gatzka CD, Cameron JD, Kingwell BA (2001). Women exhibit a greater age-related increase in proximal aortic stiffness than men. *J Hypertens.* 19: 1–8.

57. Cheung YF, Brogan PA, Pilla CB, Dillon MJ, Redington AN (2002). Arterial distensibility in children and teenagers: normal evolution and the effect of childhood vasculitis. *Arch Dis Child.* 87: 348–51.

58. Kingwell BA, Medley TL, Waddell TK, Cole TJ, Dart AM, Jennings GL (2001). Large artery stiffness: structural and genetic aspects. *Clin Exp Pharm Physiol.* 28: 1040–3.

59. Avolio AP, Clyde KM, Beard TC, Cooke HM, Ho KK, O'Rourke MF (1986). Improved arterial distensibility in normotensive subjects on a low salt diet. *Arteriosclerosis* 6(2): 166–9.

60. Lehmann ED, Gosling RG, Sonksen PH (1992). Arterial wall compliance in diabetes. *Diabetic Med.* 9: 114–19.

61. Lehmann ED, Watts GF, Fatemi-Langroudi B, Gosling RG (1992). Aortic compliance in young patients with heterozygous familial hypercholesterolaemia. *Clin Sci.* 83: 717–21.

62. Kingwell BA, Cameron JD, Gillies KJ, Jennings GL, Dart AM (1995). Arterial compliance as a determinant of baroreflex function in athletes and hypertensives. *Am J Physiol.* 268 (*Heart Circ. Physiol.* 37): H411–18.

63. Cameron JD, Jennings GL, Dart AM (1996). Conduit arterial compliance is decreased in newly diagnosed patients with coronary heart disease: implications for prediction of risk. *J Cardiovasc Risk* 3: 495–500.

64. Kelly R, Haywood C, Avolio A, O'Rourke MF (1989). Non-invasive determination of age-related changes in the human arterial pulse. *Circulation* 80: 1652–9.

65. Rajkumar C (2000). Hypertension in the elderly. In: Bulpitt CJ, ed. *Handbook of Hypertension*, vol. 20. *Epidemiology of Hypertension*. Amsterdam: Elsevier.

66. Bulpitt CJ, Rajkumar C, Cameron JD (1999). Vascular compliance as a measure of biological age. *J Am Geriatr Soc.* 47: 657–63.

67. Cameron JD (1995). Systemic arterial compliance and its association with aerobic fitness and coronary heart disease. (MD thesis, Monash University).

68. Benetos A, Laurent S, Hoeks AP, Boutouyrie PH, Safar ME (1993). Arterial alterations with aging and high blood pressure. A noninvasive study of carotid and femoral arteries. *Arteriosclerosis Thromb.* 13(1): 90–7.

69. Boutouyrie P, Laurent S, Benetos A, Girerd XJ, Hoeks AP, Safar ME (1992). Opposing effects of ageing on distal and proximal large arteries in hypertensives. *J Hypertens.* 10(6): S87–91.

70. Port S, Cobb FR, Colema E, Jones RH (1980). Effect of age on the response of the left ventricular ejection fraction to exercise. *N Eng J Med.* 303: 1133.

Aging of the Gastrointestinal System

L. Drozdowski, M. Keelan, M.T. Clandinin and A.B.R. Thomson

519 Newton Research Building, University of Alberta, Edmonton, Alberta T6G 2C2, Canada

Introduction

All multicellular organisms undergo change with time. This concept of aging may be considered in different ways. The maximum life span and the average life span are two commonly used definitions. The process of aging must take into account social, biological, physiological, emotional, socioeconomic, as well as cellular and molecular changes. Aging is a multifactorial conditional with environmental and genetic factors. While chronological age is easy to measure, biological age is a better marker of health status that allows for the concept of the introduction of processes to ameliorate the aging process, thereby compressing morbidity, improving competence, well being, and quality of life [1].

The medical model of successful aging is based on the principle of adding life to years, and not just years to life [2]. The Healthy Active Life Expectancy (HALE) measures the number of years that a particular age can expect to live in a healthy, active state, free from ill health, as well as Quality Adjusted Life Years (QALY), which gives some indication of the handicap that individuals may suffer in their remaining life [3]. The theories of aging relate to cellular, physiological, organ-based and genetic considerations [1]. Cellular theories of aging include considerations of

---

List of Abbreviations: ASA, acetylsalicylic acid; CNS, central nervous system; DNA, deoxyribonucleic acid; EHT, endoscopic hemostatic therapy; ERCP, endoscopic retrograde cholangiopancreatography; GI, gastrointestinal; H. pylori, Helicobacter pylori; HALE, Healthy Active Life Expectancy; HAV, hepatitis A virus; HCV, hepatitis C virus; IBD, inflammatory bowel disease; IBS, irritable bowel syndrome; LOH, loss of heterogozity; NSAID, nonsteroidal anti-inflammatory drug; PPI, proton pump inhibitor; QALY, Quality Adjusted Life Years

R. Aspinall (ed.), Aging of Organs and Systems, 153–177.
© *2003 Kluwer Academic Publishers. Printed in Great Britain.*

programmed cell death, waste product theory, as well as cross-linked and cross-linkage theory. Organ-based theories of aging were related to immunological, calorie-restriction and neuroendocrine considerations, while the genetic theories of aging include programmed aging, modifier genes, pleiotropic genes, gene redundancy, programmed aging, somatic mutation, DNA repair, free radicals and telomeres. Research must examine the theories of aging to better identify those extrinsic aspects which can be modified, in order to achieve elongation of life span and better quality of life amongst our senior citizens.

The gastrointestinal tract is subject to aging

The digestive tract has considerable adaptive capacity, and maintains much of its normal physiological function during the aging process [4]. The most common gastroenterological diagnoses seen in persons over the age of 65 include functional bowel disorders, peptic ulcer disease and neoplasm, seen in 31%, 16%, and 15% of patients, respectively [5]. These are followed in frequency by diverticular disease (7%), gastroesophageal reflux disease and hiatal hernia (6%), cholelithiasis (5%), and liver disease (4%).

There are special considerations which a physician must take into account when taking a history from the elderly, including the possibility of poor hearing, poor recall, minimization or exaggeration of symptoms, memory loss and confusion and social as well as economic influences such as death of a spouse and loss of independence or a change in living conditions [6]. The physician must beware: upper abdominal discomfort or pain in the elderly may represent coronary artery disease, eating disorders may be due to poor dentition and reduced chewing capacity, and caregivers must pay special attention to the older individual's eating habits and malfunction. Malnutrition is common in the elderly and may be due to reduced intake, often affected by psychosocial issues, economic considerations leading to poor choices, and reduced food intake. As a result, malnutrition in the geriatric population is unfortunately often underdiagnosed [7, 8]. Age-related medical, psychological, social and economic problems increase the likelihood of a poor nutritional state [9, 10]. The five most prevalent factors for undernutrition in the elderly include immobility, masticatory problems, problems preparing food, social isolation, and loss of appetite [11, 12]. Diarrhea may be associated with malabsorption, and is also often associated with decreased food intake as an avoidance strategy.

Physiological changes with aging

There are number of physiological changes that occur with aging that have the potential to influence drug disposition and metabolism, and may thereby influence gastrointestinal function (Table 1).

Since many drugs are eliminated by metabolic pathways in the liver, and since the use of several drugs (polypharmacy) is common in the elderly, liver function has to be considered as an additional factor modifying their drug response. As the number of drugs given increases with aging, so does the probability of a drug interaction [13].

Table 1. *Physiological changes that occur with aging and have the potential to influence drug disposition and metabolism*

System	Change
General	Reduced total body mass
	Reduced basal metabolic rate
	Reduced proportion of body water
	Increased proportion of body fat
Circulatory	Decreased cardiac output
	Altered relative tissue perfusion
	Decreased plasma protein binding
Gastrointestinal	Reduced gastric acid production
	Reduced gastric emptying rate
	Reduced gut motility
	Reduced gut blood flow
	Reduced absorption surface
	Intestinal uptake/transport?
	Intestinal metabolism (?)
Hepatic-biliary	Reduced liver mass
	Reduced liver blood flow
	Reduced albumin synthesis
	Hepatic and biliary uptake/transport?

Herrlinger and Klotz [Ref.16]

Age-related changes in metabolism may increase the risk of adverse effects. Liver mass and hepatic blood flow both decrease with age [14–16]. Pharmacokinetic interactions include those influencing drug absorption, alterations of motility, damage of the gastrointestinal mucosa, and interactions affecting drug distribution [16]. Age significantly affects the pharmacokinetics of lansoprasole and clarithromycin, both commonly used for eradication of *H. pylori* [17]. Enzyme induction or inhibition may increase or decrease the efficacy of administered medications. For example, the imidazole derivative cimetidine and ketoconazole bind to the hemoprotein of cytochrome P450 (CYP450) enzyme via their Fe atom, thereby nonspecifically inhibiting the oxymetabolism of a variety of drugs [18]. Omeprazole inhibits the hepatic transformation of diazepam, since both agents are substrates of CYP2C-19, an enzyme for the metabolism of proton pump inhibitors [19].

Physiology and pathology of the esophagus in the elderly

There are physiological processes that occur with aging that may be associated with an increased prevalence or a higher complication rate in certain disease processes, including malignancy. The care of the geriatric patient may be more complex because of multisystem diseases, and psychosocial circumstances may alter the older person's perception, reporting or tolerance of symptoms. The oral phase of swallowing may be hampered by impairment or absence of dentition, xerostomia, and reduced perception of food by the tongue [20]. Aging is associated with a gradual decrease in upper esophageal sphincter pressure [21], reduced upper esophageal sphincter opening [22], and delayed upper esophageal sphincter relaxation after swallowing [23]. Coordination between glottal and upper esophageal sphincter function may be abnormal in the elderly [24]. In the older person there is more prevalent occurrence of simultaneous contraction or a failure of contractions in the distal esophagus after deglutition [25]. The amplitude of the peristaltic waves may be decreased with aging [26]. However, other authors have reported increased distal esophageal pressure amplitude and duration with age [27]. It remains controversial whether the lower esophageal sphincter pressure changes with age [26]. The duration of reflux episodes may be longer in the elderly, possibly due to impaired esophageal clearance [28] which is common in the elderly [29], and may be due to gastroesophageal reflux disease. Elderly subjects tend to have more severe acid reflux, and more severe esophageal lesions as compared with younger persons [30]. Although the prevalence of *Helicobacter pylori* (*Hp*) infection and esophagitis is higher in the elderly populations, *Hp* eradication therapy does not influence the healing rate of esophagitis [31].

Although the onset of achalasia is usually in persons between 20 and 40 years of age, a second peak occurs in the elderly, and older patients tend to experience more chest pain [32]. Although the symptomatic benefit of the reaction to botulinum ingestion may only be temporary, it may prove to be useful, especially for elderly patients [33]. The toxin binds to a presynaptic receptor, and irreversibly inhibits axonal acetylcholine release [34]. The chest pain from diffuse esophageal spasm must be differentiated from cardiac disease. The treatment of both diffuse esophageal spasm and nutcracker esophagus require alleviation of symptoms by way of decreasing smooth muscle contractility, and by the use of smooth muscle relaxants such as calcium channel blockers and nitrates, although the results of this therapeutic approach are not uniform [35].

A high percentage of elderly patients experience relapse of their esophagitis, especially if not treated with antisecretory drugs [36]. Risk factors for relapse include presence of typical symptoms, hiatal hernia, and a severe grade of esophagitis. Maintenance therapy with antisecretory drugs, such as proton pump inhibitors (PPIs) most effectively minimizes the occurrence of relapse.

The prevalence of Barrett's esophagus increases with age. While esophageal resection may be the treatment of choice in those with high grade dysplasia, the operative mortality rate is high in older persons [37], and palliative options such as photodynamic therapy or recanalization of the obstructed esophagus by laser or argon plasma coagulation need to be considered [38]. Most patients with gastro-esophageal reflux disease can be well controlled with PPIs, but it is unclear whether their use can cause regression of Barrett's epithelium.

Stomach and duodenum: ulcer disease

Peptic ulcer disease is one of the more common gastrointestinal conditions seen in the elderly, partly because of their high prevalence of *H. pylori* infection and partly because of their frequent use of non-steroidal anti-inflammatory drugs (NSAIDs) [39]. Upper gastrointestinal non-variceal bleeding is common; because of comorbid conditions, these bleeding episodes bring a significant risk of morbidity or mortality. Indeed, persons over the age of 65 account for over 80% of the mortality rate from peptic ulcer disease [40].

For patients 70 years of age and over, *H. pylori* infection is a risk factor for the development of dyspepsia, while no association is clearly observed in younger age groups [41]. Non-ulcer dyspepsia symptoms in the elderly may be improved two months following eradication of *H. pylori* [42]. There is a steady rise in the prevalence of *H. pylori* infection with age of approximately 1% per year for the overall population in the US [43], although it has been suggested that most of the peptic ulcer disease in the elderly occurs in the absence of *H. pylori* [44]. Patients with an *H. pylori* infection secrete less acid in the acute stage, but they then enter a hypersecre-tory phase followed by atrophy and hyposecretion [45]. It remains controversial whether gastric acid secretion declines with age in the absence of an *H. pylori* infection [46, 47]. However, aging is associated with a decrease in gastric mucosal prostaglandin content [48, 49], thereby possibly contributing to a loss of mucosa defense mechanisms. While there have been no human studies of age-related changes in mucus composition, mucosal proliferation or gastric blood flow, animal studies demonstrate that these contributors to the mucosal defense mechanisms may also be impaired with aging.

Peptic ulcers in the elderly tend to be large [50], gastric ulcers are more prevalent in the proximal stomach as compared with younger persons [51], and it remains controversial whether ulcer pain is less common in the elderly. A second risk factor for the development of peptic ulcer disease in the elderly is the high prescribing use of NSAIDs [52]. Patients on NSAIDs have approximately 3 times the relative risk of developing an ulcer, as compared with patients under the age of 60 [53]. A case-controlled study in the elderly did not show synergism in the risk of ulcer bleeding, and confirmed the independent risks of *H. pylori* and NSAIDs, although this topic remains controversial. The elderly may have multiple risk factors in developing complications of NSAIDs and ulcers such as their age, use of ASA or NSAIDs, use of anticoagulants, past history of ulcer disease or ulcer bleeding, and comorbid conditions such as cardiac disease.

Triple therapy is effective in the eradication of *H. pylori* infection in the elderly, but failed eradication primarily due to low compliance and antibiotic resistance is associated with a higher incidence of ulcer recurrence in the elderly as compared to younger persons (41.6% versus 2.2%) in the first year [54–56]. If the elderly patient develops an NSAID-associated ulcer, it is best if the NSAID can be stopped. However, if the NSAID must be continued, administration of a proton pump inhibitor will allow for ulcer healing [57]. It is unknown if switching from an NSAID to a Cox-2 will allow for ulcer healing. While the use of PPIs as cotherapy with NSAIDs has been proven to be useful to reduce the risk of ulcer occurrence [58], it is unknown whether the PPIs protect against the lesser likelihood of the development of a peptic ulcer in the patient on a Cox-2 inhibitor. A recent consensus discussion suggested that those individuals on a Cox-2 inhibitor who are at risk of ulcer complications should be cotreated with PPI (Hunt *et al.*, personal communication, 2002). Increasing proportions of patients with peptic ulcers are not on ASA/NSAIDs, and are not infected with *H. pylori*. Healing of a duodenal ulcer with H_2-receptor antagonists is slower in the elderly [59], but data is not available for the PPIs.

Endoscopic investigation and treatment of GI bleeding in the elderly

The incidence of upper GI bleeding is approximately 100 per 100 000 people per year. The incidence is higher with older age and about 27% of patients with acute upper gastrointestinal bleeding are over 80 years of age [60]. The incidence of lower gastrointestinal bleeding is 21 / 100 000 per year [61], with the incidence rising sharply with advancing age.

Peptic ulcer bleeding is more serious in the elderly, there is more extensive or frequent rebleeding, and more requirement for blood transfusions [62]. The mortality rate is also higher [60, 62–64]. The severity of the initial hemorrhage, ulcer size, comorbidity and impaired liver function contribute to the risk of rebleeding and death from an upper gastrointestinal bleed in the elderly [65–67]. Intravenously administered PPI improves the outcome of patients with non-variceal upper gastro-intestinal bleeding, with outcomes varying among studies in the beneficial effect on rebleeding, the number units of blood transfused, the length of hospital stay, or the need for surgery. A trial involving 333 patients over the age of 60 compared intravenous omeprazole with placebo and demonstrated the reduced need for surgery [68].

The most common causes of upper gastrointestinal bleeding in the elderly include gastric and duodenal ulcers, and esophagitis [69]. The most common causes of lower gastrointestinal bleeding in the elderly include diverticulosis, colitis (including idiopathic inflammatory bowel disease, radiation colitis, ischemic colitis, infectious colitis, and vasculitis), neoplasia and angiodysplasia [69].

Interestingly, older patients are no different than younger patients in terms of their need for endoscopy, surgery, transfusions or in the duration of their hospital stay [70]. However, the mortality resulting from operations for peptic ulcers is 5–10 times higher in the elderly than in the young [71–74]. For example, the mortality rate from upper GI bleeding in persons under the age of 60 is less than 10%, while over 1 in 3 die when the bleeding occurs in a person over the age of 80 [65]. The mortality rate of lower GI bleeding is also higher in the elderly [61].

While endoscopy is the preferred investigation for both upper and lower gastro-intestinal bleeding, elderly individuals are at a greater risk of complications (0.2–

4.9% versus 0.03–0.13%) [75] possibly because of additional age-related difficulties such as interactions with medications or cardiopulmonary events.

Esophagogastroduodenoscopy has a diagnostic yield of 94% of detecting the source of bleeding, and this yield is increased if the procedure is done soon after the onset of bleeding [74, 76]. The endoscopic nature of the lesion may be useful for stratification of risk, and may thereby be useful to predict outcome [77]. Endoscopic hemostatic therapy (EHT) includes injection sclerotherapy, band ligation of varices, mono- or multi-polar electrocautery laser, argon plasmic coagulation, and clipping [78]. Meta-analyses have shown that EHT reduces the rate of rebleeding, the need for surgery, and the mortality rate [79].

Studies with small numbers of patients report beneficial effects from endoscopic treatment of diverticular hemorrhage [80, 81]. Post-polypectomy hemorrhage is uncommon, and can be treated by performing endoscopic maneuvers [82]. In patients with high-risk bleeding peptic ulcers (active bleeding, visible vessel, adherent clot), EHT plus intravenous infusion of omeprazole leads to the reduction of recurrent bleeding from 23% to 7%, and reduces both hospital stay, and the number of blood transfusions required [83].

If surgery is necessary, ideally it should be done under elective circumstances: early elective surgery is associated with a mortality rate of 13%, versus 25% for emergency surgery performed in persons over the age of 60 [84]. At a mean age of 65 years, EHT for treatment of peptic ulcer bleeding is associated with fewer complications than surgery [85].

Gastrointestinal neoplasms

Gastric cancer

The WHO classification of tumors of the digestive system replaced the term "dysplasia" with "intra-epithelial neoplasia" [86]. In the new Vienna classification of gastrointestinal neoplasms, high-grade dysplasia, non-invasive carcinoma, and suspected invasive carcinoma are considered to be similar [87]. Meining and co-workers (2001) have suggested that "as an alternative to the atrophy-metaplasia-dysplasia-carcinoma sequence hypothesis, the concept of gastritis of the carcinoma phenomena" has been developed.

Gastric carcinomas include an intestinal and a diffuse type, which are often found in non-atrophic mucosa with little intestinal dysplasia [89]. It is no longer clear that the atrophy-metaplasia sequence is a reality [88]. Intestinal metaplasia often arises in non-atrophic mucosa, and can then lead to focal atrophy. A small (14%) proportion of patients with intestinal metaplasia show a loss of heterogozity (LOH) and mutation of p53 [90]. It remains controversial whether intestinal metaplasia is actually a precancerous condition [88].

Gastric atrophy and intestinal metaplasia are associated with *H. pylori* infections in the elderly [91]. These strains are cytotoxin-associated gene (cag A) positive, a potential virulence factor for the development of adenocarcinoma of the stomach [92, 93]. Accordingly, *H. pylori* eradication is recommended for elderly patients [55].

Colon cancer and polyps

Colorectal cancer and adenous polyps are more common with advancing age. Prevention of colonic cancer by way of screening procedures such as colonoscopy, sigmoidoscopy plus barium enema, or fecal occult blood testing are well accepted, although not widely practiced [94].

Nutrient malabsorption

A key determinant of morbidity and mortality in the elderly is poor nutritional status [95–100]. Depending on level of independence, the prevalence of involuntary weight loss in the elderly is estimated to be between 13% and 38% [101–103]. Similarly, nutritional deficiencies are observed more frequently in functionally dependent groups when compared to healthy, independent, free living elderly (104–106]. Indeed, almost two thirds of elderly patients admitted to a tertiary care facility were found to be malnourished, or at high risk for malnourishment [107].

There are many factors that contribute to this malnourishment such as decreased appetite, reduced sense of taste and smell, reduced food intake, poor dentition, psychological issues including depression, social factors, socioeconomic status, medications and malabsorption. When the absorption of carbohydrates is assessed with hydrogen breath testing, one third of subjects over the age of 65 were considered to have malabsorption [108]. Bacterial overgrowth associated with hypochlorhydria occurs in the elderly, and may also reduce carbohydrate absorption [109]. Furthermore, the brush border membrane activity of enzymes such as lactase and sucrase fall with aging, and therefore contribute to the reductions in absorption [110]. In laboratory animals, dietary manipulation may help to partially normalize the abnormal absorption that occurs with aging, and this approach of nutritional intervention may prove to be useful in the elderly [111].

Consuming more than 300 grams of fat/day does not induce steatorrhea in undernourished elderly patients [112]. Bile acids are absorbed less efficiently in the elderly [113], but in a study of the healthy elderly there was no correlation between age and 72 hour fecal fat excretion [114]. In a recent review of malabsorption in the elderly, Holt [115] concluded that no important changes in lipid absorption with age have been described.

The effect of age on intestinal amino acid transport has not been extensively studied. Despite similar protein intakes, the elderly have reduced serum albumin concentrations when compared to younger subjects, suggesting possible peptide and amino acid malabsorption [116].

The elderly may also develop deficiencies in micronutrients. Age results in reductions in calcium absorption [117]. Diminution of vitamin D metabolite concentration may explain in part this reduced intestinal absorption [118]. However decreases in the renal production and intestinal response to 1,25 hydroxycholecalciferol [119], and reductions in vitamin D receptors in the intestinal mucosa [120] may also be contributing factors. The absorption of 5-methyltetrahydrofolate vitamin D is normal in the elderly [121–123]. Vitamin B12 deficiency seen in the elderly may be due to atrophic gastritis or bacterial overgrowth, rather than being due to malabsorption, or a lack of intrinsic factor [124]. Treatment of this deficiency and the underlying causes is recommended, as vitamin B12 deficiency may result in increased homocysteine levels and increased risk of vascular and neurological disease [125–127].

In humans, there is no difference between the young and old in terms of the height of enterocytes and the jejunal surface area-to-volume ratios [128]. In humans over the age of 65, there is an increased expression of the proliferation cell nuclear antigen in the duodenal villi and crypts, as compared to younger persons [128]. It is suggested that the maintenance of mucosal architecture throughout the aging process may be due to increased rates of small bowel enterocyte proliferation and differentiation [129]. The height of the intestinal villi may decline with aging [130], but other studies have demonstrated no significant effect of age on intestinal surface area [131].

No changes in the passive permeability of the intestine are observed with aging [132, 133]. Small intestinal motility is independent on age [134] and orocecal transit times are similar in older and younger persons [135].

Diarrhea may be life-threatening in infants, in persons with immunodeficiency, and in the elderly [136]. Infections are the most common cause of diarrhea, but the elderly also develop diarrhea because of their use of multiple medications, associated small intestine bacterial overgrowth, and fecal impaction. Colonic carcinoma and diarrhea associated with diabetes need to be excluded. Both celiac disease and inflammatory bowel disease (Crohn's disease and ulcerative colitis) may present for the first time in the elderly. In those individuals treated with antibiotics, *Clostridium difficile* infection needs to be excluded [137]. A practical approach to diarrhea has been proposed for the elderly [137] (Figure 1).

Some authors suggest that malabsorption is uncommon among elderly patients [138, 139], whereas others suggest the contrary [140–142]. In the elderly, malabsorption presents less often with diarrhea and more often with weight loss or failure to regain weight after a recurrent illness, or with signs of malabsorption such as anemia or metabolic bone disease. The elderly may be more likely on a PPI, have had previous gastric surgery, have a motility disorder due to diabetes or hypothyroidism, or have renal impairment resulting in vitamin D deficiency. Medications such as beta blockers or anticholingerics may slow intestinal motility, and thereby create the potential for developing bacterial overgrowth syndrome. Chemotherapy or antibiotics may predispose to *Clostridium difficile* infection, and non-steroidal anti-inflammatory drugs (NSAIDs) may lead to small intestinal ulceration or to strictures [143].

Among 490 geriatric patients, 11% were malnourished, and of these, 44% had occult malabsorption and 71% had small intestinal bacterial overgrowth [144]. The treatment of bacterial overgrowth may lead to weight gain, or to an increase in abnormal blood parameters such as calcium and protein [145].

Celiac disease is common in our community, and in the elderly the presenting symptoms may be extraintestinal in nature [146]. Indeed, about a quarter of the patients being diagnosed with celiac disease are now in their seventh decade [147, 148]. The usual symptoms of celiac disease in the elderly are iron deficiency, anemia, hypoalbuminemia, hypocalcemia/hypomagnesemia, and autoimmune thyroid disease associated with hypothyroidism [149, 150]. Some elderly patients with celiac

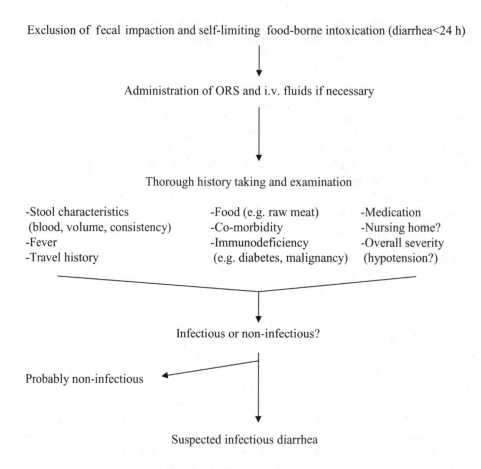

Exclusion of fecal impaction and self-limiting food-borne intoxication (diarrhea<24 h)

Administration of ORS and i.v. fluids if necessary

Thorough history taking and examination

-Stool characteristics
 (blood, volume, consistency)
-Fever
-Travel history

-Food (e.g. raw meat)
-Co-morbidity
-Immunodeficiency
 (e.g. diabetes, malignancy)

-Medication
-Nursing home?
-Overall severity
 (hypotension?)

Infectious or non-infectious?

Probably non-infectious

Suspected infectious diarrhea

Figure 1. Practical approach to diarrhea. Hoffmann and Zeit [Ref. 137].

disease may develop collagenous sprue or lymphocytic colitis [151–153]. Small intestinal non-Hodgkins lymphoma may also develop [154, 155]. Osteoporosis is common in patients with celiac disease [156], and age-adjusted bone mineral density is inversely correlated with age [157].

Gastrointestinal ischemia

The true incidence of ischemic colitis is unknown [158]. Ischemic colitis is the most common form of all gastrointestinal ischemic syndromes, and probably accounts for about 1 in 2000 acute admissions to hospital [159]. Contributing factors include the arteriosclerosis associated with age, colateral circulation, and the patient's general hemodynamic state. Obstruction of the arterial supply is usually from a thrombus or an embolus, or an occlusion on the venous side from hypercoaguable states. Non-occlusive disease will occur without any obvious precipitating event. Most patients present with abdominal pain, diarrhea, abdominal distention, and nausea or vomiting [160]. Rectal bleeding is usually mild and self-limited. While some patients who present with pain from acute mesentery ischemia are hemodynamically stable, about 1 in 3 presents with shock [161]. Elevated serum D-lactate concentrations may suggest the presence of ischemia [162, 163], but other laboratory investigations are of limited value. Endoscopic assessment is both sensitive and specific [164], and CT imaging demonstrating segmental bowel wall thickening may provide evidence of the extent of ischemic colitis [165, 166]. Doppler sonography with or without colour may help to differentiate between inflammatory and ischemic changes, and may have prognostic significance [167, 168]. The roles of magnetic resonance imaging and Indium-III white blood cell scans remain to be proven [169, 170]. Patients with ischemic bowel disease are treated with supportive care, as well as surgery where necessary for gangrenous disease. Most patients resolve spontaneously [171]. Those with associated diabetes, hemodynamic instability and ileus are more likely to require surgery [172]. Unfortunately, mortality from surgery is approximately 50% [173]. About 1 patient in 5 with ischemic colitis develops chronic changes, including stricturing [174].

Irritable bowel syndrome

Irritable bowel syndrome (IBS) is very common in young people, and there is always caution about making this diagnosis in the elderly for fear of missing an organic disease such as colorectal cancer. Using the Rome II criteria, IBS is defined as "12 weeks or more in the prior year of abdominal discomfort or pain associated with 2 or more of the following: relief by defecation or associated with a change in stool frequency or stool form (hard or loose)" [175]. IBS may be diarrhea-predominant, or constipation-predominant, or there may be alternating diarrhea and constipation. In 70 year old individuals, the prevalence of IBS is 6–18%, depending upon the definition used; 5 years later, 50–79% of the IBS sufferers no longer have symptoms [176].

In Americans age 65–93 years, the prevalence of both chronic constipation and chronic diarrhea is 24% [177]. The perception of pain may differ in older versus younger individuals, so older persons with IBS need to be considered for other possible diagnoses including extraintestinal diseases as well as mesentery ischemia and colon cancer [178]. IBS symptoms may be associated with urinary symptoms such as urgency and nocturia [179]. If the diagnosis of IBS is confirmed, then therapeutic agents that may prove useful include smooth muscle relaxants, antidiarrheal agents such as loperamide, and tricyclic anti-depressants [180, 181]. Fiber intake may need to be optimized, and patients need to be reassured.

Inflammatory bowel disease

There is a later age peak in the prevalence of inflammatory bowel disease (IBD), which needs to be distinguished from ischemia or diverticulitis [182]. Diarrhea and weight loss may be more predominant in the presentation of IBD in the elderly, while abdominal pain and rectal bleeding may be less common [183]. Although colitis occurs less commonly in the elderly [184], severe first attacks may be more common [185]. Recurrent attacks are less frequent in the elderly [185]. Complications may be more prevalent [186], although more recent studies suggest that the complication rates are similar in the young and in the old [187].

The distribution of Crohn's disease in older individuals more commonly involves the colon [188]. Misdiagnosis is more common in the young [189], leading to a delay in diagnosis and the initiation of appropriate therapy. Intestinal obstruction may be less common in the elderly, possibly because of fewer patients having small intestinal involvement [190, 191].

The principles of treatment of IBD in the elderly are similar to those in younger individuals, and include the use of mesalamine, antibiotics, corticosteroids (attention paid to the possibility of associated osteoporosis), immune suppression and surgery. Ileal pouch-anal anastomosis can be safely used in selected older patients with ulcerative colitis [192].

Constipation and fecal incontinence

The Rome II Committee defined functional constipation as "a group of functional disorders which present as persistent to the infrequent, or seemingly incomplete defecation" [193]. The two physiological causes of constipation include slow colonic transit and disordered defecation [194]. Constipation may be caused by central nervous system diseases such as Parkinsonism; endocrine causes such as diabetes, hypothyroidism, Addison's disease or pheochromocytoma; as well as anatomical complications such as dysfunction of the anal sphincter, rectocele, intussusception, or pelvic floor descent. Anti-cholinergic dopaminergic medications aggravate constipation, as does the administration of opiates.

Fecal continence is "the ability to perceive, retain and evacuate rectal contents at the time and place of choice," while fecal incontinence is defined as "any loss of continence beyond the age of 4 years" [194]. Incontinence may be due to diseases of the central nervous system, spinal cord or colon, or from conditions causing diarrhea or impairment of sphincter contraction.

Constipation is common in the elderly [195], possibly due to slow colonic transit or low fluid intake, and fecal incontinence is seen in about 1% of individuals over the age of 60 [196]. The incontinent elderly are often bed-ridden, and have associated urinary incontinence [197]. The sensitivity of the anal sphincter is maintained in the elderly, although anal sphincter pressures may be reduced [198]. Rectal prolapse or fecal impaction need to be diagnosed and treated appropriately.

Diverticular disease

Diverticulosis increases with age [199], but most people with diverticulosis do not have symptoms or complications [200]. Nevertheless, rectal bleeding may be common [201]. Emergency angiography may be necessary in those individuals with brisk bleeding, and those with a negative angiogram may benefit from a diagnostic colonoscopy [202]. Diverticulitis occurs in 10–25% of patients with diverticulosis [201], and treatment involves supportive therapy plus the use of appropriate antibiotics. Only 1 patient in 4 with diverticulosis will have recurrent episodes of diverticulitis, and in selected patients surgical intervention may be necessary [203].

Liver disease in the elderly

The weight of the liver is reduced by about 7% in elderly men and 14% in elderly women [204]. Total hepatic blood flow, as well as functional hepatic blood flow, decreases with age [205]. It is this fall in blood flow, rather than the alteration in the microsomal mono-oxygenase [206], which is responsible for the impaired oxidative metabolism of some drugs in the elderly, such as propranolol, amitryptaline, and morphine [207]. The metabolism of methoclopranide and of paracetamol is impaired in frail but not in healthy elderly subjects [208].

Treatment of hepatitis C (HCV) in the elderly is the same as in younger persons, using interferon and ribovarin. Hepatitis in older individuals often presents as cirrhosis or hepatocellular carcinoma, and unlike hepatitis C, hepatitis B-associated cirrhosis is a contraindication to interferon therapy [209].

The prevalence of anti-hepatitis A virus is 80% in American nursing homes [210]. Vaccination against HAV infection is recommended in older persons visiting developing countries, particularly if they have chronic liver disease [209]. Mortality from liver-related causes in patients with HCV is increased if there is associated alcoholism [211].

The immediate stay in an intensive care unit or in a hospital is no different following liver transplantation in younger versus older individuals [212]. Although the quality of life in older persons is better than younger individuals 1 year after liver transplantation, the number of older persons who survive 1 year is lower than in younger persons, and 5 year survival is also unfortunately lower [213].

Pancreatic and biliary disease

About a third of individuals who develop acute pancreatitis are over the age of 65 years [214]. Common causes include alcohol abuse and gallstone disease. In the elderly, only 8% had their first episode of acute pancreatitis caused by alcohol [215], and in those over 85 years of age, gallstone pancreatitis as an etiology of acute pancreatitis is as prevalent as 75% [216]. Often in the elderly the cause of chronic pancreatitis will be idiopathic, or may be due to periampullary tumor carcinoma of the head of the pancreatitis. Pancreatic duct stones are also a cause of pancreatic insufficiency in the elderly [217]. Acute pancreatitis may follow cardiac surgery, presumably due to ischemia [218]. Attention to fluid unbalance is important in the elderly and early admission to an intensive care unit may be important [219, 220].

While endoscopic retrograde cholangiopancreatography (ERCP) is both sensitive and specific in detecting abnormalities in the hepatobiliary system [221], aging itself may result in changes in the pancreatic ducts which in turn may lead to the misdiagnosis of chronic pancreatitis [222].

The most common type of pancreatic exocrine tumor is an adenocarcinoma of the duct. Most patients with pancreatic cancer are over the age of 65 years [223]. Older patients tolerate pancreaticoduodenectomy as well as younger individuals [224], but older patients may have a longer hospital stay and some authors suggest a higher complication rate [225]. Thus, age by itself is not a reason for referring a surgical intervention from an individual patient [226].

References

1. Balcolm NR, Sinclair A (2001). Ageing: definitions, mechanisms and the magnitude of the problem [Review]. *Best Pract Res Clin Gastroenterol.* 15: 835–49.
2. Fries JF (1980). Ageing, natural death and the compression of morbidity. *N Engl J Med.* 303: 130–6.
3. Jagger C (1997). Using healthy active life expectancy to measure population outcomes. In: Philip I, ed. *Outcomes Assessment for Healthcare in Elderly People.* London: Farrand Press, pp. 67–76.
4. Russell RM (1992). Associated changes in gastrointestinal function attributed to aging *Am J Nutr.* 56: 1203–7.
5. Sklar M (1983). Gastrointestinal diseases in the aged. In: Reichel W, ed. *Clinical Aspects of Aging.* Baltimore: Williams & Wilkins, pp. 205–17.
6. Sklar M, Kirsner JB (2001). Assessing and interviewing the elderly interpretation of signs and symptoms. *Best Pract Res Clin Gastroenterol.* 15: 851–67.
7. Tierney AJ (1996). Undernutrition and elderly hospital patients: a review. *J Adv Nurs.* 23: 228–36.
8. Beck AM, Ovesen L, Osler M (1999). The 'Mini Nutritional Assessment' (MNA) and the 'Determine Your Nutritional Health' Checklist (NSI Checklist) as predictors of morbidity and mortality in an elderly Danish population. *Br J Nutr.* 81: 31–6.
9. Lehmann AB, Bassey EJ, Morgan K, Dalosso HM (1991). Normal values for weight, skeletal size and body mass indicies in 890 men and women aged over 65 years. *Clin Nutr.* 10: 18–22.
10. Morley JE (1997). Anorexia of ageing: physiologic and pathologic. *Am J Clin Nutr.* 66: 760–73.
11. Volkert D, Hubsch S, Oster P, Schlierf G (1996). Nutritional support and functional status in undernourished geriatric patients during hospitalization and 6-month follow-up. *Aging (Milano)* 8: 386–95.
12. Pirlich M, Lochs H (2001). Nutrition in the elderly. *Best Pract Res Clin Gastroenterol.* 15: 869–84.
13. Kondo JJL, Blaschke TF (1989). Drug-drug interactions in geriatric patients. In: Platt D, ed. *Gerontology* Berlin: Springer Verlag, pp. 257–69.
14. Le Couteur DG, McLean AJ (1998). The aging liver. Drug clearance and an oxygen diffusion barrier hypothesis. *Clin Pharmcokinet* 34: 359–73.

15. Schmucker DL (1998). Aging and the liver: an update. *J Gerontol: Biol Sci* 53A: B315–20.

16. Herrlinger C, Klotz U (2001). Drug metabolism and drug interactions in the elderly. *Best Pract Res Clin Gastroenterol*. 15: 897–918.

17. Ammon S, Treiber G, Kees F, Klotz U (2000). Influence of age on the steady state disposition of drugs commonly used for the eradication of *Helicobacter pylori*. *Aliment Pharmaco Ther*. 14: 759–66.

18. Feely J, Pereira L, Guy E, Hockings N (1984). Factors affecting the response to inhibition of drug metabolism by cimetidine dose response and sensitivity of elderly and induced subjects. *Br J Clin Pharmacol*. 17: 77–81.

19. Andersson T, Andren K, Cederberg C (1990). Effect of omeprazole and cimetidine on plasma diazepam levels. *Eur J Clin Pharmacol*. 39: 51–4.

20. Budtz-Jorgensen E, Chung J-P, Rapin CH (2001). Nutrition and oral health. *Best Pract Res Clin Gastroenterol*. 15: 885–96.

21. Shaker R, Ren J, Podvrsan B (1993). Effect of aging and bolus variables on pharyngeal and upper esophageal sphincter motor function. *Am J Physiol*. 264: G427–32.

22. Kern M, Bardan E, Arndorfer R, Hofmann C, Ren J, Shaker R (1999). Comparison of upper esophageal sphincter opening in healthy asymptomatic young and elderly volunteers. *Ann Otol Rhinol Laryngol*. 108: 982–9.

23. Shaw DW, Cook IJ, Gabb M (1995). Influence of normal aging on oral-pharyngeal and upper esophageal sphincter function during swallowing. *Am J Physiol*. 268: G389–96.

24. Ekberg O, Feinberg MJ (1991). Altered swallowing function in elderly patients without dysphagia: radiologic findings in 56 cases. *Am J Roentgenol*. 156: 1181–4.

25. Tack J, Vantrappen G (1997). The aging oesophagus. *Gut* 41: 422–4.

26. Grande L, Lacima G, Ros E, *et al.* (1999). Deterioration of esophageal motility with age: a manometric study of 79 healthy subjects. *Am J Gastroenterol*. 94: 1795–801.

27. Adamek RJ, Wegener M, Weinbeck M, Gielen B (1994). Long-term esophageal manometry in healthy subjects. Evaluation of normal values and influence of age. *Dig Dis Sci*. 39: 2069–73.

28. Ferriolli E, Oliveira RB, Matsuda NM (1998). Aging, esophageal motility, and gastroesophageal reflux. *J Am Geriatr Soc*. 46: 1534–7.

29. Lock G (2001). Physiology and pathology of the oesophagus in the elderly patient. *Best Pract Res Clin Gastroenterol*. 15: 919–41.

30. Collen MJ, Abdulian JD, Chen YK (1995). Gastroesophageal reflux disease in the elderly: more severe disease that requires aggressive therapy. *Am J Gastroenterol*. 90: 1053–7.

31. Pilotto A, Franceschi M, Leandro G, *et al.* (2002a). Influence of *Helicobacter pylori* infection on severity of oesophagitis and response to therapy in the elderly. *Dig Liver Dis*. 34: 328–31.

32. Simmons DB, Schuman BM, Griffin JW (1997). Achalasia in patients over 65. *J Florida Med Assoc*. 84: 101–3.

33. Pasricha PJ, Ravich WJ, Hendrix TR (1995). Intrasphincteric botulinum toxin for the treatment of achalasia. *N Engl J Med*. 332: 774–8.

34. Hallett M (1999). One man's poison clinical applications of botulinum toxin. *N Engl J Med*. 341: 118–20.

35. Pandolfino JE, Howden CW, Kahrilas PJ (2000). Motility-modifying agents and management of disorders of gastrointestinal motility. *Gastroenterology* 118: S32–47.

36. Pilotto A, Franceshi M, Leandro G, Novello R, Di Mario F, Valerio G (2002b). Long-term clinical outcome of elderly patients with relfux esophagitis: a six-month to three-year follow-up study. *Am J Ther*. 9: 295–300.

37. Wong J (1981). Management of carcinoma of the esophagus. *Ann Royal Coll Surg Engl*. 26: 138–49.

38. de la Mora G, Marcon NE (2001). Endoscopy in the elderly patient. *Best Pract Res Clin Gastroenterol*. 15: 999–1012.

39. Seinela L, Ahvenainen J (2000). Peptic Ulcer in the very old patients. *Gerontology* 4: 271–5.

40. Phillips AC, Polisson RP, Simon LS (1997). NSAIDs and the elderly. Toxicity and economic implications. *Drugs Aging* 10: 119–30.

41. Bode G, Brenner H, Adler G, Rothenbacher D (2002). Dyspeptic symptoms in middle-aged to old adults: the role of *Helicobacter pylori* infection, and various demographic and lifestyle factors. *J Intern Med*. 252: 41–7.

42. Pilotto A, Franceschi M, Leandro G, et al. (1999). Efficacy of 7 day lansoprazole-based triple therapy for Helicobacter infection in elderly patients. *J Gastroenterol Hepatol*. 14: 468–75.

43. Graham DY, Malaty HM, Evans DG, Evans DJ, Klein PD, Adam E (1991). Epidemiology of *Helicobacter pylori* in an asymptomatic population in the United States. Effect of age, race and socioeconomic status. *Gastroenterology* 100: 1495–501.

44. Kemppainen H, Raiha I, Kujari H, Sourander L (1998). Characteristics of *Helicobacter pylori*-negative and -positive peptic ulcer disease. *Age Ageing* 27: 427–31.

45. Jones JIW, Hawkey CJ (2001). Physiology and organ-related pathology of the elderly: stomach ulcers. *Best Pract Res Clin Gastroenterol*. 15: 943–61.

46. Haruma K, Kamada T, Kawaguchi H (2000). Effect of age and *Helicobacter pylori* infection on gastric acid secretion. *J Gastroenterol Hepatol*. 15: 277–83.

47. Kinoshita Y, Kawanami C, Kishi K (1997). *Helicobacter pylori* independent chronological change in gastric acid secretion in the Japanese. *Gut*, 41: 452–8.

48. Cryer B, Redfern JS, Goldschmiedt M (1992). Effect of aging on gastric and duodenal mucosal prostaglandin concentrations in humans. *Gastroenterology* 102: 1118–23.

49. Goto H, Sugiyama S, Ohara A (1992). Age-associated decreases in prostaglandin contents in human gastric mucosa. *Biochem Biophys Res Comm*. 186: 1443–8.

50. Del Vecchio B, Domenico T (1993). Italian Multicenter Study Group for gastric ulcer. Can advanced age influence the characteristics of peptic gastric ulcer? *Gastrointest Endo*. 39: 50–3.

51. Kemppainen H, Raiha I, Sourander L (1997). Clinical presentation of peptic ulcer in the elderly. *Gerontology* 43: 283–8.

52. Hawkey CJ (2001). Gastrointestinal toxicity of non-steroidal anti-inflammatory drugs. In: Vane JR, Botting RM, eds. *Therapeutic Roles of Selective COX-2 Inhibitors*. London: William Harvey Press, pp. 355–94.

53. Gabriel SE, Jaakkimainen L, Bombardier C (1991). Risk for serious gastrointestinal complications related to use of nonsteroidal anti-inflammatory drugs. A meta-analysis. *Ann Intern Med*. 115: 787–96.

54. Pilotto A, Franceschi M, Di Mario F (1998a). The long-term clinical outcome of elderly patients with *Helicobacter pylori*-associated peptic ulcer disease. *Gerontology* 44: 153–8.

55. Pilotto A, Di Mario F, Franceschi M (2000a). Treatment of *Helicobacter pylori* infection in elderly subjects. *Age Ageing* 29: 103–9.

56. Pilotto A, Malfertheiner P (2002). An approach to *Helicobacter pylori* infection in the elderly. *Aliment Pharmaco Ther*. 16: 683–91.

57. Hawkey CJ, Karrasch JA, Szczepanski (1998). Omeprazole versus misoprostol for NSAID-induced ulcer management (OMNIUM) study group. Omeprazole compared with misoprostol for ulcers associated with nonsteroidal antiinflammatory drugs. *N Engl J Med.* 338: 727–34.

58. Pilotto A, Di Mario F, Franceschi M, *et al.* (2000b). Pantoprazole versus one-week *Helicobacter pylori* eradication therapy for the prevention of acute NSAID-related gastroduodenal damage in elderly subjects. *Aliment Pharmacol Ther.* 14: 1077–82.

59. Koop H, Arnold R, Classen M (1992). The RUDER Study Group. Healing and relapse of duodenal ulcer during ranitidine therapy in the elderly. *J Clin Gastroenterol.* 15: 291–5.

60. Rockall TA, Logan RF, Devlin HB, Northfield TC (1996). Risk assessment after acute upper gastrointestinal haemorrhage. *Gut* 38: 316–21.

61. Longstreth GF (1997). Epidemiology and outcome of patients hospitalized with acute lower gastrointestinal haemorrhage: a population-based study. *Am J Gastroenterol.* 92: 419–24.

62. Cooper BT, Weston CF, Neumann CS (1988). Acute upper gastrointestinal haemorrhage in patients aged 80 years or more. *Q J Med.* 68: 765–74.

63. Hasselgren G, Blomqvist A, Eriksson S (1998). Short and long-term course of elderly patients with peptic ulcer bleeding analysis of factors influencing fatal outcome. *Eur J Surg.* 164: 685–91.

64. Hudson N, Faulkner G, Smith SJ (1995). Late mortality in elderly patients surviving acute peptic ulcer in bleeding. *Gut* 37: 177–81.

65. Chow LWC, Gertsch P, Poon RTP, Branicki FJ (1998). Risk factors for rebleeding and death from peptic ulcer in the very elderly. *Br J Surg.* 85: 121–4.

66. Battaglia G, Di Mario F, Dotto P (1993). markers of slow-healing peptic ulcer in the elderly. A study on 1,051 ranitidine-treated patients. *Dig Dis Sci.* 38: 1414–21.

67. Watson RJ, Hooper TL, Ingram G (1985). Duodenal ulcer disease in the elderly: a retrospective study. *Age Ageing* 14: 225–9.

68. Hasselgren G, Lind T, Lundell L (1997). Continuous intravenous infusion of omeprazole in elderly patients with peptic ulcer bleeding. Results of a placebo-controlled multicenter study. *Scand J Gastroenterol.* 32: 328–33.

69. Lingenfelser T, Ell C (2001). Gastrointestinal bleeding in the elderly. *Best Pract Res Clin Gastroenterol.* 15: 963–82.

70. Segal WN, Cello JP (1997). Haemorrhage in the upper gastrointestinal tract in the older patient. *Am J Gastroenterol.* 92: 42–6.

71. O'Riordain DS, O'Dwyer PJ, O'Higgins NJ (1990). Perforated duodenal ulcer in elderly patients. *J Royal Coll Surg Edinburgh* 35: 93–4.

72. Kane E, Fried G, McSherry CK (1981). Perforated peptic ulcer in the elderly. *J Am Geriatr Soc.* 29: 224–7.

73. Kum CK, Chong YS, Koo CC, Rauff A (1993). Elderly patients with perforated peptic ulcers: factors affecting morbidity and mortality. *J Royal Coll Surg Edinburg* 38: 344–7.

74. Peter DJ, Dougherty JM (1999). Evaluation of the patient with gastrointestinal bleeding: an evidence based approach. *Emerg Med Clin N Am* 17: 239–61.

75. Farrell JJ, Friedman LS (2000). Gastrointestinal bleeding in older people. *Gastroenterol Clin N Am.* 29: 1–36.

76. Vreeburg EM, Snel P, de Bruijne JW (1997). Acute upper gastrointestinal bleeding in the Amsterdam area: incidence, diagnosis, and clinical outcome. *Am J Gastroenterol.* 1997: 236–43.

77. Terdiman JP (1998). Update on upper gastrointestinal bleeding. *Postgrad Med.* 103: 43–63.

78. Cappell MS, Abdullah M (2000). Management of gastrointestinal bleeding induced by gastrointestinal endoscopy. *Gastroenterol Clin N Am.* 29: 125–67.

79. Simoens M, Gevers AM, Rutgeerts P (1999). Endoscopic therapy for upper gastrointestinal haemorrhage: a state of the art. *Hepatogastroenterology* 46: 737–45.

80. Stollman NH, Raskin JB (1999). Diverticular disease of the colon. *J Clin Gastroenterol.* 29: 241–52.

81. Foutch PG (1995). Diverticular bleeding: are nonsteroidal anti-inflammatory drugs risk factors for haemorrhage and can colonoscopy predict outcome for patients? *Am J Gastroenterol.* 90: 1779–84.

82. Sorbi D, Norton I, Conio M (2000). Postopolypectomy lower GI bleeding: descriptive analysis. *Gastrointest Endosc.* 51, 690–6.

83. Lau JYW, Sung JJY, Lee KKC (2000). Effect of intravenous omeprazole on recurrent bleeding after endoscopic treatment of bleeding peptic ulcer. *N Engl J Med.* 343: 310–16.

84. Rosen A (1999). Gastrointestinal bleeding in the elderly. *Clin Geriatr Med.* 15: 511–25.

85. Lau JYW, Sung JJY, Lam YJ (1999). Endoscopic retreatment compared with surgery in patients with recurrent bleeding after initial endoscopic control of bleeding ulcers. *N Engl J Med.* 340: 751–6.

86. Hamilton SR, Aoltonen LA (2000). *Tumours of the Digestive System*. World Health Organization Classification of Tumours. Lyon: IARC Press.

87. Schlemper RJ, Riddell RH, Kato Y (2000). The Vienna classification of gastrointestinal epithelial neoplasia. *Gut* 47: 251–5.

88. Meining A, Morgner A, Miehlke S, Bayerdorffer E, Stolte M (2001). Atrophy-metaplasia-dysplasia-carcinoma sequence in the stomach: a reality or merely an hypothesis? *Best Pract Res Clin Gastroenterol.* 15: 983–98.

89. Sugano H, Nakamura K, Kato Y (1982). Pathological studies on human gastric cancer. *Acta Pathologica Japan* 32(Suppl 2): 329–47.

90. Sasaki I, Yao T, Nawate H, Tsuneyoshi M (1999). Minute gastric carcinoma of differentiated type with special reference to the significance of intestinal metaplasia, proliferative zone, and p53 protein during tumor development. *Cancer* 85: 1719–29.

91. Pilotto A, Rassu M, Bozzola L, *et al.* (1998b). Cag-A positive *Helicobacter pylori* infection in the elderly. Association with gastric atrophy and intestinal metaplasia. *J Clin Gastroenterol.* 26: 18–22.

92. Blaser MJ, Perez-Perez GI, Kleantoustl, *et al.* (1995). Infection with *Helicobacter pylori* possessing cag A is associated with an increased risk of developing adenocarcinoma of the stomach. *Cancer Res.* 55: 2111–5.

93. Parsonnet J, Friedman GD, Orestrich N, *et al.* (1997). Risk of gastric cancer in people with Cag A positive or Cag A negative *Helicobacter pylori* infection. *Gut* 40: 297–301.

94. Schulmann K, Reiser M, Schmiegel W (2002). colonic cancer polyps. *Best Pract Res Clin Gastroenterol.* 16: 91–114.

95. Rudman D (1989). Nutrition and fitness in elderly people. *Am J Clin Nutr.* 49: 1090–8.

96. Kerstetter JE, Holthausen BA, Fitz PA (1992). Malnutrition in the institutionalized elderly adult. *J Am Diet Assoc.* 92: 1109–16.

97. Abbasi AA, Rudman D (1993). Observations on the prevalence of protein-calorie undernutrition in VA nursing homes. *J Am Geriatr Soc.* 41: 117–21.

98. Mowe M, Bohmer T, Kindt E (1994). Reduced nutritional status in an elderly population (>70 y) is probable before disease and possibly contributes to the development of disease. *Am J Clin Nutr.* 59: 317–24.

99.   Sullivan D, Walls RC (1994). Impact of nutritional status on morbidity in a population of geriatric rehabilitation patients. *J Am Geriatr Soc.* 42: 471–7.

100. Payette H, Coulombe C, Boutier V, Gray-Donald K (1999). Weight loss and mortality among the free-living frail elderly: a prospective study. *J Gerontol: Med Sci.* 54A: M440–5.

101. Payette H, Gray-Donald K (1994). Risk of malnutrition in an elderly population receiving home care services. *Facts and Research in Gastroenterology: Nutrition.* Suppl 2: 71–85.

102. Sullivan DH, Patch GA, Walls RC, Lipschitz DA (1990). Impact of nutrition status on morbidity and mortality in a select population of geriatric rehabilitation patients *Am J Clin Nutr.* 51: 749–58.

103. Wallace JI, Schwartz RS, LaCroix AZ, *et al.* (1995). Involuntary weight loss in older outpatients: incidence and clinical significance. *J Am Geriatr Soc.* 43: 329–37.

104. Payette H, Gray-Donald K (1991). Dietary intake and biochemical indices of nutritional status in an elderly population, with estimates of the precision of the 7-d food record. *J Am College Nutr.* 54: 478–88.

105. McGandy RB, Russell RM, Hartz SC, *et al.* (1986). Nutritional status survey of healthy noninstitutionalized elderly: energy and nutrient intakes from three-day records and nutrient supplements. *Nutr Res.* 6: 785–98.

106. De Groot CPGM, Van Staveren WA, Dirren H, Hautvast JGAJ, eds. (1996). SENECA: Nutrition and the elderly in Europe. Follow-up study and longitudinal analysis. *Eur J Clin Nutr.* 50(Suppl 2): 1–124.

107. Azad N, Murphy J, Amos SS, Toppan J (1999). Nutrition survey in an elderly population following admission to a tertiary care hospital. *Can Med Assoc J.* 161: 511–15.

108. Feibusch JM, Holt PR (1982). Impaired absorptive capacity for carbohydrate in the aging human. *Dig Dis Sci.* 27: 1095–100.

109. Riepe SP, Goldstein J, Alpers DH (1980). Effect of secreted *Bacteroides* proteases on human intestinal brush border hydrolases. *J Clin Invest.* 66: 314–22.

110. Lee MF, Russell RM, Montgomery RK, Krasinski SD (1997). Total intestinal lactase and sucrase activities are reduced in aged rats. *J Nutr.* 127: 1382–7.

111. Woudstra T, Thomson ABR (2002). Nutrient absorption and intestinal adaptation with ageing. *Best Pract Res Clin Gastroenterol.* 16: 1–15.

112. Simko V, Michael S (1989). Absorptive capacity for dietary fat in elderly patients with debilitating disorders. *Arch Intern Med.* 149: 557–60.

113. Salemans JM, Nagengast FM, Tangerman A (1993). Effect of ageing on postprandial conjugated and unconjugated serum bile acid levels in healthy subjects. *Eur J Clin Invest.* 23: 192–8.

114. Arora S, Kassarjian Z, Krasinski SD (1989). Effect of age on tests of intestinal and hepatic function in healthy humans. *Gastroenterology* 96: 1560–5.

115. Holt PR (2001). Diarrhea and malabsorption in the elderly. *Gastroenterol Clin N Am.* 30: 427–44.

116. Gersovitz M, Munro HN, Udall J, Young VR (1980). Albumin synthesis in young and elderly subjects using a new stable isotope methodology: response to level of protein intake. *Metabolism* 29: 1075–86.

117. Bullamore JR, Wilkinson R, Gallagher JC, *et al.* (1970). Effect of age on calcium absorption. *Lancet* 2: 535.

118. Morris HA, Nordin BEC, Fraser V (1985). Calcium absorption and serum 1,25 dihydroxy vitamin D levels in normal and osteoporotic women. *Gastroenterology* A1508.

119. Armbrecht HJ, Zenser TV, Bruns ME, Davis BB (1979). Effect of age on intestinal calcium absorption and adaptation to dietary calcium. *Am J Physiol.* 236: E769–74.

120. Ebeling PR, Sandgren ME, DiMagno EP, Lane AW, DeLuca HF, Riggs BL (1992). Evidence of an age-related decrease in intestinal responsiveness to vitamin D: relationship between serum 1,25 dihydroxyvitamin D-3 and intestinal vitamin D receptor concentrations in normal women. *J Clin Endocrinol Metab.* 75: 176–82.

121. Said HM, Hollander D (1984). Does aging affect the intestinal transport of 5-methyltetrahydrofolate? *Digestion* 30: 231–5.

122. Nilsson-Ehle H, Landahl S, Lindstedt G (1989). Low serum cobalamin levels in a population study of 70- and 75-year old subjects. Gastrointestinal causes and hematological effects. *Dig Dis Sci.* 34: 716–23.

123. Kinyamu HK, Gallagher JC, Knezetic JA, DeLuca HF, Prahl JM, Lanspa SJ (1997). Effect of vitamin D receptor genotypes on calcium absorption, duodenal vitamin D receptor concentration, and serum 1,25 dihydroxyvitamin D levels in normal women. *Calcif Tiss Int.* 60: 491–5.

124. Kassarjian Z, Russell RM (1989). Hypochlorhydria: a factor in nutrition. *Annu Rev Nutr.* 9: 271–85

125. Stampfer MJ, Malinow MR, Willet WC, *et al.* (1992). A prospective study of plasma homocysteine and risk of myocardial infarction in US physicians. *J Am Med Assoc.* 268: 877–81.

126. Selhub J, Jaques PF, Bostom AG, *et al.* (1995). Association between plasma homocysteine concentrations and extracranial carotid-artey stenosis. *N Engl J Med.* 332: 226–91.

127. Riggs KM, Spiro A III, Tucker K, Rush D (1996). Relations of vitamins B-12, vitamin B-6, folate, and homocysteine to cognitive performance in the normative ageing study. *Am J Clin Nutr.* 63: 306–14.

128. Corazza GR, Ginaldi L, Ouaglione G (1998). Proliferation cell nuclear antigen expression is increased I small bowel epithelium in the ederly. *Mech Ageing Dev.* 104: 1–9.

129. Ciccocioppo R, Di Sabatino A, Luinetti O, *et al.* (2002). Small bowel enterocyte apoptosis and proliferation are increased in the elderly. *Gerontology* 48(4): 204–8.

130. Warren PM, Pepperman MA, Montgomery RD (1978). Age changes in small-intestinal mucosa. *Lancet* 2: 849–50.

131. Corazza GR, Frazzoni M, Gatto MRA, *et al.* (1986). Ageing and small bowel mucosa: a morphometric study. *Gerontology* 32: 60.

132. Beaumont DM, Cobden I, Sheldon WL, Laker MF, James OF (1987). Passive and active carbohydrate absorption by the ageing gut. *Age Aging* 16: 294–300.

133. Saltzman JR, Kowdley KV, Perone G, *et al.* (1995). Changes in small intestine permeability with ageing. *J Am Geriatr Soc.* 43: 160

134. Husebye E, Engedal K (1992). The patterns of motility are maintained in the human small intestine throughout the process of ageing. *Scand J Gastroenterol.* 27: 397–404.

135. Madsen J (1992). Effects of gender, age, and body mass index on gastrointestinal transit times. *Dig Dis Sci.* 37: 1548.

136. Kawanishi H, Kiely J (1989). Immune-related alterations in aged gut-associated lymphoid tissues in mice. *Dig Dis Sci.* 34: 175–84.

137. Hoffmann JC, Zeitz M (2002). Small bowel disease in the elderly: diarrhoea and malabsorption. *Best Pract Res Clin Gastroenterol.* 16: 17–36.
138. Buzby JC, Roberts T (1997). Economic costs and trade impacts of microbial foodborne illness. *World Health Stat Q.* 50: 57–66.
139. Nagar A, Roberts IM (1999). Small bowel diseases in the elderly. *Clin Geriatr Med.* 15: 473–86.
140. Brasitus TA, Sitrin MD (1990). Intestinal malabsorption syndromes. *Ann Rev Med,* 41: 339–47.
141. Montgomery RD, Haeney MR, Ross IN (1978). The ageing gut: a study of intestinal absorption in relation to nutrition in the elderly. *Q J Med.* 47: 197–224.
142. Montgomery RD, Haboubi NY, Mike NH, Chesner IM, Asquith P (1986). Causes of malabsorption in the elderly. *Age Ageing* 15: 235–40.
143. Madhok R, MacKenzie JA, Lee FD, Bruckner FE, Terry TR, Sturrock RD (1986). Small bowel ulceration in patients receiving non-steroidal anti-inflammatory drugs for rheumatoid arthritis. *Q J Med.* 58: 53–8.
144. McEvoy A, Dutton J, James OF (1983). Bacterial contamination of the small intestine is an important cause of occult malabsorption in the elderly. *Br Med J.* 287: 789–93.
145. Haboubi NY, Montgomery RD (1992). Small-bowel bacterial overgrowth in elderly people: clinical significance and response to treatment. *Age Ageing* 21: 13–19.
146. Freeman H, Lemoyne M, Pare P (2002). Coeliac disease. *Best Pract Res Clin Gastroenterol.* 16: 37–49.
147. Swinson CM, Levi AJ (1980). Is coeliac disease underdiagnosed? *Br Med J.* 281: 1258–60.
148. Kirby J, Fielding JF (1984). Very adult coeliac disease! The need for jejunal biopsy in the middle aged and elderly. *Irish Med J.* 77: 35–6.
149. Freeman HJ (1995a). Celiac associated autoimmune thyroid disease. A study of 16 patients with overt hypothyroidism. *Can J Gastroenterol.* 9: 242–6.
150. Freeman HJ (1995b). Clinical spectrum of biopsy-defined celiac disease in the elderly. *Can J Gastroenterol.* 9: 42–6.
151. Weinstein WM, Saunders DR, Tytgat GN, Rubin CE (1970). Collagenous sprue – an unrecognized type of malabsorption. *N Engl J Med.* 283: 1297–301.
152. Robert ME, Ament ME, Weinstein WM (2000). The histologic spectrum and clinical outcome of refractory and unclassified sprue. *Am J Surg Pathol.* 24: 676–87.
153. Cellier C, Delabesse E, Helmer C, et al. (2000). Refractory sprue, coeliac disease, and enteropathy-associated T-cell lymphoma. French Coeliac Disease Study Group. *Lancet* 356: 203–8.
154. Freeman HJ, Weinstein WM, Shnitka TK, Piercey JR, Wensel RH (1977). Primary abdominal lymphoma presenting manifestation of celiac sprue or complicating dermatitis herpetiformis. *Am J Med.* 63: 585–94.
155. Freeman HJ (1996). Neoplastic disorders in 100 patients with celiac disease. *Can J Gastroenterol.* 10: 163–6.
156. Meyer D, Stavropolous S, Diamond B, Shane E, Green PHR (2001). Osteoporosis in a North American adult population with celiac disease. *Am J Gastroenterol.* 96: 112–19.
157. Valdimarsson T, Lofman O, Toss G (1996). Reversal of osteopenia with diet in adult coeliac disease. *Gut* 38: 322–7.
158. MacDonald PH (2002). Ischaemic colitis. *Best Pract Res Clin Gastroenterol.* 16: 51–61.
159. Brandt LJ, Boley SJ (1992). Colonic ischemia. *Surg Clin N Am.* 72: 203–29.
160. Bower TC (1993). Ischemic colitis. *Surg Clin N Am.* 73: 1037–53.

161. Guttormson NL, Burrick MP (1998). Mortality from ischemic colitis. *Dis Colon Rectum* 32: 469–72.

162. Poeze M, Froon AHM, Greve JWM, Ramsay G (1998). D-Lactase as an early marker of intestinal ischemia after ruptured abdominal aortic aneurysm repair. *Br J Surg.* 85: 1221–4.

163. Murray MJ, Gonze MD, Nowak LR, Cobb CF (1994). Serum D-lactase levels as an aid to diagnosing acute intestinal ischemia. *Am J Surg.* 167: 575–8.

164. Scowcroft CW, Sanowski RA, Kozarek RA (1981). Colonoscopy in ischemic colitis. *Gastrointest Endo.* 27: 145–61.

165. Balthazar EJ, Yen BC, Gordon RB (1999). Ischemic colitis: CT evaluation of 54 cases. *Radiology* 211: 381–8.

166. Bharucha AE, Tremaine WJ, Johnson D, Batts KP (1996). Ischemic proctosigmoiditis. *Am J Gastroenterol.* 91: 2305–9.

167. Danse EM, Van Beers BE, Jamart J (2000). Prognosis of ischemic colitis: comparison of color Doppler sonography with early clinical and laboratory findings. *Am J Roenterol.* 175: 1151–4.

168. Teefey SA, Roarke MC, Brink JA (1996). Bowel wall thickening: differentiation of inflammation from ischemia with color Doppler and duplex US. *Radioloogy* 198: 547–51.

169. Wilkerson DK, Mezrick R, Drake C (1990). MR imaging of acute occlusive intestinal ischemia. *J Vasc Surg.* 11: 567–71.

170. Moallem AG, Gerard PS, Japanwalla M (1995). Positive in-111 WBC scan in a patients with ischemic ileocolitis and negative colonoscopies. *Clin Nucl Med.* 20: 483–5.

171. Gandhi SK, Hanson MM, Vernava AM, Kaminski DL, Longo WE (1996). Ischemic colitis. *Dis Colon Rectum* 39: 88–100.

172. Longo WE, Ward D, Vernava AM, Kaminski DL (1997). Outcome of patients with total colonic ischemia. *Dis Colon Rectum* 40: 1448–54.

173. Abel ME, Russell TR (1983). Ischemic colitis: comparison of surgical and non-operative management. *Dis Colon Rectum* 26: 113–15.

174. Capella G, Guillaumes S, Galera MJ, Rius X (1988). Ischemic colitis. Review of 20 cases. *Rev Esp Enferm Apar Dig.* 74: 32–6.

175. Drossman Da, Coruzziari IE, Talley NJ, Thompson WG, Whitehead RWE, eds. (2000). *Rome II – Functional Gastrointestinal Disorders*, 2nd edn. McLean, VA: Degnon Associates.

176. Kay L (1994). Prevalence, incidence and prognosis of gastrointestinal symptoms in a random sample of an elderly population. *Age Ageing* 23: 146–9.

177. Ragnarsson G, Bodemar G (1998). Pain is temporally related to eating but not to defaecation in the irritable bowel syndrome. *Eur J Gastroenterol Hepatol.* 10: 415–21.

178. Halar EN, Hammond MC, La Cavae (1987). Sensory perception thresholds measurement. *Arch Phys Med Rehab.* 68: 499–507.

179. Monga AK, Marrero JM, Stanton S (1997). Is there an irritable bladder in the irritable bowel syndrome. *Br J Obstet Gynaecol.* 104: 1409–12.

180. Jailwala J, Imperiale TF, Kroenke K (2000). Pharmalogical treatment of the irritable bowel syndrome: a systematic review of randomized, controlled trials. *Annu Intern Med.* 13: 136–47.

181. Jackson JF, O'Malley PG, Tomkins G (2000). Treatment of functional gastrointestinal disorders with antidepressant medications: a meta-analysis. *Am J Med.* 108: 65–72.

182. Gurudu S, Fiocchi C, Katz JA (2002). Inflammatory bowel disease. *Best Pract Res Clin Gastroenterol.* 16: 77–90.

183. Langholz E, Munkholm P, Haagen Nelson O (1991). Incidence and prevalence of ulcerative colitis in the Copenhagen County from 1962 to 1987. *Scand J Gastroenterol.* 26: 1247–55.

184. Softley A, Myren J, Clamp S (1988). Inflammatory bowel disease in the elderly patient. *Scand J Gastroenterol.* 23: 27–30.

185. Sinclair T, Brunt P, Ashley N (1983). Nonspecific proctocolitis in Northeastern Scotland: a community study. *Gastroenterology* 76: 7–11.

186. Edwards RC, Truelove SC (1963). The course and prognosis of ulcerative colitis. *Gut* 4: 299–315.

187. Gupta S, Saverymuttu SH, Keshavarzian A, Hodgson HJ (1985). Is the pattern of inflammatory bowel disease different in the elderly? *Age Ageing* 14: 366–70.

188. Polito J, Childs B, Mellits E (1996). Crohn's disease: influence of age at diagnosis on site and clinical type of disease. *Gastroenterology* 111: 580–6.

189. Foxworthy D, Wilson J (1985). Crohn's disease in the elderly: prolonged delay in diagnosis. *J Am Geriatr Soc.* 33: 492–5.

190. Harper P, McAuliffe T, Beeken W (1986). Crohn's disease in the elderly: a statistical comparison with younger patients matched for sex duration of disease. *Arch Intern Med.* 146: 753–5.

191. Stalnikowicz R, Eliakim R, Diab R (1989). Crohn's disease in the elderly. *J Clin Gastroenterol.* 11: 411–15.

192. Church JM (2000). Functional outcome and quality of life in an elderly patient with an ileal pouch-anal anastomosis: a 10 year follow-up. *Aust NZ J Surg.* 70: 906–7.

193. Thompson WG, Longstreth GF, Drossman DA (1999). Functional bowel disorders and functional abdominal pain. *Gut* 45: 1143–7.

194. Muller-Lissner S (2002). General geriatrics and gastroenterology: constipation and faecal incontinence. *Best Pract Res Clin Gastroenterol.* 16: 115–33.

195. Everhart JE, Go VLW, Johannes RS (1989). A longitudinal survey of self-reported bowel habits in the United States. *Dig Dis Sci.* 34: 1153–62.

196. Thomas TM, Egan M, Walgrove A, Meade TW (1984). The prevalence of fecal and double incontinence. *Comm Med.* 6: 216–20.

197. Sonnenberg A (1989). Epidemiologie der analen Inkontinenz. In: Muller-Lissner SA, Akkermans LMA, eds. *Chronische Obstipation und Stuhlinkontinenz.* Berlin Heidelberg, New York: Springer, pp. 157–62.

198. McHugh SM, Diamant NE (1987). Effect of age, gender and parity on anal canal pressures. *Dig Dis Sci.* 32: 726–36.

199. Parks TG (1975). Natural history of diverticular disease of the colon. *Clin Gastroenterol.* 4: 53–69.

200. Thompson WG (1968). Do colonic diverticula cause symptoms? *Am J Gastroenterol.* 81: 613.

201. Young-Fadok TM, Roberts PL, Spencer MP, Wolff BG (2000). Colonic diverticular disease. *Curr Prob Surg.* 37: 457–514.

202. Forde KA (1981). Colonoscopy in acute rectal bleeding. *Gastrointest Endo.* 27: 219–20.

203. Rege RV, Nahrwald DL (1989). Diverticular disease. *Curr Prob Surg.* 26: 138–89.

204. Popper H (1986). Aging and the liver. *Prog Liver Dis.* 8: 659–83.

205. Martin SP, Ulrich CD II (1999). Pancreatic disease in the elderly. *Clin Geriatr Med.* 15: 579–605.

206. Williams D, Woodhouse K (1996). Age-related changes in O-deethylase and aldrin epoxidase activity in mouse skin ad liver mirosomes. *Age Ageing* 25: 377–80.

207. Altman DF (1990). Changes in gastrointestinal, pancreatic, biliary, and hepatic function with aging. *Gastroenterol Clin N Am.* 19: 227–34.
208. Wynne HA, Cope LH, Herd B (1990). The association of age and frailty with paracetamol conjugation I man. *Age Ageing* 19: 419–24.
209. Jansen PLM (2002). Liver disease in the elderly. *Best Pract Res Clin Gastroenterol.* 16: 149–58.
210. Chien NT, Dundoo G, Horani MH (1999). Seroprevalence of viral hepatitis in an older nursing home population. *J Am Geriatr Soc.* 47: 1110–3.
211. Harris DR, Gonin R, Alter HJ (2001). The relationship of acute transfusion-associated hepatitis to the development of cirrhosis in the presence of alcohol abuse. *Annu Intern Med.* 134: 120–4.
212. Zetterman RK, Belle SH, Hoofnagle (1998). Age and liver transplantation: a report of the liver transplantation database. *Transplantation* 66: 500–6.
213. Collins BH, Pirsch JD, Becker YT (2000). Long-term results of liver transplantation in older patients 60 years of age and older. *Transplantation* 70: 780–3.
214. Go VLW (1994). Etiology and epidemiology of pancreatitis in the United States. In: Bradley EL, ed. *Acute Pancreatitis: Diagnosis and Therapy.* New York: Raven Press, pp. 235–41.
215. Scholhamer CF Jr, Spiro HM (1979). The first attack of acute pancreatitis: a clinical study. *J Clin Gastroenterol.* 1: 325–9.
216. Park J, Fromkes J, Cooperman M (1986). Acute pancreatitis in elderly patients. *Am J Surg.* 152: 638–42.
217. Nagai H, Ohtsubo K (1984). Pancreatic lithiasis in the aged. *Gastroenterology* 86: 331–8.
218. Fernandez-del Castillo C, Harringer W, Warshaw AL (1991). Risk factors for pancreatic cellular injury after cardiopulmonary bypass. *N Engl J Med.* 325: 382–7.
219. Mutinga M, Rosenbluth A, Tenner SM (2000). Does mortality occur early or late in acute pancreatitis? *Int J Pancreatol.* 28: 91–5.
220. Lowham A, Lavelle J, Leese T (1999). Mortality from acute pancreatitis. Late septic deaths can be avoided but some early deaths still occur. *Int J Pancreatol.* 25: 103–6.
221. Clain JE, Pearson RK (1999). Diagnosis of chronic pancreatitis. Is a gold standard necessary? *Surg Clin N Am.* 79: 829–45.
222. Schmitz-Moormann P, Himmelmann GW, Brandes JW (1985). Comparative radiological and morphological study of human pancreas. Pancreatitis like changes in post-mortem ductograms and their morphological pattern. Possible implication for ERCP. *Gut* 26: 406–14.
223. Niederhuber JE, Brennan MF, Menck HR (1995). The National Cancer Data Base report on pancreatic cancer. *Cancer* 76: 1671–7.
224. Hodul P, Tansey J, Golts E (2001). Age is not a contraindication to pancreaticoduode-nectomy. *Am Surg.* 67: 270–5.
225. Sohn TA, Yeo CJ, Cameron (1998). Should pancreaticoduodenetomy be performed in octogenarians? *J Gastrointestal Surg.* 2: 207–16.
226. Gloor B, Ahmed Z, Uhl W, Buchler MW (2002). Pancreatic disease in the elderly. *Best Pract Res Clin Gastroenterol.* 16: 159–70.

References not cited in the text

Balcombe NR, Sinclair A (2001). Ageing: definitions, mechanisms and the magnitude of the problem. *Best Pract Res Clin Gastroenterol.* 15: 835–49.

Becker GH, Meyer J, Necheles H (1950). Fat absorption in young and old age. *Gastroenterology* 14: 80–90.

Bennett G, Talley NJ (2002). Irritable bowel syndrome in the elderly. *Best Pract Res Clin Gastroenterol*. 16: 63–76.

Borum ML (1999). Peptic ulcer disease in the elderly. *Clin Geriatr Med*. 15: 457–71.

Cappell MS (1998). Intestinal (mesenteric) vasculopathy II. Ischemic colitis and chronic mesenteric ischemia. *Gastroenterol Clin N Am*. 27: 827–60.

Cellier C, Delabesse E, Helmer J (1984). Refractory sprue, coeliac disease, and enteropathy-associated T-cell lymphoma. *Lancet* 356: 203–8.

Cullen DJ, Hawkey GM, Greenwood DC (1997). Peptic ulcer bleeding in the elderly: relative roles of *Helicobacter pylori* and non-steroidal anti-inflammatory drugs. *Gut* 41: 459–62.

Garry PJ, Goodwin JS, Hunt WC, Hooper EM, Leonard AG (1982). Nutritional status in a healthy elderly population: dietary and supplemental intakes. *J Am Coll Nutr*. 36: 319–31.

Gupta S, Savermuttu S, Keshavarzian A (1985). Is the pattern of inflammatory bowel disease different in the elderly? *Age Ageing* 14: 366–70.

Kassarjian Z, Russell RM (1989). Hupochlorhydria: a factor in nutrition. *Annu Rev Nutr*. 9: 271–85.

Kern M, Bardan E, Arndorfer R (1999). Comparison of upper esophageal sphincter opening in healthy asymptomatic young and elderly volunteers. *Ann Otol Rhinol Laryngol*. 108: 982–9.

Khaghan N, Holt PR (2002). Peptic disease in elderly patients. *Can J Gastroenterol*. 14: 922–8.

Kinyamu HK, Gallagher JC, Balhorn KE, Petranick KM, Rafferty KA (1997). Serum vitamin D metabolites and calcium absorption in normal young an elderly free-living women and in women living in nursing homes. *Am J Clin Nutr*. 65: 790–7.

Longo WE, Ballantyne GH, Gusberg RJ (1992). Ischemic colitis: patterns and prognosis. *Dis Colon Rectum* 35: 726–30.

Pilotto A (2001) *Helicobacter pylori*-associated peptic ulcer disease in older patients: current management strategies. *Drugs Aging* 18: 487–94.

Place RJ, Simmang CL (2002). Diverticular disease. *Best Pract Res Clin Gastroenterol*. 16: 135–48.

Schlemper RJ, Riddell RH, Kato Y (2000). The Vienna classification of gastrointestinal epithelial neoplasma. *Gut* 47: 251–5.

Scholhamer CF Jr, Spiro HM (1979). The first attack of acute pancreatitis: a clinical study. *J Clin Gastroenterol*. 1: 325–9.

Tierney AJ (1996). Undernutrition and elderly hospital patients: a review. *J Adv Nurs*. 23: 228–46.

Vetta F, ronzoni S, Tagliere G, Bollea MR (1999). The impact of malnutrition on the quality of life in the elderly. *Clin Nutr*. 18: 259–67.

Volkery D (1996). Ernahrungsprobleme in der Geriatrie Mangelernahrung bei geriatrischen patienten. *Aktuelle Ernahrungsmedizin* 21: 200–2.

Waye JD, Kahn O, Auerbach M (1996). Complications of colonoscopy and flexible sigmoidoscopy. *Gastrointest Endosc Clin N Am*. 6: 343–77.

Weinstein WM, Saunders DR, Tytgat GN (1970). Collagenous sprue- an unrecognized type of malabsorption. *N Engl J Med*. 283: 1297–301.

Wynne HA, Yelland C, Cope LH (1993). The association of age and frailty with the pharmacokinetics and pharmacodynamics of metoclopramide. *Age Ageing* 22: 354–9.

Zoli M, Magalotti D, Bianchi G (1999). Total and functional hepatic blood flow decrease in parallel with ageing. *Age Ageing* 28: 29–33.

Aging of the Eye

John J. Harding

Nuffield Laboratory of Ophthalmology, University of Oxford, UK

The first aging change noticed by many of us before reaching our 48th birthdays was the inadequate length of arms no longer able to hold books far enough away to allow us to focus on the print. At an age when walking, running, and cycling were all possible the lens had called a halt to its adjustable focusing powers, and we were doomed to a life with reading spectacles. Of course the loss of adjustment had started in infancy (Figure 1) but its effect kicked in abruptly before many of the expected age-related changes. The expected changes include slow loss of various physical and mental functions followed by the brutal appearance of age-related disease. The eye suffers age-related disease in all areas so that eye hospitals are full of elderly patients.

All the different tissues of the eye decline in function but each has different weaknesses (Figure 2). The retina has a high metabolic activity and has to rapidly replace the photoreceptor membranes. Lens and cornea are transparent thus limiting the content of pigments and particles, and in turn limiting the metabolic activity. These weaknesses have dictated which features decline in each tissue and so this review will cover the eye by tissue. The literature is dominated by changes to the long-lived proteins of the lens and the accumulation of lipofuscin, "the age pigment," in retina so this review will emphasise those same areas. Both feature slow accumulation of damaged material in keeping with the slow progress of aging; and both types of change result ultimately in age-related diseases: cataract and age-related macular degeneration respectively. Interest in these diseases dictates that this review will concentrate on aging changes in human eyes.

Aging of the eye has been covered in an excellent book [2]. Weale has related 19 age-related changes to life span [3].

Aging of the lens

The transparent lens is an avascular tissue suspended between aqueous humour and vitreous body (Figure 2). It is contained within the lens capsule, a basement

R. Aspinall (ed.), Aging of Organs and Systems, 179–200.
© 2003 *Kluwer Academic Publishers. Printed in Great Britain.*

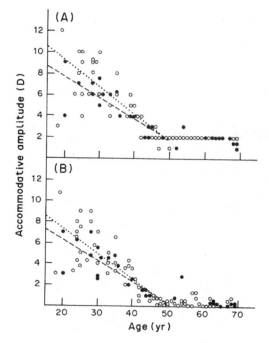

Figure 1. *The loss of lens accommodation (adjustment of focus) with age in humans. Open symbols, female; closed symbols, male. (A) left eye; (B) dilated right eye [1].*

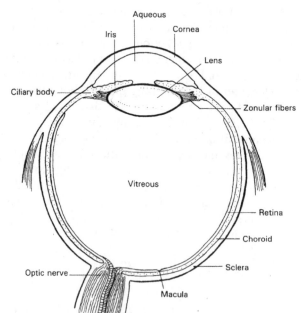

Figure 2. *A section through the eye.*

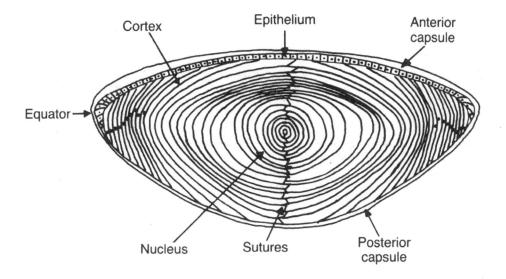

Figure 3. *A section through the lens, from a drawing by John Cronin.*

membrane, and never loses cells (Figure 3). There is a single layer of epithelial cells covering the inside of the anterior capsule. These divide, move towards the equator and then elongate to form the long thin fibre cells that fill the lens. During elongation the cells lose their organelles including cell nuclei and are therefore unable to synthesise protein to replace that which is damaged. The cells at the centre of the lens, in the lens nucleus, have been there since birth and most of the protein in this region has a similar longevity. Thus these proteins are certain to accumulate post-translational modifications and have therefore attracted the attention of scientists interested in the role of post-translational modification in the aging of proteins.

Counteracting the apparent weaknesses of the lens is a formidable array of defense mechanisms from reducing compounds (anti-oxidants) to molecular chaperones. Part of the story of aging in the lens concerns the undermining of these defenses.

Because lens cells divide and are never shed the lens continues to grow throughout life (Figure 4), and lens weight has been used as a means of determining the age of wildlife species. Lens thickness also increases with age as does its yellow color and fluorescence in man [5]. Light scattering increases with age.

Accommodation

Accommodation is the ability to adjust the physiological lens to provide near vision. It is achieved by ciliary muscle contraction, relaxing the tension in the zonular fibres and allowing the lens to become more globular. It decreases with age (Figure 1) due to decreased elasticity of both capsule and the lens substance, as well as changes in

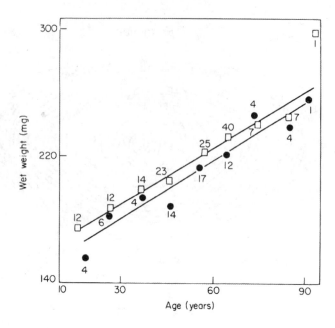

Figure 4. *The increase of human lens weight with age. Open symbols, male; closed symbols, female* [4].

the ciliary muscle and in the curvature of the lens [5, 6]. Various contributing factors may account for the relatively early onset of presbyopia, loss of accommodation, with age [7].

Physical properties

The human lens is almost colorless at birth but becomes increasingly yellow with age. Its transparency is at first surprising in a tissue with a very high protein content (34% in human lens) and with no crystal-like regularity in the arrangement of its proteins. Transparency is achieved by the short-range order of these proteins, especially alpha-crystallin [8]. Aging brings accumulation of pigment decreasing transparency. Aggregation of proteins is responsible for the concomitant increase in light scattering (Figure 5) [9]. An increasing fluorescence with increased coloration may be due to the same processes.

Metabolites

The lens has most of the metabolic pathways found in other tissues but some are restricted to the more superficial cells of the epithelium and peripheral cortex because the others lack the cell organelles necessary for the citric acid cycle, oxidative

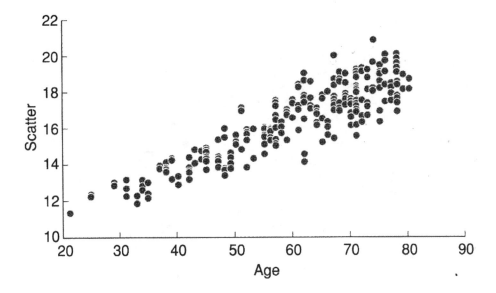

Figure 5. *The increase in central nuclear light scatter from the lens with age. Taken from ref. 9.*

phosphorylation and protein synthesis [10]. Nutrients enter the lens from the aqueous humour and waste products return there. Many metabolites and enzymes have been studied in lens and many age-related changes, usually decreases, have been reported.

Glutathione, γ-glutamylcysteinylglycine, is a key protecting molecule in the lens, which decreases steadily with age (Figure 6). Some of it becomes bound to lens proteins as a mixed disulphide. The enzymes for the synthesis of glutathione also fall. Nevertheless the concentrations in the older lenses remain much higher than those attained for other compounds in experiments designed to augment the reducing capacity of lens. Ascorbic acid also decreases with age in human lens.

The pigments of the human lens have fascinated lens biochemists since the first was identified by Ruth van Heyningen (Figure 7) [12]. The hydroxykynurenine glucoside and a more recent addition decrease with age (Figure 8) [13]. O-glucoside links are rare in mammals.

Enzymes

Without mitochondria most lens cells depend on glycolysis as the source of energy and this is the main route for sugar metabolism. The pentose phosphate pathway plays a subsidiary role and both pathways are controlled as in other tissues [10]. The specific activities of many enzymes decline significantly during adult life [14, 15]. More subtle changes like a change in heat lability have also been found. Inactive enzyme protein that cross-reacts with antibodies represents some of the lost activity.

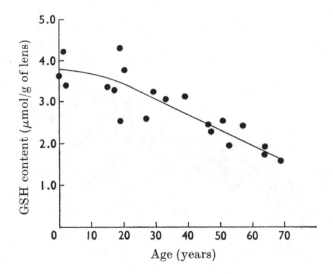

Figure 6. *The decrease of glutathione with age in human lens. Taken from ref. 11*

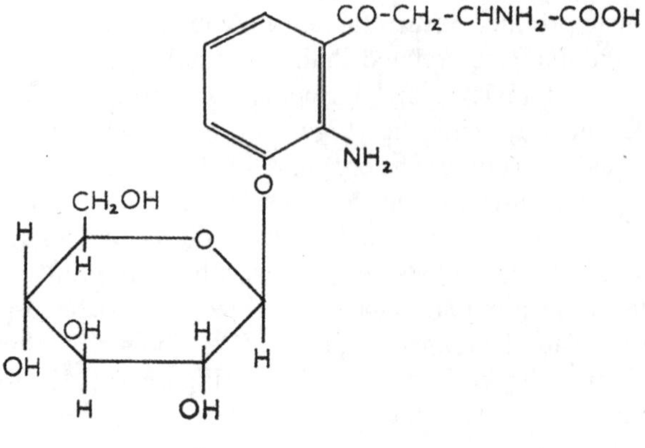

Figure 7. *The O-beta-D-glucoside of L-3-hydroxykynurenine identified by van Heyningen [12].*

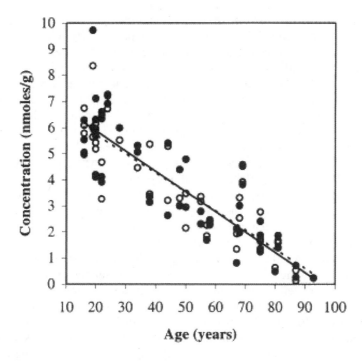

Figure 8. *Decreasing concentration of 4-(2-amino-3-hydroxyphenyl)-4-oxobutanoic acid O-diglucoside with age in human lens. Nucleus: solid circles; cortex: open circles. Taken from ref. 13.*

One can imagine that enzymes could pass through several stages as their function was compromised during ageing. The native enzyme could first become heat-labile as happened to glutathione reductase, then lose activity while retaining antigenicity as found with superoxide dismutase, before final loss of antigenicity [15]. Loss of enzyme activity does not necessarily undermine the defenses of the lens because there may be an excess of the enzyme. For example human lens has 20–30 times the specific activity of glutathione reductase that is found in most laboratory and domestic animals. More attention must be given to identification of the weak links in the armour protecting the lens.

Crystallins

The lens has a very high protein content and more than 90% of this protein can be separated into three groups of proteins, the α-, β- and γ-crystallins, that provide the high refractive index of the lens (Figure 9). These were long regarded as structural proteins but alpha-crystallin is in addition a molecular chaperone protecting the other proteins of the lens against conformational change and its consequences. It appears that the crystallins have been selected as particularly tough proteins both physically and in terms of resistance to chemical attack, nevertheless many changes to crystallins occur with aging including increased fuzziness and anodic shifts on electrophoresis. When the lens is homogenized an increasing proportion of the protein appears in soluble (Figure 9) and insoluble aggregated fractions. To an extent this may be an artefact of preparation but it serves as an indicator that the protein conformations are altered with age. The aggregated protein from human lens consists largely of α-crystallin with lesser amounts of other proteins.

The protective chaperone function of the lens decreases with age as a combination of two effects. Firstly more of the α-crystallin is present in an aggregated form which has a severely diminished chaperone function, and secondly even the α-crystallin of normal size isolated from the lens nucleus has decreased chaperone function [16]. Therefore whatever damaging effects impinge on the proteins the ability to defend against them decreases steadily during adult life.

There are many possibilities for damaging attacks on the lens proteins left vulnerable by the absence of both turnover and blood supply, exposed to whatever compounds enter the lens from without or are produced within. Oxidation is often discussed as a major stress but there is little evidence of oxidative damage during aging perhaps due to a combination of excellent defenses and a very low oxygen tension. Very little protein-protein disulphide has formed and markers of more drastic oxidative damage are not detected although a limited unfolding of protein

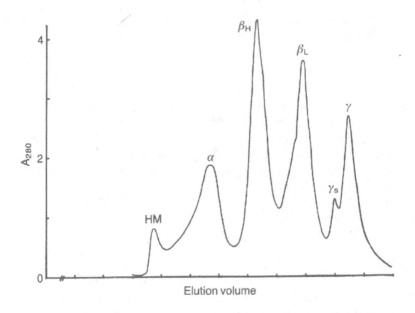

Figure 9. *Lens proteins separated by size-exclusion chromatography. Aggregated protein is followed by α-, β_H-, β_L-, γ_S- and γ-crystallins (Beswick and Harding, unpublished).*

has occurred [15]. Other chemical aging changes to proteins have been detected including truncation of βB$_1$-crystallin [17], glycation, deamidation and racemization of aspartyl residues [15]. The truncated version of βB$_1$-crystallin showed anomalous association behavior to the extent that it formed good crystals [17]. Use of mass spectra has confirmed post-translational modification of many different crystallins, but βB$_2$-crystallin appeared to be relatively resistant to such change [18].

Membranes/other changes with age

There is a loss of membrane potential and increases in calcium and sodium with age [15]. The major membrane protein (MIP), like some crystallins, undergoes truncation with age, but deamidation was found to be complete before the age of seven years [19].

There is a moderate shortening of telomeres in lens epithelium from older rats, which may or may not be enough to limit cell division [20].

Cataract

Cataract is opacification of the lens sufficient to impair vision, and is a distinctly age-related disease. Patients presenting before the age of 50 are few in number and are usually found to have some powerfully predisposing factor such as diabetes. The prevalence of cataract with age in several countries is shown in Figure 10. In all countries there is a strong age-dependence but it can be said that the onset is about 20

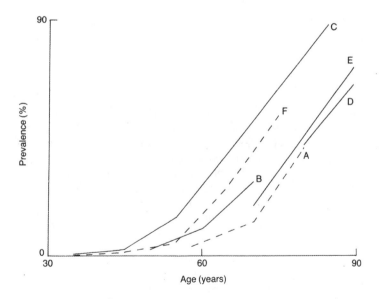

Figure 10. Prevalence of cataract in different countries: A and B, USA; C, India; D, England; E, New Zealand; F, Tibet. Taken from ref. 15.

years earlier in India and Tibet than in Western countries. The association between cataract and mortality continues to intrigue [21] and indicates that cataract, which can be easily observed *in vivo*, can be a surrogate indicator for the deterioration of a variety of organs with age.

The age-related changes discussed above are generally much exaggerated in cataract. The cataract lenses are usually more yellow than age-matched counterparts, and more fluorescent, and they certainly scatter more light. The glutathione levels have fallen further, as mixed disulphides increased. Ascorbate and inositol also fall as sodium, calcium and other components increase (Table 1). Enzyme activities fall to a greater extent in the cataract lenses (Table 2).

Table 1. Changes in levels of some small molecules in human cataract lenses. Adapted from ref. 15

Increases	Decreases
Sodium	Glutathione
Calcium	Inositol
Lead	Ascorbate
Cadmium	Tetrahydropterin
Hydrogen peroxide	
"Malondialdehyde"	
Fluorophors	

Table 2. Enzyme activities lowered in some human cataracts. Adapted from ref 15

Glucose 6-phosphate dehydrogenase
6-Phosphogluconate dehydrogenase
Lactate dehydrogenase
Glyceraldehyde 3-phosphate dehydrogenase
Polyol dehydrogenase
Na,K-ATPase
Glutathione reductase
Glutathione S-transferase
Glutathione peroxidase
Catalase
Superoxide dismutase
Quinonoid dihydropterin reductase

Even changes to the proteins are seen in a more extreme form in cataract lenses. The chemical post-translational modifications are more pronounced and probably lead to the changes of conformation characteristic of cataract.

Prevention of cataract might be expected to provide clues to the mechanism of the age-related changes. A variety of agents have been studied for the ability to prevent or delay cataractogenesis [22]. Aspirin-like drugs have been tested in the greatest range of studies and remain promising. Their action may well be to prevent post-translational modification of proteins. Other agents, such as aminoguanidine, were introduced to prevent specific modifications like the late glycation changes. Vitamins, including the antioxidant vitamins, were tested extensively with equivocal results leading to the conclusion that they will not be successful anti-cataract agents. One could go further to suggest that their lack of efficacy indicates that oxidation is not very important in the etiology of cataract.

Aging of the retina

The retina is the light-responsive part of the eye and lines the globe behind the lens (Figure 2). It is essentially part of the central nervous system which is accessible and visible from without, serving as a window into the brain [23]. It has six neural cell types organized in ten layers (Figure 11). Unlike the lens it has a good blood supply.

Retinal pigment epithelium and photoreceptors

The cells most frequently discussed in relation to aging of the retina are the photoreceptor cells (rods and cones) and the underlying retinal pigment epithelium (RPE) (Figures 11 and 12). The human eye has about 100 million rods and 6 million cones. Photoreceptor density decreases with increasing age, in a similar way there is loss of ganglion cells and RPE [25]. The visual pigment opsin is a transmembrane protein found in the many discs stacked in the outer segment of photoreceptors. There can be as many as 1700 discs in an outer segment each with more than a million molecules of rhodopsin. New discs are added continuously to the basal end and at the other end the tips of the outer segments are phagocytozed by RPE cells. The RPE cells recognize the tips, ingest them and then digest away all the membrane within a few hours. Each RPE cell may digest thousands or tens of thousands of discs a day to be efficiently degraded within hours by a variety of enzymes [26].

The RPE consists of a single layer of epithelial cells between the neural retina and choroid, and is partly responsible for the red color of the fundus [27]. The cells do not divide during adult life. In addition to its role in phagocytosis it provides the blood-retinal barrier controlling the passage of metabolites into and out of the retina; it absorbs light that has passed through the photoreceptors; and probably protects against the effects of high oxygen levels and light. RPE also serves to control and preserve retinol, an essential component of rhodopsin, within the retina. It accepts it from the bloodstream, esterifies it for storage, and re-esterifies that which passes back from photoreceptors after bleaching [28].

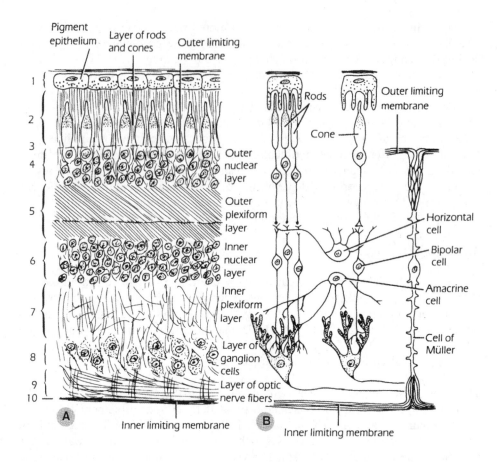

Figure 11. (A) The ten layers of the retina. (B) The arrangement of the nerve cells. The cell processes of the supporting cells of Mueller form the outer and inner limiting membranes. Taken from ref. 24 with permission.

The retina may be at particular risk from oxidation because it has a good oxygen supply, a high level of polyunsaturated fatty acids in the photoreceptor membranes, abundant mitochondria and is exposed to light [27]. In compensation it is well endowed with antioxidant defenses, with RPE in particular having high levels of vitamin E, superoxide dismutase, catalase, glutathione peroxidase, and melanin. However the light-absorbing pigment, melanin is depleted with age, while the age-related pigment lipofuscin accumulates. A second component of melanin with a different fluorescence emission spectrum accumulates with age.

Lipofuscin was thought of as a lipid aggregate but in fact is mostly protein in nature [28]. It forms fluorescent granules in the RPE that represent incompletely digested photoreceptor outer segments, possibly crosslinked by malondialdehyde and other peroxidation products (Figure 13), although attempts to generate lipofuscin

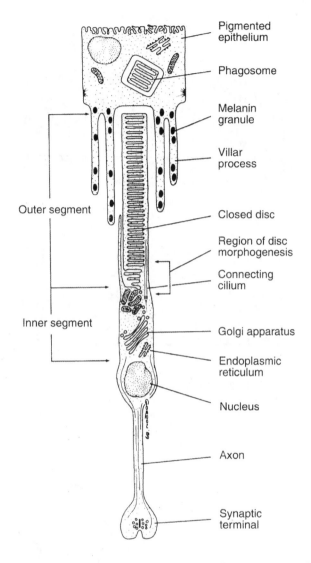

Pigmented
epithelium

Phagosome

Melanin
granule

Villar
process

Outer segment

Closed disc

Region of disc
morphogenesis

Connecting
cilium

Inner segment

Golgi apparatus

Endoplasmic
reticulum

Nucleus

Axon

Synaptic
terminal

Figure 12. *A retinal pigment epithelial cell at the top and a rod photoreceptor cell. Taken from ref.
23.*

from malondialdehyde, lipid and protein *in vitro* have failed [29]. This lipofuscin
differs from that in other tissues. The aggregates accumulate although some
lysosomal enzymes increase in activity with age [27]. RPE lipofuscin contains many
fluorophores, with a retinal derivative being a major contributor [29]. The only
lipofuscin fluorophores to be structurally characterized are A2E and *iso*-A2E, each
formed from two retinal molecules (Figure 14) [30]. They are pyridinium bisretinoids
that accumulate in RPE with age.

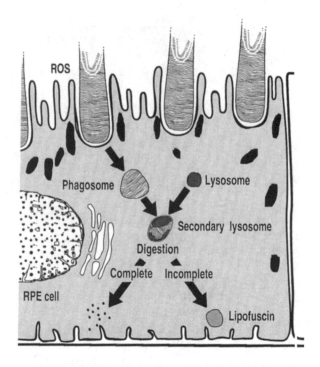

Figure 13. *RPE cell digesting rod outer segments (ROS) and producing lipofuscin. Initially the RPE cell recognises and binds ROS, whose tip is then engulfed to form a phagosome, which fuses with a lysosome. Digestion takes place in the secondary lysosome producing products small enough to be released into the cytoplasm. Incomplete digestion produces lipofuscin. Taken from ref. 29 with permission.*

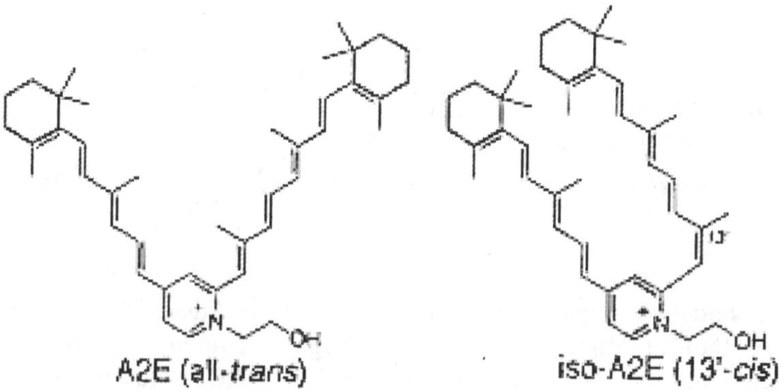

A2E (all-*trans*)   iso-A2E (13'-*cis*)

Figure 14. *Structures of A2E and iso-A2E [30].*

When exposed to blue light lipofuscin generates reactive oxygen species causing lipid peroxidation, protein oxidation, lysosomal damage, loss of glutathione, inactivation of enzymes, membrane blebbing, and ultimately cell death at least in cell culture [28, 31]. Various factors influence lipofuscin formation (Table 3). The role of light is supported by finding that whites, whose fundi reflect 5% of incident light, accumulate more lipofuscin that blacks who reflect only 1% 29. The role of the visual pigments in lipofuscin formation is emphasized in a recent model (Figure 15) [32]. Interference with the visual cycle affects lipofuscin formation. It is of interest that the formation of Schiff bases from the reaction of aldehydes with proteins and the subsequence formation of fluorophores are common features of both lipofuscin formation in RPE and of protein aggregation in the centre of the lens, as well as in tissues affected by diabetic complications.

Advanced glycation endproducts also accumulate with age in RPE cells and are associated with lipofuscin [28]. These reactive entities could possibly inactivate lysosomal enzymes and alter growth factor production.

Table 3. Factors influencing the formation of lipofuscin. Modified from ref. 29

Promoting formation	Retarding formation
Age	Melanin
Protease inhibitors	Vitamin C
Oxygen	Vitamin E
Light	Superoxide dismutase
Pro-oxidants (e.g., Fe)	Glutathione peroxidase
Vitamin A	Catalase
Low food intake	

Bruch's membrane

Bruch's membrane serves as a filter between the choriocapillaris, the primary capillary bed of the choroid, and photoreceptors allowing nutrients to pass into the photoreceptors and for degradation products to pass out to the blood. Bruch's membrane has five layers: two basement membranes, two collagenous and one elastic layer. Each exhibits its own age-related changes which include a proliferation of fibres in both collagen and elastin layers. The collagenous layers contain collagen of types I, III, and V, while the basement membranes contain type IV [35]. The collagen of Bruch's membrane, like most other collagens, becomes more crosslinked with age. Some crosslinking may be by glycation products. A long-space collagen also accumulates.

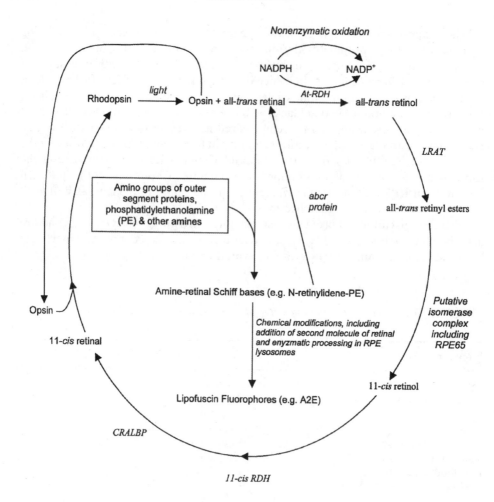

Figure 15. *Model for RPE lipofuscin formation. Light bleaches the visual pigments in the photoreceptor cells to release all-trans- retinal. Most of the retinal is reduced to retinol and then recycled to regenerate the visual pigments. However a fraction of the retinal reacts with amines to produce RPE lipofuscin chromophores. Taken from ref. 32 with permission.*

Beyond middle age characteristic lipid-rich accumulations appear on Bruch's membrane restricting the flow and are recognized during ocular examination as "drusen" increasing with age but especially associated with age-related maculopathy and macular degeneration [27, 33]. Careful observation shows that drusen can disappear. The lipids of drusen include glycolipids and cholesterol, but other glycoconjugates and proteins are present [34]. Among the proteins identified are ubiquitin, integrins, a metalloproteinase inhibitor, advanced glycation endproducts, amyloid components, apolipoprotein E, factor X, matrix metalloproteinases, vitronectin, Ig lambda, and complement components. It has been pointed out that many of these proteins are associated with immune and/or other immune related processes [34]. Drusen may originate by deposition of material from RPE or photoreceptors; and/or by transformation of RPE cells; and/or by immune-related processes following damage to RPE.

Other deposits appear in the collagenous layers. The deposits may represent RPE activity but appear not to be simply remnants of outer segments because they do not react with rhodopsin antibodies [33].

Other retinal cells

Although RPE and photoreceptors have been emphasized in studies of aging there are changes to other cells. The number of astrocytes declines as they accumulate glial fibrillary acidic protein, GFAP, and lipofuscin [35].

Choriocapillaris

The area and density of choriocapillaris decrease with age [33]. It is not possible to say whether decreased removal of debris leads to the deposits on Bruch's membrane or whether the deposits on Bruch's membrane lead to the changes in choriocapillaris.

Macular pigment

The central area of the retina, called the macula, contains two hydroxycarotenoid pigments, lutein and zeaxanthin, which screen damaging light and serve as anti-oxidants. Their concentration peaks at the centre of the macula but declines with age and further decline is associated with macular degeneration [36]. Their total concentration was not closely related to dietary intake of the two pigments.

Age-related maculopathy and age-related macular degeneration

Age-related maculopathy (ARM) is a degenerative disorder of the macula often associated with visual impairment which is more common after 65 years of age [37]. It is characterized by soft drusen, areas of hyperpigmentation and areas of hypopigmentation. Drusen, like cataracts, tend to be similar in appearance in both eyes. Late stages of ARM are called age-related macular degeneration (AMD) which is characterized by areas of depigmentation with absence of RPE, detachment of RPE, subretinal or sub-RPE neovascular membranes, sub-retinal hemorrhages, and hard exudates (lipids) in the macular area. Growth of new vessels through Bruch's membrane towards the retina, detachment of the RPE, and geographic atrophy of outer retina and choriocapillaris cause the loss of central vision [33]. The basement membrane of the retinal capillaries thickens [35].

AMD is invariably associated with loss of vision but early ARM may not be. Both increase markedly with age [38]; the increases in both major features of AMD are shown in Figure 16.

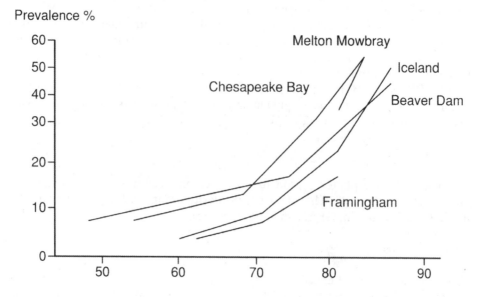

Figure 16. Age-related prevalence of late stage geographic atrophy and neovascular AMD. Taken from ref. 38 with permission.

AMD is the major cause of non-preventable blindness in Western countries, but the major molecular pathways leading to AMD have yet to be elucidated. Pigment epithelium detachments may be caused by fluid from RPE itself being unable to penetrate Bruch's membrane due to the accumulation of hydrophobic material [33]. This is consistent with the observation that the more hydrophobic drusen, as assessed by fluorescein, are more likely to lead to detachment of the pigment epithelium. The same barrier would limit metabolic exchange between choroid and RPE compromising photoreceptor function. The neovascularization may be precipitated by changed levels of growth factors. However the role of oxidation in AMD has been widely investigated. The availability of oxygen, polyunsaturated fatty acids, and light make widespread oxidation almost inevitable but the evidence provided by prevention studies has been mixed [39].

Lutein and zeaxanthin the only carotenoids to be found in retinal pigment appeared promising but Beatty et al. concluded that firm evidence of a causal link between oxidation and AMD was still lacking [39]. Publication of results from a clinical trial of the anti-oxidant vitamins C and E plus beta-carotene and zinc may encourage the view that oxidation is important [40]. A combination of these compounds decreased the risk of patients with earlier retinal changes progressing to advanced AMD. However the dominance of an ageing effect is supported by recognition that many risk factors for cataract [15] are also risk factors for AMD [41]. These include smoking and lack of education, which may be a reflection of relative poverty. The appearance of lens opacities as a risk factor for AMD may simply indicate that in part the two diseases have a common etiology.

Aging of the cornea and sclera

The cornea has a layered structure with the stroma, which is largely acellular, making up almost 90% of the thickness. Bowman's layer and the epithelial basement membrane lie anterior to the stroma with Descemet's membrane behind. The outermost layers of the cornea are the cellular layers with epithelial layers in front and a single layer of endothelium behind facing the aqueous humour. The stroma has many layers containing parallel fine collagen fibres; with their direction sharply altered from layer to layer. The sclera has a less regular arrangement but still has few cells in the main collagenous layer. The sclera becomes thinner and yellower with age, and less elastic as lipids and calcium accumulate. Cornea and sclera form the outer coat of the eye and must be strong enough to resist the intraocular pressure, and collagen provides this strength. The cornea must be transparent to allow light to pass through to the retina, but sclera, notwithstanding its similar composition, must be opaque. The greatest refraction of light by the eye occurs at the air/cornea interface. Light scattering by the cornea increases with age after 70 years and light transmittance decreases as fluorescence increases [2].

Collagen fibres are separated by proteoglycans and in sclera the larger proteoglycans, aggrecan, accumulate with age [42]. The cell density in the corneal endothelium decreases with age possibly because glycation of Descemet's membrane decreases the adhesion and spreading of endothelial cells [2, 43].

The front of the cornea is protected by the tears flowing over it, providing nutrition, and lubrication, but even here aging takes its toll. The nature of the tear producing cells alters and total protein output from the lachrymal glands falls [44]. The concentrations of lipocalin and lysozyme decrease after the age of 60 years, as the time before break-up of the tear film diminishes [45]. In many older people this trend continues to the stage of "dry eye" and the need to administer artificial tears.

Aging of the vitreous body

The vitreous body is a thin gel with more than 98% water occupying the posterior part of the globe (Figure 2). It is virtually acellular with a network of thin collagen fibrils kept apart by the glycosaminoglycan, hyaluronan [46]. The function of the vitreous is hard to pin down especially as surgical removal has few consequences beyond accelerating the development of cataract. The intraocular pressure acts though it to maintain the shape of the eye, and growth factors pass through it from retina to lens. The layman's awareness of vitreous occurs when "floaters" are noted moving in the field of vision when looking at the ceiling or a clear sky. Floaters increase with age but are of no concern unless they suddenly increase. Collagen is responsible for the gel structure [47]. In man the gel gradually liquefies with age, and the collagen fibres concentrate in the remaining gel. The predominant collagen if the vitreous body is type II collagen, unlike the type I collagen that predominates in cornea and sclera, but other types are present and form mixed fibres of type V/XI [48]. It is the latter that aggregate during liquefaction [46].

Vision

With all the age-related changes in tissues throughout the eye it is not surprising that vision itself declines with age after the age of 50 years. This slow loss is usually attributed to the changes in lens and retina.

The future

The purpose in studying aging of the eye and age-related eye has usually been seen in terms of retarding aging and especially disease. Few have expected to reverse either, but recently a more optimistic view of aging was proposed with a list of potentially reversible steps [49] which we might consider in relation to the eye. Some putative interventions, such as dispersion of intracellular and extracellular aggregates have a particular resonance with changes we have seen in lens and retina, in cataract and AMD respectively. Splitting covalent crosslinks from late glycation products could also be important. But if post-translational modifications that had caused conformational change were reversed would the native conformation re-emerge? This is possible especially with active molecular chaperone available to assist the process.

Acknowledgments

I am grateful to Professor Neville Osborne for useful comments on an earlier version of this review, and to Dr Diyaa Rachdan for help with the figures.

References

1. Koretz JF, Kaufman PL, Neider MW, Goeckner PA (1989). Accommodation and presbyopia in the human eye – aging of the anterior segment. *Vision Res.* 29: 1685–92.
2. Weale RA (1982). *A Biography of the Eye.* London: HK Lewis and Co.
3. Weale RA (1995). Why does the human visual system age in the way it does? *Exp Eye Res.* 60: 49–55.
4. Harding JJ, Rixon KC, Marriott FHC (1977). Men have heavier lenses than women of the same age. *Exp Eye Res.* 25: 651.
5. Bron AJ, Vrensen GFJM, Koretz J, Maraini G, Harding JJ (2000). The ageing lens. *Ophthalmologica* 214: 86–104.
6. Davson H (1990). *Physiology of the Eye.* London: Macmillan.
7. Weale RA (2000). Why we need reading glasses before a zimmer frame. *Vision Res.* 40: 2233–40.
8. Delaye M, Tardieu A (1983). Short-range order of crystallin proteins accounts for eye lens transparency. *Nature* 302: 415–17.
9. Sparrow JM, Frost NA (1999). Physical aspects of lens clarity and image degradation. In: Easty D, Sparrow JM, eds. *Oxford Textbook of Ophthalmology*, Vol 1, Oxford: Oxford University Press, pp. 463–5.

10. Harding JJ (1997). Lens. In: Harding JJ, ed. *Biochemistry of the Eye*. London: Chapman and Hall, pp. 94–134.
11. Harding JJ (1970). Free and protein-bound glutathione in normal and cataractous human lenses. *Biochem J.* 117: 957–60.
12. van Heyningen R (1971). Fluorescent glucoside in the human lens. *Nature* 230: 393–4.
13. Taylor LM, Aquilina JA, Willis RH, Jamie JF, Truscott RJW (2001). Identification of a new human lens UV filter compound. *FEBS Lett.* 509: 6–10.
14. Hockwin O, Ohrloff C (1981). Enzymes in normal, ageing and cataractous lenses. In: Bloemendal H, ed. *Molecular and Cell Biology of the Eye Lens*. New York: Wiley, pp. 367–413.
15. Harding JJ (1991). *Cataract: Biochemistry, Epidemiology and Pharmacology*. London: Chapman and Hall.
16. Derham BK, Harding JJ (1997). Effect of aging on the chaperone-like function of human α-crystallin assessed by three methods. *Biochem J.* 328: 763–8.
17. Bateman O, Lubsen NH, Slingsby C (2001). Association behaviour of human βB_1-crystallin. *Exp Eye Res.* 73: 321–31.
18. Zhang Z, David LL, Smith DL, Smith JB (2001). Resistance of human βB_2-crystallin to *in vivo* modification. *Exp Eye Res.* 73: 203–11.
19. Schey KL, Little M, Fowler JG, Crouch RK (2000). Characterization of human lens major intrinsic protein structure. *Invest Ophthalmol Vis Sci.* 41: 175–82.
20. Pendergrass WR, Penn PE, Li J, Wolf NS (2001). Agerelated telomere shortening accurs in lens epithelium from old rats and is slowed by caloric restriction. *Exp Eye Res.* 73: 221–8.
21. Wang JJ, Mitchell P, Simpson JM, Cumming RG, Smith W (2001). Visual impairment, age-related cataract and mortality. *Arch Ophthalmol.* 119: 1186–90.
22. Harding JJ (2001). Can drugs or micronutrients prevent cataract. *Drugs Aging* 18: 473–86.
23. Gordon WG, Bazan NG (1997). Retina. In: Harding JJ, ed. *Biochemistry of the Eye*. London: Chapman and Hall, pp. 144–275.
24. Snell RS, Lemp MA (1998). *Clinical Anatomy of the Eye*. Oxford: Blackwell Science.
25. Panda-Jonas S, Jonas JB, Jakobczyk-Zmija M (1995). Retinal photoreceptor density decreases with age. *Ophthalmology* 102: 1853–9.
26. Berman ER (1991). *Biochemistry of the Eye*. New York: Plenum Press.
27. Boulton M (1991). Ageing of the retinal pigment epithelium. *Prog Ret Res.* 11: 125–51.
28. Davies S, Elliott MH, Floor E, *et al.* (2001) Photocytotoxicity of lipofuscin in retinal pigment epithelial cells. *Free Rad Biol Med.* 31: 256–65.
29. Kennedy CJ, Rakoczy PE, Constable IJ (1995). Lipofuscin of the retinal pigment epithelium: a review. *Eye* 9: 763–71.
30. Ben-Shabat S, Parish CA, Vollmer HR, Itagaki Y, Fishkin N, Sparrow JR (2002). Biosynthetic studies of A2E, a major fluorophore of retinal pigment epithelial lipofuscin. *J Biol Chem.* 277: 7183–90.
31. Shamsi FA, Boulton M (2001). Inhibition of RPE lysosomal and antioxidant activity by the age pigment lipofuscin. *Invest Ophthalmol Vis Sci.* 42: 3041–6.
32. Katz ML (2002). Bernard Strehler – inspiration for basic research into the mechanisms of aging. *Mech Ageing Dev.* 123: 831–40.
33. Guymer R, Luthert P, Bird A (1998). Changes in Bruch's membrane and related structures with age. *Prog Ret Eye Res.* 18: 59–90.
34. Hagemann GS, Luthert PJ, Chong V, Johnson LV, Anderson DH, Mullins RF (2001) An integrated hypothesis that considers drusen as biomarkers of immune-mediated processes at the RPE-Bruch's membrane interface in aging and age-related macular degeneration. *Prog Ret Eye Res.* 20: 705–32.

35. Ramirez JM, Ramirez AI, Salazar JJ, de Hoz R, Trivino A (2001). Changes of astrocytes in retinal ageing and age-related macular degeneration. *Exp Eye Res.* 73: 601–15.
36. Beatty S, Murray IJ, Henson DB, Carden D, Koh H-H, Boulton ME (2001). Macular pigment and risk for age-related macular degeneration in subjects from a northern European population. *Invest Ophthalmol Vis Sci.* 42: 439–46.
37. Bird AC, Bressler NM, Bressler SB, *et al.* (1995) An international classification and grading system for age-related maculopathy and age-related macular degeneration. *Survey Ophthalmol.* 39: 367–74.
38. Fletcher A (2001) Epidemiological considerations: how common is AMD and what causes it? In: Pratt S, ed. *Age-Related Macular Degeneration (AMD) and Lutein: Assessing the Evidence.* London: Royal Society of Medicine Press, pp. 7–17.
39. Beatty S, Koh H-H, Henson D, Boulton M (2000). The role of oxidative stress in the pathogenesis of age-related macular degeneration. *Survey Ophthalmol.* 45: 115–34.
40. Age-related Eye Disease Study Group (2001). A randomized, placebo-controlled, clinical trial of high-dose supplementation with Vitamins C and E, beta-carotene, and zinc for age-related macular degeneration and vision loss. *Arch Ophthalmol.* 119: 1417–36.
41. Age-Related Eye Disease Study Research Group. (2000). Risk factors associated with age-related macular degeneration. A case-control study in the age-related eye disease study. *Ophthalmology* 107: 2224–32.
42. Rada JA, Achen VR, Penugonda S, Schmidt RW, Mount BA (2000). Proteoglycan composition in the human sclera during growth and aging. *Invest Ophthalmol Vis Sci.* 41: 1639–48.
43. Kaji Y, Amano S, Usui T, *et al.* (2001) Advanced glycation end-products in Descemet's membrane and their effect on corneal endothelial cell. *Curr Eye Res.* 23: 469–77.
44. Draper CE, Adeghate EA, Singh J, Pallot DJ (1999). Evidence to suggest morphological and physiological alterations of lacrimal gland acini with ageing. *Exp Eye Res.* 68: 265–76.
45. Tiffany JM, Gouveia SM (2002). Age-related changes in human tear composition and stability. *Adv Exp Med Biol.* 506: 587–91.
46. Bishop PN (2000). Structural molecules and supramolecular organisation of the vitreous gel. *Prog Ret Eye Res.* 19: 323–44.
47. Pirie A, Schmidt G, Waters JW (1948). Ox vitreous humour. I – The residual protein. *Br J Ophthalmol.* 32: 321–39.
48. Mayne R, Brewton RG, Ren Z-X (1997). Vitreous body and zonular apparatus. In: Harding JJ, ed. *Biochemistry of the Eye.* London: Chapman and Hall, pp. 135–43.
49. de Grey ADNJ, Baynes JW, Berd D, Heward CB, Pawelec G, Stock G (2002). Is human aging still mysterious to be left only to scientists? *BioEssays* 24: 667–76.

Aging and the Oral Cavity

Dirk Bister

Department of Orthodontics, GKT Dental Institute, London

Introduction

The oral cavity is not an organ system but because of the anatomical unity of its various functions it is sometimes viewed as one. It is the beginning of the alimentary system and the function of the oral cavity for food digestion is paramount: taste (and inseparably connected to it smell), mastication and lubrication take place here, the sterilization of the food bolus is initialized here. The oral cavity also enables us to articulate and it helps with the humidification of air for ventilation.

Aging has been described as the process of growing old or to mature. It is therefore logical to look for changes related to the aging process in the elderly. However as so often in nature variation is the norm rather than an "average" and therefore on closer inspection findings are not always conclusive. Differentiation between physiological changes due to aging and disease is often difficult. Some changes are virtually only found in the elderly but not exclusively. The involutionary characteristics of some aging processes led previous investigators to describe these changes (now associated with aging) as disease particularly when associated with loss of structure and/or function. Loss of periodontal attachment, for example, is common in the elderly but it is not a "disease" of the elderly *per se*. Loss of periodontal attachment is, on average, more likely to be due to prolonged exposure to environmental factors and this can be interpreted as an expression of aging rather than disease.

However age changes of the oral cavity are varied and some of these changes, previously thought to be related to aging, are now thought to be secondary to disease or medication for instance.

Research has long sought to differentiate between changes mainly due to environmental influences and pure genotypic aging by controling external influences in animal experimentation. However these attempts are often frustrated by the fact that the aging process expresses great variation between individuals. The following chapter gives a short summary of the anatomy and physiology of the relevant

R. Aspinall (ed.), Aging of Organs and Systems, 201–223.
© 2003 *Kluwer Academic Publishers. Printed in Great Britain.*

structures and tries to summarize the changes of the oral cavity with aging. Epidemiological studies are mentioned where relevant.

The teeth

General anatomy and physiology (Figure 1; refs 1, 2)
A tooth consists essentially of two parts (crown and the root) and three layers. The innermost part, the cavitas dentis is occupied by the pulp, which in turn consists of vessels, nerves, fibers and fibrous tissue. The pulp connects to the outside via the foramen apicis dentis. The pulp cavity is surrounded by dentine, which forms the bulk of the tooth. It consists of organic and inorganic material. However it also comprises vital tissue: its immermost layer (the most peripheral layer of the pulp) is formed by odontoblasts, and their processes lie in the tubules, which radiate towards its outermost layer: the mantle dentin. The odontoblastic processes are thought to be responsible for the sensitivity of teeth although the extent of the neural innervation has always been controversial. The coronal part of the tooth, which is visible intraorally, is covered by enamel. This is the most densely mineralized and hardest tissue of the human body, consisting of an organic and inorganic matrix. The inorganic matrix consists of soluble and insoluble components and its largest part is on the inner third of the enamel mantle. The crystalline part of the enamel consists of hydroxylapatite (calcium and phosphorus: $Ca_{10}[PO^4]_6[OH]_2$). The enamel is developmentally of epithelial origin, whilst the dentine and pulp are of mesenchymal origin. There is mutual dependency during development.

Age changes of the pulp
Age related changes of the pulpal tissues which diminish its capacity to react appropriately to stimuli include: reduction of pulpal space (due to secondary dentine deposits), decreasing numbers of fibroblasts and blood vessels in the pulpal tissue, an increase of collagen fibers in the root canal, formation of denticles (s.b.) [3]. Collagen synthesis appears to remain constant throughout life [4]. Most studies investigated differences in morphology on a cross sectional basis. Atrophy previously thought to be due to aging are now thought to be due to autolytic artefacts during tissue preparation [5]. Whether pulp-stones are a pysiological characteristic of aging remains controversial. Although most pulpal calcifications are a reaction of the tissue due to pathological processes (up to 75%), there is a 1:10 chance of finding these in older patients when compared to younger ones. However diffuse mineralization has been confirmed to increase with age.

Animal experiments show age related changes of neural innervation of the pulp: the number of myelinated and unmyelinated fibers decrease as degeneration of the pulpal tissue increased.

Markers of immune reactivity such as calcitonin-gene related peptide and substance P-like reactivity decrease with age and thus reflect the diminished ability of the aged pulp to react to pathological stimuli in the same way a young pulp does.

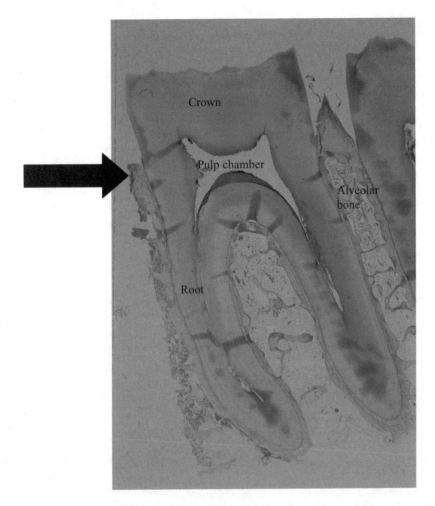

Figure 1. *Decalcified saggital section of human molars showing crown, root and periodontal attachment (arrow), (by kind permission of Stalin P.A. Kariyawasam, Craniofacial Development, GKT Dental Institute, London).*

Age changes of the dentine and root

Primary dentine is produced until the root formation is complete. Further physiological dentine additions are called secondary dentine; therefore apposition of secondary dentine can be interpreted as aging process [6]. Subsequently the space available to the pulpal tissue will diminish. Once tooth formation is complete morphological changes of the dentine are most often due to external stimuli such as demineralization (carious challenges) or sudden exposure to the environment of the oral cavity (such as trauma) [7]. These changes are more commonly found in the elderly. Often they express past readiness of the dental tissues to respond to stimuli rather than age related changes, but a variety of the changes can occur in both aging and past physiological reaction to stimuli, however the latter are usually locally confined. Age related changes of the dentin, which usually appear generalized, are: increase of sclerotic dentine (starting apically moving towards the coronal part of the tooth later in life [8]), addition of peritubular dentin leading to obliteration of the tubules (however usually less than 50% become obliterated in the coronal part [9]). The sclerosing process has also been observed in teeth, which were impacted and thus never exposed to the to the oral cavity.

The examination of teeth of individuals can be used to determine their age. This is often used for identification of individuals in forensic medicine. The most reliable criterion is the transparency of root dentin [10]. But other methods such as identification of the thickness of cementum and secondary dentin, are also used and has been found to be reliable for victims the over 20 years of age. Racemization of aspartic acid in dentin appears to be a very reliable method of age determination for all ages [11].

Root length appears to diminish with age. Idiopathic root shortening (see Figure 2) occurs in up to 48% of the population although usually in a minor form. As it is a process that proceeds with age some authors refer to it as a physiological process. It is more severe in the deciduous dentition (even in the absence of succedaneous teeth) than in the permanent dentition. Teeth most commonly affected are the maxillary central incisors, second premolars, maxillary lateral incisors [12].

Age changes of the enamel
Changes of the enamel, once the crown is exposed in the oral cavity, can only come from direct contact to the environment (saliva or food). Enamel defects caused by abrasion (wearing away of the tooth surface such as due to tooth-brushing or pipe-smoking etc., see Figure 3) and attrition (wearing away of the tooth surface in bruxism and mastication, see Figure 4) will not remineralize sufficiently to regenerate the original shape and form of the tooth. Anatomical details such as perikymata and imbrication lines are lost and the enamel surface appears smooth and polished with age, thus changing its reflection characteristics. So-called "wear facets" appear on the occlusal surfaces and interproximal "contact points" become "contact areas." Occasionally this can lead to exposure of dentin, which normally mineralize sufficiently to protect the pulp (sclerosis). Also so-called tertiary dentin can be formed on the outermost layer of the pulp, thus increasing the thickness of the hard tissue. Depending on the dietary habit, masticatory load parafunctional habits and brushing habits of an individual the amount of attrition can give clues to their age. In modern societies however loss of tooth surface due to caries are far more common than the changes due to attrition. Demineralization and remineralization of the enamel is a physiological phenomenon, which occurs on a daily basis. Thus deposits as well the general changes of the surface characteristics (s.a.) and the changes of the dentine over time (s.a.) will change the color of teeth over time: teeth become more yellow and less translucent, duller. Acids either taken directly (fizzy drinks, juices) or produced by the intra-oral flora as a by-product of its sugar-metabolism demineralize the enamel readily. This process however changes with age and the difference in

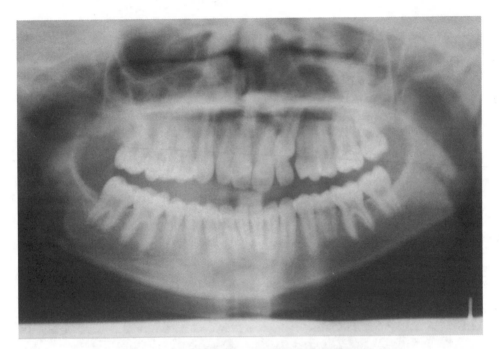

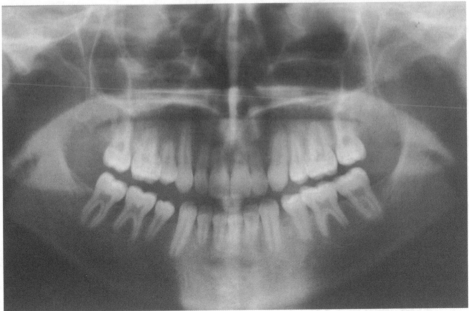

Figure 2. Shortening of roots of various teeth, in this case probably due to orthodontic treatment. Radiographs of the same patient before and after treatment.

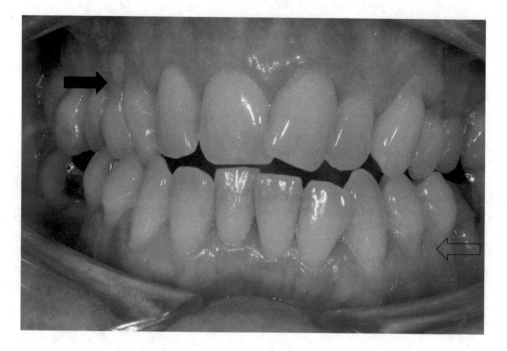

Figure 3. *Abrasion on buccal surfaces of the upper and lower teeth, some of the damage has already been repaired using composite filling material (transparent arrow).*

response can either be due to changes in the composition of the saliva or changes of the composition of the enamel. Immature enamel of a recently emerged tooth is more likely to absorb ions such as fluoride just as demineralized areas are and this is being used for therapeutic fluoride application to prevent later demineralization. Once fluoride had been incorporated it is less soluble than the OH groups it replaces and is also thought to have bacteriostatic effects.

The inorganic content of the enamel matrix increases with age [13], as water is lost. Subsequently crystalline structure of the enamel becomes denser. This process is often referred to posteruptive maturation. Its chemical composition changes because some of the original OH groups are replaced by other elements. Subsequently it becomes less soluble but also more brittle. This in turn leads to increased vulnerability to chemical challenges (it is more easily etched) and temperature changes (more cracks appearing later in life; 60% of all permanent teeth in adults are affected [14]). However density of the human enamel does not appear to change significantly with age, although fluoride content increases as does the nitrogen content, the latter only remaining stable after age 60. The surface layers of the enamel are affected more by these changes substantiating the view that these are due to ion exchange mechanisms which changes the surface composition with time. Enamel-prisms increase in size when exposed to the oral environment and can therefore be interpreted as age related changes.

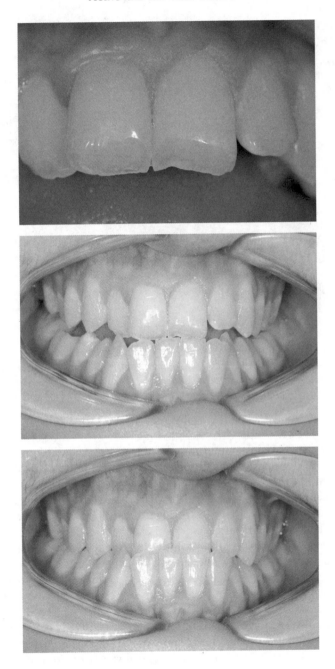

Figure 4. *Attrition of the upper incisors. This is due to displacement: the upper incisors are contacting the lower incisors "edge to edge" on closure of the mandible. The lower jaw then slides forewards to maximal intercuspation to give the patient a comfortable bite.*

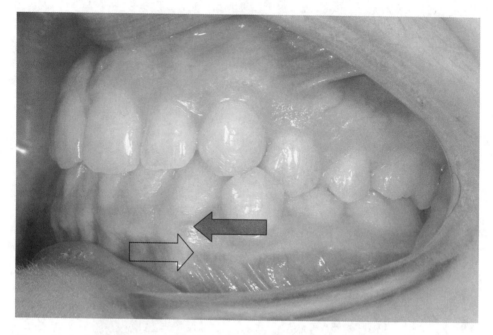

Figure 5*. Healthy anatomy showing clinical crown, free gingival groove (transparent arrow), attached gingiva, muco-gingival junction (tranparent arrow) and alveolar mucosa. The only anomalies visible are: the lower first premolar is in crossbite with the opposing arch and the second deciduous molar is still present.*

Maturation of the dentition

The first part of the visible maturation of the dental apparatus in humans is the development of the deciduous dentition. These first teeth are subsequently replaced by succedaneous teeth (incisors, canines and bicuspids). These together with the accessional teeth (first, second and third molars) constitute the permanent dentition. Once the root formation of the teeth (regardless of which dentition) is complete (approximately three years after emergence of permanent teeth) the tooth has finished growing. This is in contrast to other mammals, which are also diphyodont (no more than two sets of teeth), where teeth continue to grow (i.e., rodents).

Maturation of the dental complex is however by no means complete once the roots are fully formed. The facial skeleton continues to grow well into our 20's and there is mounting evidence that this growth continues, albeit at a slower rate, well into adulthood (see also section on bone [15]).

The position of the permanent teeth in the orofacial skeleton depends by and large on five factors: (1) Pressure of the cheeks and lips; (2) Pressure of the tongue; (3) Position of the basal bones of maxilla and mandible; (4) Pressure from the posterior dentition to the anterior dentition (the so-called "mesial migration"or occlusal drift; this also includes development of the third molars and attrition changes of the dentition); (5) Periodontal attachment of the teeth in the alveolus (see section on periodontal ligament).

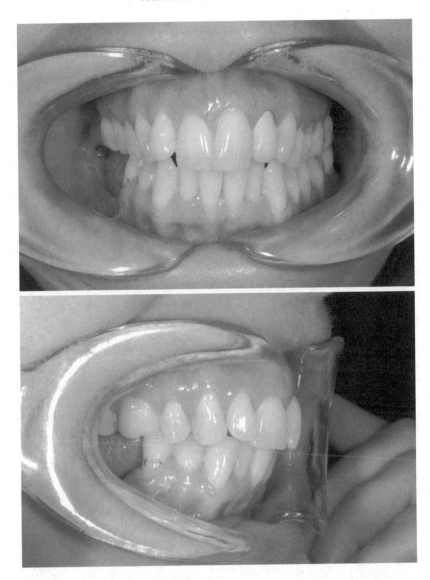

Figure 6. *Loss of teeth many years ago on the right mandible lead to loss of bone: the height of the basal bone is now visible. Interestingly the upper teeth, although no longer in contact with the lower teeth, did not over-erupt (the patient did not wear a denture).*

All the above factors may undergo changes with maturation and aging of the body. It has bee recognized for sometime that, teeth continue to change position throughout life and this can therefore be interpreted as an aging process. Particularly the position of the lower front teeth is unstable. More recent evidence suggest that late onset lower incisor crowding does not happen to all patients, but that generally dental arch width and length decreases until early adulthood and beyond and that females appeared to be more affected than males. Findings are similar for patients who underwent orthodontic therapy [16].

Disuse atrophy occurs once teeth have lost their occlusal contacts and these changes are similar to the ones seen in teeth undergoing general aging processes. The size of the pulp chamber reduces as does vascularity and innervation. Deticular canals reduce in diameter and may obliterate. Whether teeth show unlimited over-eruption once they lost their occlusal contact remains controversial (see Figure 6).

Epidemiology of the dental status of the elderly
Dental diseases are thought to be on the decrease at least in the population in North America but populations in northern Europe are thought to behave in a similar fashion. Total tooth loss was once thought to be synonymous with aging. This however is changing fast: 50% of North Americans were edentulous by 65 in 1958; this figure is now reached at 75+ years and will continue to decline [17]. It has been predicted that by 2024 in the age group between 65 and 74 only 10% of Americans will be without teeth [18]. This increase in number can most likely be attributed to better prophylaxis and earlier treatment of carious and periodontal lesions. However these data vary widely depending more on the socioeconomic data rather than age *per se*. This may reflect variations in accessibility to dental care or attitude of patients to tooth retention [19]. It has also been postulated that tooth loss early in life will predispose patients to edentulousness later in life.

One index used to measure dental decay is the DMF, which is an acronym for decayed, missing (because of caries) and filled (teeth). The DMF rates are lower in areas with higher fluoride content in the water. Although this index has been criticised for its inaccuracy it is widely used. One must assume that all teeth which carry fillings do so because of previous decay (teeth may be filled or crowned for esthetic reasons) and that missing teeth have been removed because of caries (teeth may be removed because of periodontal disease). Generally the DMF shows an interesting trend particularly for the USA; high DMF rates are more prevalent amongst the well educated and patients from households with higher incomes. However this population also has the fewest carious and unfilled teeth, thus reflecting treatment choice rather than reflecting disease prevalence as it is difficult to explain the difference by variations of dietary habits alone [20].

There are two underlying tendencies influencing the DMF index in recent years: one leading to an increase in the rate is the growing number of elderly patients who are more predisposed to develop caries, root caries in particular [21]. On the other hand the younger population currently has a lower DMF. This however leaves a longer period of time to develop decay on previously unfilled teeth. Overall we can therefore expect to see an increase of DMF with time.

Root caries is a disease exclusive to patients who have lost periodontal support thereby exposing root surfaces of teeth. Histologically root caries differs from caries of the enamel and its etiology is also different; it is often associated with a dry mouth [22]. Although root caries is not a phenomenon of recent times reliable data have only been available since the late 1980's probably because of the reduced awareness beforehand which in turn was due to the small patient numbers of a higher age group with teeth present [23]. Although nearly exclusive to the aged it is thought mostly to be of secondary nature: diminishing rate of saliva flow which in turn is no longer thought to be purely age related, but rather due to medication taken to combat other mostly age related diseases. Xerostomia (dry mouth) is thought to have its etiology rooted in multiple medications. Root caries also correlates negatively with higher socio-economic status, oral hygiene levels, preventive behavior, and use of dental services and interestingly number of remaining teeth.

The periodontal ligament

Anatomy and physiology (Figures 1 and 5)
The periodontal ligament connects the teeth with the surrounding bone and defends the attachment from bacterial and mechanical assaults.

The cementum covering the root surface constitutes the inner boundary of the periodontal ligament. Cementum is avascular, and largely mineralized. However it contains cementocytes in the outer layer. Cementum allows the Sharpey's fibers of the ligament to attach to the teeth, thereby translating the vertical (pressure) forces of occlusion into tension. Cementum itself is largely resistant to pressure-resorption. Intact cementum with vital cells is paramount to the successful transplantation and reimplantation of teeth. The thickness of the cementum layer increases with age as it is continuously deposited by the cementocytes throughput life and its thickness triples between the ages of 10 to 75 years [24]. The chemical composition of cementum changes with age and flouride and magnesium content increase.

The cemento-enamel junction is the most coronal part of the cementum and three forms of connections between the cementum and the enamel have been reported: edge to edge, cementum overlapping emanmel, and a "gap" junction where dentin is exposed.

The attached gingiva covers the alveolar processes and the cervical region of the teeth. It comprises three anatomically distinct types of epithelium: junctional, sulcular and the oral epithelium. The depth between the cemento-enamel junction and the junctional/sulcular epithelium is the so-called probing-depth and should measure 1–2 mm. The most coronal part the so-called "free margin" stretches from the tooth to the keratinized gingiva, the demarcation is the free gingival groove. The keratinized gingiva is demarcated apically by the mucogingival junction and this in turn connects gingiva to oral mucosa (alveolar mucosa). The attached gingiva has also be defined as the tissue extending from the free margin of the tooth to the muco-gigival line minus the pocket depth: in other words the part of the gingiva which has to withstand masticatory forces. It measures between 0–9 mm. The height necessary to maintain gingival health is controversial. Interestingly, the height of the attached gingiva appears to increase with age.

The structural elements of the periodontal ligament comprise cells, connective tissue components, matrix, vessel and nerves. The cellular component is dominated by firbroblasts. Osteoblasts and osteoclasts, are found in the peripheral part of the ligament; cementoblasts are only found during active apposition of cementum and line the mixed fiber layer. Osteo- and cemento-progenitor cells are also present. Epithelial cells protect the apparatus from the oral environment cervically.

The gingival cervicular fluid is an exudate from the superficial plexus of the dento-gingival vessels. Permeability of the vessels correlates well with the amount of gingival inflammation present in the sulcus. Various constituents have therefore been used as markers of periodontal breakdown.

The connective tissue fibers consist predominantly of collagen but oxytalan is also present. The spatial arrangement of the periodontal fibers is complex and varies significantly with anatomical location. Sharpey's fibers are anatomically distinct and characterize the periodontal ligament: they are anchored in the cementum on one side and in alveolar bone on the other.

The residual matrix is a viscous gel and shares the characteristics of other connective tissues. It is composed of glucosaminoglycans (mucopolysaccharides), glycoproteins and lipids.

The periodontal ligament maintains is ability to remodel with advancing age: its progenitor cells appear to be able to respond to force application via the tooth even in older patients, albeit at a slower rate. The periodontal ligament absorbs pressure from teeth, if short lived, and it thereby protects the surrounding alveolar bone. If however the pressure is sustained for longer periods of time it results in spatial re-arrangement of the ligament allowing the teeth to move. If the pressure on the tooth is high rapid and extensive remodeling will be the consequence, which can lead necrosis of the ligament and subsequent root shortening. The latter is also noted as a physiological aging process affecting about 20% of the population, in whom 70% of teeth are affected.

The outer border of the periodontal ligament is a thin, perforated, layer of bone the cribriform plate. The collagen bundles of the ligament insert here and nerve fibers and blood vessels cross through it. The alveolar bone, which supports the teeth does not only consist of what on radiographs appears as the lamina dura; it also consists of the buccal and lingual cortical shelves and the connecting cancellus bone. The thickness of the shelves varies with age and generally diminishes with time, occasionally leaving denuded areas of root surface, the so-called fenestrations or dehiscence, if the root is exposed along the cleft like defect starting at the cemento-enamel junction. Normal alveolar bone lies about 2 mm apical to the cemento-enamel junction in adolescents and adults. Once teeth are lost the alveolar bone disappears leaving only the basal bone behind (see Figure 6).

Age-related changes of the periodontal ligament
The effects of age on the periodontal ligament have been documented extensively [25, 26]. Generally the changes are parallel to the changes of other connective tissues in the human body. The gingival tissue increases in density, which is subsequent to the reduction of the number of fibroblasts. From adolescence onwards the fibroblasts in

the ligament are still able to produce collagen and proteoglycans although the turnover diminishes substantially with time. The half live of collagen doubles with age (from 2.5 to 6.5 days to 15 days) [27]; this is due to reduced solubility and more stable supramolecular structure [28]. A subsequent increase collagen content has been observed. In humans fiber content has been reported to decrease and the interstitial compartment to increase but no change of collagen fiber orientation has been observed.

Although most studies agree that the vascularity of the ligament decreases with age [29], the effect on the innervation is not clear. The effect of age on the width of the periodontal ligament itself is controversial. The width of the periodontal ligament is decreased in non-functional teeth and an increased deposition of cementum has been suggested as causative factor [30]. However an increased width of the cementum layer does not per se mean that the width of the periodontal ligament is reduced as shown in old dogs. A consistent finding is the reduced number of epithelial cells with aging for humans and dogs.

Epidemiology of the periodontal status of the elderly
Long teeth or being "long in the tooth" has long been synonymous with being of advanced age and this common colloquialism either refers to periodontal disease or overeruption of teeth due to loss of occlusal contact (see Figure 7). The term periodontal disease is non-specific and is applied to all conditions where the attachment of the tooth to the socket is reduced. The term gingivitis is applied to a reversible superficial form of the periodontal disease. Periodontitis is an irreversible form of this condition, which has lead to permanent loss of attachment. The latter always includes inflammation of the gingiva. However gingivitis may or may not proceed to periodontal disease.

Four types of periodontal disease are generally recognized: juvenile, prepubertal, rapid progressive and adult. Some bacterial species have been identified with certain forms of the disease although a clear cause and effect relationship has not been established. There seems to be individually different susceptibility to developing the disease. Although poor periodontal condition correlates positively to the amount of dental plaque present and age, significant attachment loss is not necessarily a natural consequence of aging and is not thought to be a major cause for tooth loss. The leading cause for tooth loss, at least before the age, of 35 is caries.

Prevalence of significant attachment loss (6 mm or more in at least one site) has been reported in 7 to 15% of populations [31]. For the USA this figure goes up to 34% in the population aged over 65. However no good predictive markers have yet been identified, apart from increased probing depth itself. Periodontal disease is generally quiescent and progresses slowly with time. Increased prevalence in the aged is therefore an expression of disease-accumulation and not increased susceptibility. It is thought that susceptibility varies widely between individuals. In a longitudinal study investigating progression of periodontal disease in untreated Sri Lankans three distinct patters emerged. A group of patients with severe gum disease at the start of the observation period (8%) continued to lose attachment, effectively losing all their teeth at age 45 average loss was 1 mm per year. Eighty-one percent of the population

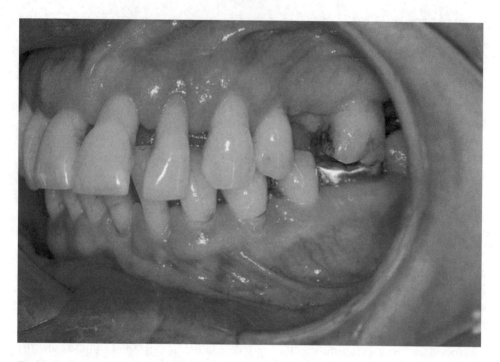

Figure 7. *Exessive lengthening of the upper front teeth. This was due to advanced periodontal disease and all upper front teeth were mobile on examination.*

with moderate progression (attachment loss of more than 2 and less than 6 mm) progressed at a slower rate; 0.3 mm when patients were in their 20's gathering momentum as patients grew older (0.5 mm per year in their 40's). The residual group with no disease at the start only progressed very slowly 0.5 mm per year not accelerating with time [32]. It is now thought that the rapid loss of attachment is due to an aggravated host-response to infection. This would explain the differences between individuals undergoing the same infection with different amount of period-ontal destruction.

Evidence therefore suggests that older patients with teeth present are less susceptible to rapid progressing periodontal disease: this would have expressed itself earlier leading to tooth loss [33]. However it is difficult to come to a decisive conclusion as treatment protocols change overtime and tooth survival might be affected.

The bones of the maxillo-facial skeleton: alveolus, maxilla and mandible

Maxilla and mandible
Aging of bone is covered elsewhere in this book, however significant differences exist between the development, anatomy and physiology of the bones of the maxilla and mandible and the residual skeleton. The skeleton of the head is of ectomesenchymal

origin, derived from neural crest cells whilst the rest of the skeleton derives directly from cells of mesenchymal origin [34]. Ossification of large parts of maxilla and mandible are via intramembranous pathways rather than endochondral ossification such as in the long bones. However both osteoblastic lineages develop via mesen-chymal cells and some markers such as alkaline phosphatase and alpha 2 collagen expression have been used to identify differentiation of cells to their eventual phenotypes [35].

Segmental development along branchial arches is another distinct feature of the development of the carnium; both maxilla and mandible are derived from the first branchial arch. Genetic control of the segmentation process is via the Homeobox gene family and the Msx gene family in particular appears to play an important role here. The name homeobox is derived from the so-called homeotic mutations first described in the fly. It is a commonly used model for the description of the patterned development of the body and was named after muscle segment homeobox (msh) gene [36]. The human equivalents are the Msx genes and at least three have been described to play a part in mammal development: Msx1, Msx2 and Msx3 [37].

Msx1 and Msx2 have been implicated in the development of eyes and teeth as well as bone development. Msx gene expression has been identified as pivotal for activity of osteocalcine gene expression in rats. However Msx2 gene expression has been reported to be present in some osteoblastic cell lines and not in others [38]. Indeed overexpression can lead to suppression of osteocalcin levels [39]. More intriguing is the finding that Msx2 is expressed in the periosteum of the mature maxilla and mandible in mice, whilst it was not expressed in the long bones [40]. Some unique regulatory mechanisms have therefore been proposed for the maxillo-facial skeleton in the aging facial skeleton [41]. This in turn might help to explain the continuous growth of the maxillo-facial skeleton well into adulthood, although this growth is only moderate. Also the regenerative capacity of the alveolar bone does not appear to be diminished with age: The healing rate of alveolar extraction sites in patients older than 40 was not different from younger patients [42]. Healing of lesions in general may be different at various ages though [43].

Apart from the binding of Msx genes to DNA these genes have also been proposed in the expression of the family of bone morphogenic proteins (BMP) genes [44].

Further research involving tissue grafting and knockout experiments will shed new light on the importance of Msx genes in the mature bone of the maxillo-facial skeleton.

The alveolar bone
Alveolar bone loss is common in the elderly especially in the edentulous patient. However quantification is mainly by measuring bone height rather than density and is therefore not as well described as in the appendicular skeleton. Reduction of the number of trabeculae and increase of the density of cortical bone have been described [45]. The amount of bone produced decreases as the size of the Haversian canals increases resulting in a more spongeous structure. Overall the amount of the amount of bone per time unit formed decreases reducing the overall amount of bone surface. Also the time from formation of osteoid to complete mineralization is prolonged in

adults, subsequent to orthodontic tooth movement [46]. Overall these changes seem to suggest the age-induced process parallel those seen in the general skeleton such as in type I and type II osteoporosis. But it is also known that the craniofacial area is generally less affected by bone loss in aging when compared to long bones and the vertebrae.

The presence of the periodontal ligament appears to be essential for the height maintenance of the alveolar bone and is independent of the bone density generally suggesting local effects override generalized factors causing generalized bone loss [47]. Loss of periodontal ligament (s.a.) will inevitably lead to vertical loss of alveolar bone height therefore suggesting disuse atrophy. Interestingly not all patients affected by tooth loss are equally affected by loss of alveolar ridge height. The question is how does the presence of teeth indirectly via the periodontal ligament maintain the presence of bone. New bone formation in areas previously affected by atrophy has been reported in the literature. However we also know that this happens only when teeth are moved along the alveolus (see Figure 8) and not when teeth are moved in the bucco-lingual direction. The response to the latter movement normally leads to recession with loss of attachment (see Figure 9). Some authors suggest that by extruding teeth the alveolar bone height increases in the direction of the extrusion [48]. This suggests that promoter cells are still active even in previously diseased areas, leading to new bone formation, but only in areas where bone had been previously present. The precise mechanism of bone cell response to orthodontic tooth movement is still not well established. Although a variety of factors have been implicated (such as cAMP, cGMP, prostaglandin E2) the precise interaction of these with each other is not well understood [49]. Cytokines as well as neurotransmitters, which are involved in general inflammation processes, are also thought to be involved. What effect aging has on the expression of the above named factors has not been investigated. Currently there is also no data linking bone turnover to speed of tooth movement, which would substantiate the currently widely held concept that orthodontic tooth movement is reduced in adults. This issue is still controversial as some authors report treatment time of adolescent and adult patients to be of similar duration, whilst the amount of tooth movement per se can be significantly faster in adolescents [50]. Lastly the prevalence of metabolic bone disease and generalized metabolic disorders are more common in the aged and this is more likely to have an effect on tooth movement.

The mucous membranes of the oral cavity

The anatomy and physiology of the mucous membranes is very differentiated. Most oral pathology occurs in older adults, but pathology is not a condition of old age *per se*. Reported changes of the oral mucosa includes thinning (up to 30%) with age and the surface is thereby more susceptible to injury. Keratinization is decreased, elasticity is lost and the surface characteristics change: the normally prevalent stippling disappears and the surface appears oedematous. The tongue loses its filiform papillae [51–54]. However other investigators have failed to verify the above findings [55, 56]. Indeed studies on rodents, which were controlled for a variety of

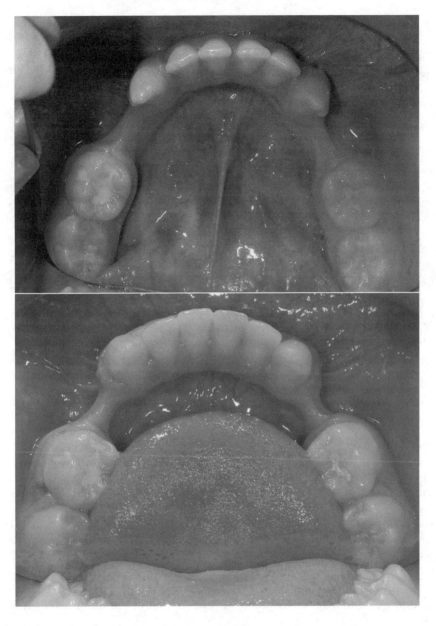

Figure 8. *Reduced distance between lower canines and first molars after orthodontic treatment: the lower molars were moved forewards into an area that had a previously reduced bone width and height. This suggests that induction of bone is possible without loss of attachment.*

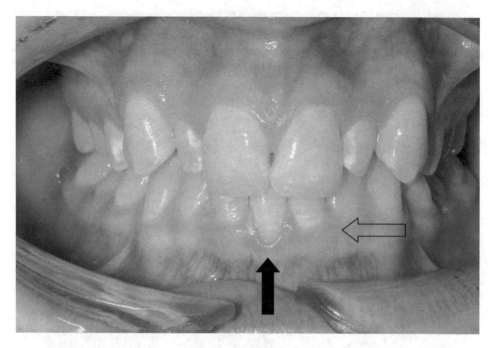

Figure 9*. Gingival recession of the lower supplemental incisor (black arrow). Loss of labial attachment is most likely to be due to crowding but not to abrasion or periodontal disease. Note the normal gingival height of adjacent teeth (transparent arrow).*

factors, failed to produce significant change with age regarding proliferative activity [57]. It may well be that some of the age-associated changes previously reported in the human mucosa were due to vitamin deficiency or a consequence of general nutritional deficiencies or other systemic influences such as reduced hormone levels (oestrogen in particular) [58]. Interestingly one study showed that patients who showed skin changes, which were due to, increased photoexposure also showed more minor oral lesions. This suggests a systemic immune effect on all epithelial tissues [59]. Generally it can be said the results of studies looking at the effects of aging on a variety of epithelial tissues are inconsistent in design and fail to produce distinct changes associated with aging. Variable tissue properties demonstrating inconsistent changes over time have been assumed for explanation of the variation of the above results.

Taste and smell

The effect of aging on taste and smell has previously been investigated by a variety of authors. Generally quantification of taste and smell are difficult even when employing modern psycho-physical methods. Inter-individual variations of strength and associations with certain tastes and smells may depend on previous experiences and preferences. However most studies agree that olfactory response decreases dramatically with age [60]. Taste and trigeminal responses such as tingle and pungency are less affected and independent of one another. As often seen with the aging processes not all are affected to the same extend, but it is generally accepted that threshold deficits tend to increase with age. Recent studies seem to suggest smaller differences in sensitivity to taste than earlier ones [61]; the differences in methodology playing a greater role than the aging process. Tastes are generally easier to test than smells as there are only four qualities: sweet, sour, bitter and salty. Usually a threshold is tested and the relevance of the test-method in daily live in not certain. Quantification of taste intensity on a sliding scale does not seem to show convincing differences between young and old in terms of the ability to differentiate between strengths. However older adults appear to give smaller estimates for the overall rating of a particular strength [62]. Looking at the performance of repeated measurements using the Interclass Correlation Coefficient not all tastes are scored in an equally robust way with increasing age: stimuli representing sweet was the only one which did not decline with age. However the elderly are more likely to exhibit localized taste losses than the younger population, this however does not mean that there is a generalized loss of taste sensation [63].

In summary taste sensation is very robust with age, more so than smell. Taste complaints of the elderly are more likely to be due to other pathology and or side effects of medication rather than aging processes.

The saliva and salivary glands

Saliva functions as lubricator and protects the oral and pharyngeal mucosa as well as the teeth. It plays an essential part of the immune system (it carries immunoglobins as well as other antimicrobial agents: lysozymes, lactoferrins, peroxidases and histatins), and it is saturated with calcium and phosphorus ions thereby aiding remineralization of teeth. Saliva has the capacity to buffer acids, which are primarily a product of intraoral bacteria digesting carbohydrates (see above section on decay). Saliva aids taste sensation and has digestive as well as humidifying properties. Healthy adults produce approximately 1.5 litres per day (see relevant textbooks for further reference).

Saliva is produced by three paired major salivary glands: parotid (serous), submandibular and sublingual glands both (both mixed: mucous and serous). The minor salivary glands are distributed throughout the oral cavity. A gland is composed of saliva producing elements (acinar) and modifying elements (ductal). The latter secret K^+ and also add a variety of proteins and resorb Na^+ and Cl^-. Reduced saliva output had long been thought of as inevitable consequence of gland atrophy, which occurs with aging [64]. The weight of the major glands decreases steadily with age. Histological studies reveal a reduction of acinar elements, which are replaced by fatty of fibrous tissue. Duct volume and vascularity increase with age. However the major glands age at a different rate, the submandibular gland overtaking the parotid and minor salivary glands [65].

Although it appears plausible that salivary output diminishes with age [66], various studies reveal controversial results [67]. This might be due to variations of examination techniques (production measured at rest, stimulated: gustatory or mechanical stimulation; whole saliva or individual glands). Longitudinal studies are rare and generally it is thought that flow rates do not decline at a significant rate and that not all glands reduce output at the same rate (output of the parotid gland in particular being particularly stable [68]). However Na^+ concentration decreases whilst K^+ remains unchanged with increasing age [69]. This would suggest that a differential aging process is at work influencing absorption and secretion differently.

The amount of immunoglobin in saliva samples of the aged is also controversial. Some investigators found decrease of IgA, whilst others demonstrated an increase. Other protein-constituents associated with remineralization of the teeth appear to be unchanged or increased (lactoferrin and lysozyme). However one study suggest that the antimicrobial effectiveness of saliva decreases with age [70].

It is now thought that significant reduction of salivary output is often iatrogenic. Medication and or radiotherapy often significantly reduce saliva output and function and some older studies did not exclude patients from the above cohorts. Also a patient complaining of oral dryness may not suffer from primarily decreased saliva flow; dehydration and systemic disease such as Sjögren's syndrome and diabetes mellitus are also often responsible. Salivary output must be decreased by more than 50% before patients become symptomatic, so tumors destroying one gland are not thought to be a common cause of xerostomia. Often psychological factors play a major role in the complaint of oral dryness. Most causes leading to a decrease of salivary function are not primarily due to aging although they are more common in the aged as the causative pathology is more common in the aged.

Conclusion

Some changes previously associated with the aging process were probably unjustified and were often based on patients' samples where secondary changes (disease and medication) were prevalent. As so often in nature variation is the rule and the expression of aging characteristics is diverse, sometimes dominated by continuous apposition (cementum and secondary dentine) often at the detriment of another structure (reduction of root length and pulp cavity volume). Growth of some structures continues even in advanced age (ears and nose). Some tissues clearly undergo involutionary changes (salivary glands), however the resulting function of the organ is sometimes not compromised (saliva flow) thus suggesting that the residual tissue is able to compensate. Overall the functional properties of the oral cavity are different in the aged, but probably to a lesser degree than previously thought. Many conditions previously commonly found in the elderly are preventable. Easy access of the elderly to facilities enabling adequate oral healthcare is paramount to achieving better oral health.

References

1. Woelfel and Scheid (1997). *Dental Anatomy.* Baltimore: Williams & Wilkins.
2. *Oral Structural Biology* (1991). Stuttgart: Schroeder, Thieme.
3. Avery JK (1986). Pulp. In: Bhaskar SN, ed. *Orban's Histology and Embryology.* St. Louis: Mosby.
4. Nielsen CJ, *et al.* (1983). Age related changes in reducible crosslinks of human dental pulp collagen. *Arch Oral Biol.* 28: 759.
5. Stanley HR (1973) The effect of systemic diseases on the human pulp. In: Siskin M, ed. *The Biology of the Human Dental Pulp.* St. Louis: Mosby.
6. Baume LJ (1980). *The Biology of Pulp and Dentine.* Monographs in Oral Science, Vol 8, Basel: Karger.
7. Churchill HR (1932). *Human Odontography and Histology.* Philadelphia: Lea & Febinger.
8. Nalbandian J, *et al.* (1960). Sclerotic age changes in root dentine of human teeth as observed by optical, electron and X-ray microscopy. *J Dent Res.* 39: 598.
9. Weber DF (1974). Human dentine sclerosis: a microradiographic survey. *Arch Oral Biol.* 19: 163.
10. Bang G, Ramm E (1970). Deteremination of age in humans from root dentine transparency. *Acta Odont Scand.* 24: 3.
11. Helfman PM, Bada JL (1976). Aspartic acid racemisation in dentine as a measure of ageing. *Nature* 262: 279.
12. Hotz R (1967). Wurzelresorptionen an bleibenden Zaehnen. *Fortschr Kieferorthop.* 28: 217.
13. Goldberg M, *et al.* (1979). Maturation de tarditive de l'email dentaire humain. *J Biol Bucc.* 7: 353.
14. Zachrisson BU, *et al.* (1980). Enalmel cracks in debonded, debanded and orthodontically untreated teeth. *Am J Orthod.* 77: 307.
15. Behrents RG (1989). The consequences of adult craniofacial growth. In: Carlson, ed. *Orthodontics in the Aging Society.* Craniofacial Growth Series. Center for Human Growth and Development. Ann Arbor: University of Michigan.
16. Sinclair P, Little R (1983) Maturation of untreated normal occlusions. *Am J Orthod.* 83: 114.
17. Burt BA, *et al.* (1992). *Dentistry, Dental Practice and the Comminity.* Philadelphia: WB Saunders.
18. Weintraub JA, Burt BA (1985). Tooth loss in the United States. *J Dent Educ.* 49: 368.
19. Bouma J, *et al.* (1986). Caries status at the moment of total tooth extraction in arural and an urban area in the Netherlands. *Comm Dent Oral Epidemiol.* 14: 345.
20. National Institute of Dental Research NIH Publ No 87 – 2868. Government Printing Office (1987), Washington.
21. Hand JS, *et al.* (1988). Incidence of coronal and root caries in an older adult population. *J Public Health Dent.* 48: 14.
22. Brown LR *et al.* (1986). Quantitative comparisons of potentially cariogenic microorganisms cultured from noncarious and rarious root and coronal tooth surfaces. *Infect Immun.* 51: 765.
23. Banting DW, *et al.* (1980). Prevalence of root surface caries among institutionalized older persons. *Comm Dent Oral Epidemiol.* 8: 84.
24. Zander HA, Huerzeler B (1958). Continuous cementum apposition. *J Dent Res.* 37: 1035.
25. Reitan K (1954). Tissue reaction as related to the age factor. *Dent Rec.* 74: 271.
26. Grant D, Bernick S (1972). The periodontium of ageing humans. *J Periodontol.* 43: 66.

27. Rippin JW (1976). Collagen turnover in the periodontal ligament under normal and altered functional forces. *J Periodont Res.* 11: 101.
28. Skougaard M *et al.* (1970). Collagen metabolism in skin and periodontal membrane of the marmoset. *Scand J Dent Res.* 78: 256.
29. Norton LA (1988). The effect of ageing cellular mechanisms on tooth movement. *Dent Clin N Am.* 32: 437.
30. Coolidge ED (1937). The thickness of the human periodontal membrane. *J Am Dent Assoc.* 24: 1260.
31. Griffiths GS, *et al.* (1988). Detection of high-risk groups and individuals for periodontal diseases. Clinical assessment of the periodontium. *J Clin Periodontol.* 15: 403.
32. Loe H, *et al.* (1986). Natural history of periodontal disease in man; rapid, moderate and no loss of attachment in Sri Lankan laborers 14 to 46 years of age. *J Clin Periodontol.* 13: 431.
33. Hunt RJ, *et al.* (1990). The prevalence of periodontal attachment loss in an Iowa population aged 70 and older. *J Public Health Dent.* 50: 251.
34. Hall B (1987). Tissue interactions in the development and evolution of the vertebrate head. In: Madison, ed. *Developmental and Evolutionary Aspects of the Neural Crest.* New York: Wiley Interscience.
35. Zernik JH, *et al.* (1990). Regulation of alkaline phosphatase and alpha2(I) procollagen expression during early intramambraneous bone formation in the rat mandible. *Differentiation* 44: 207.
36. Walldorf U, *et al.* (1989). Comparison of homeobox-containing genes of the honeybee and *Drosophila. Proc Nat Acad Sci USA* 86: 9971.
37. Davidson D (1995). The function and evolution of Msx genes: pointers and paradoxes. *Trends Genet.* 11: 405.
38. Hoffman HM, *et al.* (1994). Transcriptional control of the tissuespecific, develpmentally regulated osteocalcin gene requires a binding motif for the Msx family of homeodomain proteins. *Proc Natl Acad Sci USA* 91: 12887.
39. Towler DA, *et al.* (1994). Msx-2/Hox 8.1: a transcriptional regulator of the rat osteocalcin promoter. *Mol Endocrinol.* 8: 1484.
40. Nowroozi N, *et al.* (1996). Site specific differences in osteoblastic phenotype in the mature skeleton defined by transgene expression in Msx2-lacZ mice. *Am J Bone Miner Res.* 11: S395.
41. Zernik JH, *et al.* (1997). Development, maturation, and aging of the alveolar bone: new insights. *Dent Clin N Am.* 41: 1.
42. Amler MH (1993). Age factor in human alveolar bone repair. *J Oral Implant.* 19: 138.
43. Holm-Petersen P, Loe H (1971). Wound healing in the gingiva of young and old individuals. *Scand J Dent Res.* 79: 40.
44. Bei M, *et al.* (1995). The homeobox gene Msx1 regulates Bmp4 expression in dental mesenchyme. *J Dent Res.* 74: 393 (Abstract).
45. Amling M, *et al.* (1994). Polyostotic heterogeneity of the spine in osteoporosis. Quantitative analysis and three-dimensional morphology. *Bone Miner.* 27: 193.
46. Melsen B (1991). *Limitations in Adult Orthodontics.* Current Controversies in Orthodontics. Quinterssence: Berlin
47. Reitan K (1951). The initial tissue reaction incident to orthodontic tooth movement as related to the influence of function. *Acta Odontol Scand.* 6: 1951.
48. van Venroy JR, Yukna RA (1985). Orthodontic extrusion of single-rooted teeth affected with advanced periodontal disease. *Am J Orthod.* 87: 67.

49. Rodan GA, *et al.* (1975). Cyclic AMP and cyclic GMP: mediators of the mechanical effects on bone remodelling. *Science* 189: 467.

50. Dyer GS, *et al.* (1991). Age effects on orthodontic treatment: adolescents contrasted with adults. *Am J Orthod.* 100: 523.

51. Bottomly WK (1979). Physiology of the oral mucosa. *Otol Clin N Am.* 12: 15.

52. Pickett HG, *et al.* (1972). Changes in denture supported tissues associated with aging. *J Prosthet Dent.* 27: 35.

53. Frantzell A, *et al.* (1945). Examination of the tongue: a clinical and photographic study. *Acta Med Scand.* 122: 207.

54. Miles AEW (1972). Sans teeth: changes in the oral tissues with advancing age. *Proc R Soc Med.* 65: 801.

55. Loe H, Karring T (1971). The three dimensional morphology of the epithelium connective tissue interface of the gingiva as related to age and sex. *Scand J Dent Res.* 79: 315.

56. Wolff A, *et al.* (1993). Oral mucosa appearance is unchanged in healthy, different aged persons. *Oral Surg Oral Med Oral Pathol.* 76: 569.

57. Cameron IL (1972). Cell proliferation and renewal in ageing mice. *J Gerontol.* 27: 162.

58. Belding JH, Tade WH (1978). Evaluation of epithelial maturity in hormonally related stomatitis. *J Oral Med.* 33: 17.

59. Engel A, *et al.* (1988). Health effects of sunlight exposure in the US. *Arch Dermatol.* 142: 72.

60. Stevens JC, *et al.* (1985). Chemical senses and aging: taste versus smell. *Chem Senses* 10: 517.

61. Weiffenbach JM (1984). Taste and smell perception in aging. *Gerodontology* 3: 137.

62. Bartoshuk LM, *et al.* (1986). Taste and aging. *J Gerontol.* 41: 51.

63. Weiffenbach JM, *et al.* (1986). Taste intensity perception in aging. *J Gerontol* 41: 460.

64. Drummond JR, Chisholm DM (1984). A qualitative and quantitative study of the ageing human labial salivary glands. *Arch Oral Biol.* 29: 151.

65. Scott J (1977). Degenerative changes in the histology of the human submandibular gland occurring with age. *J Biol Buccale* 5: 311.

66. Meyer J, *et al.* (1940). Basal secretion of digestive enzymes with old age. *Arch Int Med.* 17: 171.

67. Pravinden T, Lamas M (1982). Age dependency of stimulated salivary flow rate, pH, and lactobacillus and yeast concentrations. *J Dent Res.* 61: 1052.

68. Wu AJ, *et al.* (1993). Cross sectional and longitudinal analyses of stimulated parotid salivary constituents in healthy, different aged subjects. *J Gerontol.* 48: M219.

69. Baum BJ, *et al.* (1984). Sodium handling by aging human parotid glands is inconsistent with a two stage secretion model. *Am J Physiol.* 246: R35.

70. Ganguly R, *et al.* (1986). Defective antimicriobal functions of oral secretions in the elderly. *J Infect Dis.* 153: 163.

Aging and the Immune System

Sian M. Henson and Richard Aspinall

Department of Immunology, Faculty of Medicine, Imperial College, London

The function of the immune system declines with age leading to an increase in the frequency and severity of infectious diseases, an increase in the prevalence of cancers amongst the elderly compared with the young and a recognition that the regulation of immune function is changing from the increased prevalence of autoimmune disease with age.

Age-dependent alterations of immune functions are associated with both the adaptive and innate arms of the immune system. The involvement of the adaptive immune system during aging has been well documented, with alterations in the ratio of memory to naïve T cells [1], decline of T cell functions [2] including cytokine production [3], decreased antibody production [4] and shortened duration of protective immunity following vaccination [5] all being reported. In contrast there are far fewer reports for the involvement of the innate immune system in aging with what has been documented being contradictory.

Two key events precipitate the changes outlined above. The first is a dramatic involution of the thymus resulting in a reduced thymic output. The second a homeostatic compensatory mechanism that maintains T cell numbers, causing the peripheral expansion of mature T cells to regenerate the T cell pool preventing the development of lymphopaenia.

Age-dependent alterations to antigen presenting cells

Defects at the level of the T cell are known to be associated with aging, however recent research suggests that defects may also occur within antigen presenting cells (APC). The alterations to APC with age have been most extensively studied in monocytes and macrophages and these APC when isolated from old donor have been shown to inhibit T cell proliferative responses [6]. Monocyte numbers remain unchanged in the elderly, however the balance of cytokine production is altered with increased levels of interleukin-6 (IL-6) and transforming growth factor β (TGF-β)

R. Aspinall (ed.), Aging of Organs and Systems, 225–242.
© 2003 *Kluwer Academic Publishers. Printed in Great Britain.*

being reported. Both these cytokine are known to have immunosuppressive properties at elevated concentrations. IL-6 does not upregulate TGF-β production, but high
concentrations of IL-6 increase the percentage of cells expressing the TGF-β receptor
[7]. The production of prostaglandin E_2 (PGE_2) by monocytes/macrophages has also
been reported to increase in aged mice [8] and T cells isolated from elderly subjects
have been shown to be more sensitive to exogenous PGE_2 than cells from young
subjects [9]. This increased production of PGE_2 is thought to lower T cell function by
inhibiting the production and responsiveness to IL-2 and the expression of the IL-2
receptor [10].

Studies have also implicated dendritic cells to participate in the aging process. The
used of animal models has helped to reveal a number of degenerative characteristics
in dendritic cells with age. Dendritic cells of the lymph node of old mice show a
decreased expression of the adhesion molecule CD86. It has been speculated that
because these dendritic cells are located in the germinal center of the lymph node the
decreased expression of CD86 leads to the induction of anergy [11]. Old mice were
also unable to transport antigen to regional lymph nodes, an effect that the authors
attributed to impaired dendritic cell antigen uptake [12]. The lymph nodes of aged
rats showed that follicular dendritic cells had dense distribution in involuted follicles,
but showed a reduced antigen trapping capacity. These animal studies implied that
dendritic cells of the lymph node take part in aging of the immune system, however
their role in this process remains unclear. Experiments to determine the capacity of
APC have also been carried out in animal models, in which mice were challenged
with different microbial agents. The results of these studies demonstrate that aging is
associated with decreased antigen presenting capacity [13]. Taken together, all these
experiments in animal models imply that dendritic cell functions decrease with age.

However it should be noted that defects in APC function are not always found with
age [14, 15], also defects in T cell function are apparent in APC-free systems [16] and
in situations where defined cell lines are used as APC [17].

T cell development

Precursor T cells begin to colonize the thymic anlage early during embryonic
development. This begins from around day 14 of gestation in mice [18] and at the
7th to 8th week of gestation in humans [19]. The rate of development of T cells within
the thymus is greatest before puberty and declines with a stepwise kinetic thereafter.
This stepwise decline has been observed in mice to have distinct reductions at 1.5, 3–
7, 12–16, 18 and 22–27 months [20]. Although the number of naïve T cells decline
with age the mature T cells in the periphery do not show a similar reduction [21]. This
decline in T cell production also correlates with atrophy of the thymic stroma, a
condition known as thymic involution. Age related thymic involution is referred to as
chronic involution and is distinct from involution induced by stress, pregnancy, HIV
infection or chemotherapy. This distinction can be made on the ability of the thymus
to negatively select autoreactive T cell clones, which has been shown to be impaired in
disease states [22] but remains intact with age [23]. Several changes to the tissue
architecture can be detected in the aging involuted thymus. The perivascular space

becomes enlarged resulting in a decrease in the thymic epithelial space, which houses both the cortical and medullary components responsible for thymopoiesis [24]. The perivascular space is enlarged by lymphocytic infiltration but as aging progresses adipose tissue eventually replaces the lymphocytic perivascular space [25]. However, thymic reconstitution studies have shown that the thymus remains functional in the aged mouse [23]; and in aged humans a small thymic rudiment composed of epithelium and thymocytes remains identifiable [25].

The earliest precursor cells that arrive at the murine thymus are CD3$^-$CD4^{lo}CD8$^-$ progenitors [26] these cells do not express the T cell receptor (TCR) so are incapable of antigen specificity. This effectively triple negative population can be further subdivided on the basis of expression of CD44 and CD25, with the most immature stage being CD44$^+$CD25$^-$. It is at this stage that c-kit, the receptor for stem cell factor is expressed with its expression persisting until the cells lose expression of CD44 [27]. Differentiation and commitment to the thymocyte lineage is associated with the transient expression of CD25 and loss of CD44. The cells become CD44$^+$CD25$^+$ then CD44$^-$CD25$^+$ and it is during this period that TCRβ chain rearrangement begins [28]. The thymic signal that induces this rearrangement is though to include interleukin-7 (IL-7) and the action of recombination-activating genes 1 and 2 (RAG-1, RAG-2) [29] whose expression is only terminated after the positive selection process is completed. Expression of the TCRβ chain at the thymocyte surface requires a TCRα chain equivalent to form the pre-T cell receptor (pre-TCRα). The pre-TCRα induces expansion and differentiation of these cells so that they become TCR$\alpha\beta$ bearing CD4$^+$8$^+$ thymocytes [30]. At this double positive stage the immature thymocytes undergo either positive or negative selection. Positive selection involves an obligatory interaction with self-major histocompatibility complex (MHC) molecules in the thymus and it is a process that rescues thymocytes from programmed cell death and induces their differentiation into mature T cells expressing either CD4 or CD8. Negative selection is a mechanism designed to prevent maturation of hazardous autoreactive T cells causing the elimination of T cells with self-reactive receptors [31]. Both these processes result the loss of greater than 95% of developing thymocytes.

Recently a new technique looking at TCR rearrangement circles has been developed to quantitate thymic function in humans following T cell depletion [32]. Episomal DNA circles that are generated during excisional rearrangement of TCR genes were used to monitor the emergence of recent thymic emigrants. These DNA circles, termed TCR-rearrangement excision circles (TRECs), have been shown to be stable, are not duplicated during mitosis, and are therefore diluted out with each cellular division. A TREC is produced during the deletion of the *TCRD* locus from the *TCRA*. End-to-end ligation of the recombination signal sequences flanking the δrec locus and the ψJα locus removes the *TCRD* gene region, forming a single TREC containing a unique signal joint sequence. The recombined δrec to ψJα junction, the coding joint is retained in the germline DNA until *TCRAV* to *TCRAJ* recombination occurs, the coding joint then becomes part of a second TREC. As the δrec to ψJα recombination cannot produce a functional TCR, signal- and coding-joint sequences can only occur in TRECs in mature T cells. TREC analysis has been used to investigate the ability of the thymus to generate new T cells during aging [32]. The numbers of both coding- and signal-joint TRECs were shown to decline exponentially in both $CD4^+$ and $CD8^+$ T cells with age but were still detectable. The continued presence of TREC-containing cells with age lead to the suggestion that the thymus was still capable of producing thymocytes that actively rearrange TCR genes.

Thymic involution

Many papers have been written about the possible mechanisms causing age-associated thymic atrophy. Such reports focus on whether it is an inadequate supply progenitors or whether involution relates to changes in the microenvironment of the thymus. The first hypothesis to be put forward was that atrophy was brought about by a reduction in the number of thymocyte precursors. This theory came about mainly from transfer experiments, where bone marrow from old animals is implanted into young irradiated animals [33, 34]. These experiments showed that thymic regeneration was significantly impaired when the young animals received marrow cells from aged donors, which suggested that there was a decrease in the number of T cell progenitors found in aged animals. *In vitro* studies also supported this theory, with one report finding the frequency of T cell precursors of young mice to be $2.0 \pm 0.4 \times 10^3$ compared to $1.2 \pm 0.4 \times 10^3$ found in aged mice [35]. However, this reduction was modest and subsequent experiments have shown that using the bone marrow from young mice to repopulate the thymus of aged animals fails to rescue the histological changes associated with thymic involution [23]. Other studies showed that the numbers of early progenitor cells, $(CD44^+CD25^-)$ remained unaltered by aging, while their immediate progeny, $(CD44^+CD25^+$ and $CD44^-CD25^+)$ were reduced with age [36–38]. These reports suggested that age-associated thymic atrophy occurs between the $CD44^+CD25^-$ cells and their progeny or even in the survival of the $CD44^+CD25^+$ population. This evidence suggests that a reduction in T cell progenitors is not the primary limiting factor in thymopoiesis in the aged.

If having fewer progenitors is not the primary cause of thymic involution an alternative theory postulated that precursor thymocytes had an intrinsic defect causing their failure to differentiate. Initial experiments testing this theory used bone marrow cells to reconstitute fetal thymic lobes *in vitro*. These studies found that bone marrow cells from aged mice maintained their capacity to reconstitute fetal thymic explants, to respond to mitogens and to differentiate into the different T-cell subsets. However, when using mixtures of old and young bone marrow cells for reconstitution they saw that the cells of old mice were less efficient than those of young in their capacity to give rise to T cells. The authors concluded that bone marrow cells from aged mice could reconstitute the thymus and differentiate into T cells, although their reconstituting capacity was inferior to that of bone marrow cells from young mice [39]. The repopulating cells used in these experiments were derived from bone marrow and may not be identical to those seen in the thymus. This prompted a set of experiments to test the ability of $CD44^+CD25^-$ cells isolated from the thymus of young or aged animals to reconstitute fetal thymic organ cultures. The results of these experiments showed that there was no difference between young and old cells to cause differentiation to all the defined thymocyte subsets, suggesting that there is no intrinsic defect in the CD44$^+$CD25$^-$ population as a result of aging [40]. This hypothesis could be further discounted by experiments showing that F5 transgenic mice show none of the age-associated thymic atrophy normally associated with non-transgenic mice [37].

An alternate hypothesis for the mechanism for thymic involution relates to the gradual loss of function to the thymic microenvironment, which would result in a loss of thymic function as a secondary event. A number of extrathymic factors have been thought to cause age-related changes in thymic function. Zinc is thought to play an important role in the functioning of the immune system, it is involved in the maturation and differentiation of T cells and it is required for the biological activity of a thymic peptide, thymulin. Zinc is reduced in old age but supplementing zinc has been shown to restore thymic and NK activities as well as promoting survival in aged mice [41]. Steroid, thyroid, and pituitary hormones, have also been shown to increase thymopoiesis. Growth hormone has been shown to have a positive influence on thymic function, in animal models, administration of growth hormone was found to enhance development of the thymus [42, 43] and promote the engraftment of murine or human T cells in immunodeficient mice [44]. Melatonin has been shown to have an immune reconstituting effect in aged mice [45], an effect which is thought to be mediated by the zinc pool, via glucocorticoids [46]. Finally thyroxine when administered to aged rats increased the serum thymulin level up to 50% of age-matched controls [47]. However if the factors in the extrathymic environment were the only cause of thymic involution one would expect that neonatal thymic grafts placed into aged animals would receive similar negative regulatory influences. This is not the case, for aged thymi displayed a normal thymocyte subset distribution after receiving young bone marrow in bone marrow transplantation [48]. Thus it is thought that both intra and extrathymic factors, such as bone marrow cells, thymocytes, thymic microenvironment as well as endocrine hormones all contribute to age-related changes in thymic function.

Current thinking is that thymic involution is caused mainly by changes to the thymic microenvironment. The most noticeable of which is a reduction in the number of thymocytes. It has been suggested that a lesion in the T cell developmental pathway within an early subset may cause these reductions. To investigate this hypothesis further the phases of thymopoiesis were studied in detail in aged mice and a significant decline in the CD3$^-$CD4$^-$CD8$^-$ subset was seen by 20 months of age. The decline occurred in the CD44$^+$CD25$^+$ population and the progeny of these cells, CD44$^-$CD25$^+$ but no significant difference was seen in the CD44$^+$CD25$^-$ population, leading to the suggestion that the lesion occurred after this subset [37]. The expression of CD25 is associated with the rearrangement of TCR β-chain genes [49]. The failure of aged cells to undergo the transition from CD44$^+$CD25$^-$ to CD44$^+$CD25$^+$ lead to the suggestion that thymic involution is the product of the failure of the cells to undergo differentiation through a stage of T cell development associated with the rearrangement and expression of the β-chain of the TCR.

The inability of aged cells to efficiently rearrange the β-chain of the TCR may be a result of a local reduction in one or more factors produced by the stromal epithelium of the thymus essential for the process of rearrangement of the β-chain of the TCR. One possible candidate may be IL-7 as it has been shown to play a role in the rearrangement of the V(D)Jβ chain of the TCR [29]. Further support for a link between IL-7 and thymic involution comes from both *in vitro* and *in vivo* studies. It has been shown that IL-7 could support the survival of cultured $CD44^+CD25^-$, $CD44^+CD25^+$ and $CD44^-CD25^+$ cells and also allow the differentiation of $CD44^+CD25^-$ cells in fetal thymic organ cultures [50]. *In vivo* studies using anti-IL-7, induced thymic atrophy, that was similar to the atrophy seen in aging, but could be reversed when treatment with anti-IL-7 ended [40]. However, a recent report suggested that the administration of IL-7 to aged mice did not augment thymopoiesis, shown by a lack of difference in the number of TRECs per 100 000 thymocytes in saline-treated versus IL-7-treated young mice [51]. However, in the mouse there are 3 δrec elements and these δrec's can use many Jα's as 3' acceptor sites and many do so in preference to ψJα. In the experiment above the authors used primers specific for the human which would produce PCR products with a range of sizes due to the promiscuity of the δrec and its acceptor sites, leading to an underestimation of the number of TRECs present. A better method for measuring thymic output from the mouse is to use T cell receptor delta excision circle (δEC) analysis, which is based on the number of TCRδ locus excision circles found within an αβT cell. When this method was employed IL-7 treatment was shown to increase the level of δECs detected in IL-7-treated mice compared to saline treated controls, suggesting that IL-7 could increase thymic [52].

Other cytokines have also been shown to decline with age such as IL-2, IL-9, IL-10, IL-13 and IL-14, while leukemia inhibitory factor, oncostatin M, IL-6, stem cell factor (SCF) and macrophage-colony stimulating factor (M-CSF) were all found to increase with age [53]. *In vivo* administration of oncostatin M, IL-6, SCF and M-CSF to mice induced thymic atrophy as measured by the loss of the $CD4^+CD8^+$ cell population, lead to the speculation that thymopoiesis may be regulated by these cytokines [53]. However this effect must be indirect as c-kit the receptor for SCF is not expressed by more than 95% of the cells with in the $CD4^+CD8^+$ population [54]. Oncostatin M, IL-6, and M-CSF, all of which are members of the IL-6 family of cytokines and share the gp130 receptor component are also not expressed in this $CD4^+CD8^+$ thymic subset [55].

Decline in T cell function

The development of effectors from naive $CD4^+$ cells is thought to occur in two stages [56]. The first stage is a TCR dependent response to antigen and co-stimulatory APCs, which brings about a limited proliferation. The second stage involves an IL-2 driven expansion and differentiation of naïve T cells to become $CD4^+$ effectors. Effectors are highly differentiated, activated cells that are capable of producing large amounts of cytokine upon restimulation, even in the absence of co-stimulation. This gives effectors the ability to make cytokines in response to non-professional APCs.

What distinguishes fully differentiated effectors from activated CD4$^+$ T cells is the production of high levels of cytokines other than IL-2.

Defects in CD4$^+$ T cell function are evident during the aging process, they expand less giving rise to fewer effectors with less activated phenotypes and a reduced ability to produce cytokines. This decline in function is though to be caused by a reduction in the number of CD4$^+$ T cells producing IL-2 and a decreased expression of IL-2Rs [57]. The addition of exogenous IL-2 was found to support the expansion and differentiation of aged CD4$^+$ cells and lead to the generation of effectors that themselves are capable of IL-2 production [58]. It was also shown that other cytokines that signal through the γc chain, IL-15, IL-4, and IL-7 could increase the proliferative capacity of aged CD4$^+$ T cells. However it was thought that these effects operated through an IL-2 independent mechanism that does not involve increased IL-2 production or upregulation of IL-2Rα expression. Experiments using a TCR transgenic (Tg) mouse model have been used to investigate the reduced expansion and IL-2 production by CD4$^+$ T cells seen with age. TCR Tg cells were isolated from young and old mice, labelled with CFSE and transferred to peptide immunized young adoptive hosts. The intracellular dye CFSE distributes equally between cells upon each cell division resulting in the halving of fluorescence intensity with each round of cell division. This allowing the visualization of the proportion of cells in each population undergoing cell division. Aged cells were shown to expand less than young cells during the 5-day period of *in vitro* effector generation. The early stage of expansion, days 1–3 were similar for both the young and aged cultures, whereas later expansion, days 4 and 5 was shown to be three to five times greater in the cultures of young cells as observed by flow cytometric analysis of the CFSE label. Also, during this late expansion period aged CD4 effectors expressed decreased levels of CD25 (IL-2Rα) leading the authors to suggest that this late stage is more dependent upon IL-2, which is present in greater amounts in the young cultures. Effectors generated from Tg CD4 cells from aged mice were also shown to produced less IL-2, and the frequency of effector cells secreting IL-2 was also significantly reduced in the aged. Providing further support for the continued presence of IL-2 for efficient effector generation [58].

A further consequence of a decrease in IL-2 production is an incomplete differentiation of effector populations, which result in a decreased effector function. Potentially this could cause a less vigorous cytotoxic T lymphocyte (CTL) activity or reduced humoral response, which could influence responses to infections as well as vaccinations [59]. Studies have focused on antibody production in response to vaccination and have shown that the efficacy of vaccinations for highly contagious infections is greatly reduced in the elderly [60, 61]. It has also been shown that these antibodies do not functions as well, showing a lower affinity, diminished protection against infections and autoreactivity [62]. These results are thought to be due in part to age-related declines in CD4 function. Generating Th1 and Th2 effectors from young and old CD4$^+$ cells and transferring them into adoptive hosts assessed the effect of age on the ability of naïve CD4$^+$ T cells to develop into memory cells. Two weeks after transfer the memory cells were recovered and their responses to stimulation with antigen/APC were assessed. The aged CD4$^+$ cells were shown to expand poorly and have a decreased level of cytokine production in response to stimulation. In addition the *in vivo* function of these memory cells was examined by immunizing the adoptive host an examining primary B cell responses. While the young memory cells showed high levels of CD4 helper activity leading to vigorous B cell response aged memory cells showed poor CD4 helper activity [63]. Suggesting that the aged naïve CD4 T cells have been altered permanently and that the defect exhibited is heritable and re-expressed in their progeny. The loss of CTL activity in response to infections seen in the elderly is also though to be due to a limited expansion of $CD8^+$ T cells and to an intrinsic defect in $CD8^+$ effector function. The detection of antigen specific $CD8^+$ T cells using MHC class I peptide tetrameric complexes has greatly aided the study of antigen specific immune response to infections. Using tetramer technology it has been shown that for influenza infection the reduced $CD8^+$ T cell activity seen with age was due primarily to a decrease in the number of virus-specific cells, and not due to diminished function of the antigen-specific $CD8^+$ T cells [64].

An important aspect of differentiation is the commitment of the $CD4^+$ T cells to production of restricted or polarized pattern of cytokine production. It is well know that IL-4 can drive polarization to produce Th2 cytokines and that IL-12/IFN-γ can drive Th1 polarization [65]. Aging changes the balance between type 1 and type 2 cytokines skewing production towards Th2 cytokines. It has been shown that the elderly have a reduced capacity to produce IFN-γ and that its production also revealed different kinetics with young individuals releasing IFN-γ faster and in higher amounts than the elderly [66]. Furthermore IL-4 and IL-10 were shown to be produced in higher amounts by T cells isolated from elderly individuals when compared to those taken from the young [67]. Changes in the T cell phenotype have been suggested to underlie much of the age-related changes in the cytokine network, in particular a replacement of naïve T cells expressing $CD62L^+CD45RA^+$ on their surface by memory T cells expressing CD45RO and/or CD95. In deed the percentage of T cells expressing Th2 cytokine was shown to correlate to the *in vivo* expression of CD95 and CD45RO [68].

Expansion of memory T cells

The reduction in thymic output with age is not matched by a decline in the total number of T cells in the peripheral T cell pool. The absolute number of T cells in the peripheral pool is carefully regulated to maintain T cell levels at a desirable limit. Such maintenance occurs through the expansion of the memory T cell subset thus increasing their number with age, as well as their progression along the replicative life span and reducing their replicative capacity [69]. This replication leads to the progressive shortening of telomeres to a critical length which can lead to cell-cycle arrest and/or apoptosis [70]. Telomerase, a telomere-synthesizing reverse transcriptase, can compensate for the loss of telomere associated with cell divisions [71, 72]. Telomere shortening with age has been reported in peripheral blood mononuclear cells (PBMC) [73, 74] and in T cells [75]. The rate of telomere shortening appears to decrease rapidly from newborn to 4 years of age, with a gradual decline between 4 and 39 years and telomere shortening continuing at a stable and low rate between ages 40 to 95 years [73]. Telomerase has also been shown to decrease with age being detectable in many newborns and some children before adolescence but decrease to low or undetectable levels after age 20 [74].

The rate of T cell proliferation has been studied in aged mice and was shown to have a marked reduction in turnover at the level of memory-phenotype CD44^{hi}CD8$^+$ cells compared to young mice [76]. The longevity of memory cells appears to be based on protective signals delivered by cytokines, in particular IL-15 [77, 78]; since the number of CD44^{hi}CD8$^+$ cells has been shown to be selectively reduced in IL-15$^{-/-}$ mice [79]. However direct support for the role of IL-15 came from the finding that the background turnover of CD44^{hi}CD8$^+$ cells is inhibited by anti-CD122/IL-2Rβ antibodies, which is an important component of the IL-15R but not by anti-IL-2 antibodies [78]. How cytokines promote memory cell survival is unclear but it is thought that the upregulation of anti-apoptotic molecules such as Bcl-2 and Bcl-X$_L$ plays a role. Bcl-2 upregulation is controlled *in vitro* by γc-controlled cytokines (IL-2, IL-4, IL-7, IL-9, IL-15) [80], especially by IL-15 in CD8$^+$ T cells, whereas Bcl-X$_L$ upregulation is elicited by other cytokine, including type I interferons (IFN-I) [81]. CD8$^+$ memory cells have been shown to express high levels of Bcl-2 and to a lesser extend Bcl-X$_L$ relative to naïve T cells [82]. In contrast, CD4$^+$ memory cells show upregulation of Bcl-X$_L$ but not Bcl-2 [83]. Based on adoptive transfer experiments, the reduced rate of turnover of aged CD44^{hi}CD8$^+$ cells is thought to reflect an inhibitory influence of the aged host environment. For aged CD44^{hi}CD8$^+$ cells show poor *in vivo* responses to IL-15 and IL-15-inducing agents, but respond well to IL-15 *in vitro* [76]. This impaired *in vivo* response to IL-15 correlates with increased levels of IFN-I and was reversed by the addition of anti-IFN-I antibody [76]. Also aged CD8$^+$ memory cells show an enhanced expression of Bcl-2 [76], leading to the hypothesis that the reduced turnover of CD44^{hi}CD8$^+$ cells reflects the combined inhibitory effect of enhanced Bcl-2 expression and high IFN-I levels.

As described above the survival of memory CD44^{hi}CD8$^+$ cells is in part due to protective signals by IL-15, however the γc cytokine IL-2 is capable of inhibiting the turnover of these CD44^{hi}CD8$^+$ cells [78]. For animals lacking IL-2 or a component of its receptor molecules have more expanded T cells with an activated memory phenotype [84]. Why IL-2 is inhibitory for memory CD8$^+$ cells is unclear but it is thought that the stimulation of a suppressive population of regulatory CD4$^+$ cells may be responsible. T regulatory cells constitute only small proportion, $\sim$10% of CD4$^+$ cells, they co-express the IL-2R α-chain, CD25 and have been shown to be both hyporesponsive and suppressive [85, 86]. The number of CD4$^+$CD25$^+$ T regulatory cells found in humans increases with age, young subjects were found to have less than 10% compared to 17% in old subjects (over 60 years of age). When the population was further subdivided into CD4$^+$CD25$^+$CD45RO$^+$ cells the numbers rose further, 2.2% in young subjects compared to 17% in elderly subjects (P. Moss, unpublished observations). T regulatory cells are thought to be members of a unique lineage of T cells that are selected during the process of T cell differentiation in the thymus. It remains unclear where and when this occurs. Although one possibility is that CD4$^+$CD25$^+$ T cells acquire expression of CD25 and suppressor functions in the thymic medulla, where they recognize self-antigens that are presented on MHC class

II molecules by medullary dendritic cells in a process that is known as altered negative selection [87]. T regulatory cells can mediate their suppression either directly or indirectly. Studies using CD8+ responder T cells have helped to demonstrate the mechanism for direct suppression. CD8+ T cells, in the absence of APC were stimulated using peptide MHC class I tetramers in the presence or absence of CD4+CD25+ T cells. A marked suppression of both proliferation and IFN-γ production was seen in the presence of CD4+CD25+ T cells, leading to the hypothesis that CD4+CD25+ T cells mediate their inhibitory effects by acting on the responder T cells [88]. T regulatory cells have also been shown to act on APC to inhibit the upregulation of expression of co-stimulatory molecules that are required for the activation of CD4+CD25- T cells and so indirectly inhibiting the induction of IL-2 and the proliferation of the CD4+CD25- responder T cells [89, 90]. This indirect mechanism of T regulatory suppression may be important in aging, for the increase in CD4+CD25+ T cells numbers seen in aged subjects may provide further positive feedback signals to enhance the failure of aged cells to undergo the transition from CD44+CD25- to CD44+CD25+.

The molecular pathways that are responsible for mediating T regulatory cell suppression are still unknown. One possible candidate mechanism may be the engagement of a cell-surface molecule on CD8+CD25+ T cells that contains an immunoreceptor tyrosine-based inhibitory motif (ITIM) by a ligand on CD4+CD25+ T regulatory cells. ITIMs are structural motifs containing tyrosine residues that are found in the cytoplasmic tails of several inhibitory receptors. Activation of these inhibitory receptors results in tyrosine phosphorylation, often by SRC-family tyrosine kinases, and are thought to mediate suppression [91]. One possible cell-surface molecule could be the killer cell lectin-like receptor G1 (KLRG1) whose expression has been demonstrated on effector memory CD8+ cell [92]; that is, expression is induced in cells very near the end of their senescent life span. It has also been shown that these KLRG1+CD8+ memory cells were capable of performing effector cell functions but were shown to be severely impaired in their ability to proliferate after antigen stimulation [93]. However, no evidence has been presented yet that this mechanism is operative in CD25-mediated suppression.

Age-dependent alterations to the T cell receptor

During the aging process in mice and humans spontaneous changes occur to the receptor repertoires of T cells and these changes have been shown to be idiosyncratic, that is changes produced by aging in the peripheral T cell repertoire seems to be unique to each individual [94, 95]. The proportion of skewed repertories increases with age in both the CD4+ and CD8+ subset and in particular to CD4+ and CD8+ cells expressing specific TCR Vβ chains [96]. Distortion of the Vβ usage profile during aging is particularly prominent in the CD28- subpopulation of CD8+ cells [95]. This distortion is thought to be caused by the reduced thymic output leading to a reduction in the naïve T cell pool [97] and the accumulation of clonally expanded memory CD8+ T cells as a consequence of prolonged antigenic stimulation through-out life [98]. However the mechanisms that select and maintain such clonal expansions are not yet known. It is thought that repeated or persistent infections with viruses such as influenza, CMV, and EBV may drive responses that result in a large number of T cell clones [99, 100, 101]. CMV seropositivity has been shown to be closely associated with increases in the size of the $CD57^+CD8^+$ T cell pool, which is thought to represent a highly differentiated population of late memory cells [102]. Also CD8 T cells were analyzed from elderly CMV seropositive and CMV seronegative donors and the results indicated a higher number of clonal expansions occurred in seropositive donors, suggesting that CMV may be a factor that drives oligoclonal expansions in old age [103].

Aging leads to a decline in the ability to mount T cells responses to new antigens and to previously encountered antigens. Research has shown that T cells from old mice exhibit multiple defects early in the signal transduction pathway. These include defects in tyrosine-specific protein phosphorylation [104], calcium signal generation [105], Shc-tyrosine phosphorylation [106], and activation of the extracellular signal-related kinase, RAF [107]. However, ZAP-70 protein kinase, a key enzyme in the TCR signaling pathway was shown to have no difference in the activity between activated $CD4^+$ T cells isolated from young and old mice [108], which lead to the hypothesis that aging might alter the accessibility to ZAP-70 of one or more of its substrates. These substrates include the adapter protein LAT and the guanine-nucleotide exchange factor Vav, which has a role in the reorganization of the cytoskeleton and whose phosphorylation is crucial for T cell activation [109]. LAT and Vav also play a role in the formation of the immune synapse, which is a specialized junction between a T cell and APC and consists of a central cluster of T cell receptors surrounded by a ring of adhesion molecules [110]. LAT migration to the immune synapse has been shown to depend on actin polymerization as well as on activity of Src family kinases. However, aging has been shown to reduce the percentage of $CD4^+$ cells that redistribute F-actin to the site of APC contact. These results lead to the suggestion that defects in the ability of T cells from aged donors to move kinase substrates and coupling factors into the T cell/APC contact region may contribute to the reduction in T cell activation and in turn the protective immune response [111].

B cells

The age-related decline in the immune response has often centered around T cells because they are known to orchestrate the immune response to many antigens. For example only 56% of individuals over the age of 65 showed effective immunity following vaccination with influenza compared with 80% of a younger population a reduced function which is often associated with antibody titer [112, 113]. Furthermore analysis of the antibody response in older people shows it to contain antibodies with a lower affinity for the immunizing antigen compared with those produced in a younger person. In addition, the number of auto-antibodies produced during an immune response is greatly increased in the elderly [114].

B cell production and aging

The means of production of B cells is remarkably similar to the production of T cells in that a stem cells precursor showing none of the properties of the final mature lymphocyte, differentiating through a number of defined stages accompanied by rearrangement of segments of the genome to produce a 2 chain antigen receptor which is clonally distributed. In B cells the production of one chain (the H chain in B cells) precedes production of the second (light) chain. It is therefore not surprising to discover that there is an age-associated bottleneck in B cell production in the bone marrow, which is associated with the production of the larger chain. Output of B cells from the bone marrow to join the pool of peripheral B cells therefore declines with age. Despite this reduced production the numbers of B cells in the peripheral pool appears to be unaffected by age [115]. Thus as with the T cell pool it is possible that a decreased output of naive B cells from the marrow would result in the B cell pool being increasingly made up of long-lived B cells which have previously been exposed to antigen. However unlike T cells, B cells have the ability to undergo affinity maturation in the development of an immune response. In the initial phases of a response, B cells whose antigen receptor recognizes the challenging antigen are activated and enter the germinal center within secondary lymphoid tissue (gut, spleen, lymph node) where they proliferate and receptor genes hypermutate altering the specificity of the encoded antibody. The B cells whose receptor shows high affinity for the antigen are saved from apoptotic death and go on to produce high affinity antibody. With age it appears that there is no difference in the level of hypermutation occurring during the germinal center reaction [116, 117]. This would suggest that either the starting population is less than optimal in terms of its affinity, which is a distinct possibility in the aged where the number of naïve B cells is much reduced and the response may be made with B cells from a previous response. Alternatively the selection methods during the germinal centered phase may alter with age.

Concluding comments

The elderly are more likely to suffer from numerous infectious diseases, with a mortality rate often two to three times higher among elderly patients than among younger individuals with the same disease. This disparity between young and old individuals is brought about by the malfunction of the immune system with age. The aged immune system does however retain some residual function, for example, the thymus retains a limited thymopoietic capacity and over half of people over 65 vaccinated with influenza shown effective immunity. The extent to which these and other residual functions can be enhanced remains the focus of many researchers.

References

1.   Kurashima C, Utsuyama M, Kasai M, Ishijima SA, Konno A, Hirokawa K (1995). The role of thymus in the aging of Th cell subpopulations and age-associated alteration of cytokine production by these cells. *Int Immunol.* 7: 97–104.

2.   Utsuyama M, Varga Z, Fukami K, Homma Y, Takenawa T, Hirokawa K (1993). Influence of age on the signal transduction of T cells in mice. *Int Immunol.* 5: 1177–82.

3.   Hobbs MV, Weigle WO, Noonan DJ, *et al.* (1993). Patterns of cytokine gene expression by CD4$^+$ T cells from young and old mice. *J Immunol.* 150: 3602–14.

4.   Grubeck-Loebenstein B, Berger P, Saurwein-Teissl M, Zisterer K, Wick G (1998). No immunity for the elderly. *Nat Med.* 4: 870.

5.   Steger MM, Maczek C, Berger P, Grubeck-Loebenstein B (1996). Vaccination against tetanus in the elderly: do recommended vaccination strategies give sufficient protection. *Lancet* 348: 762.

6.   Franklin RA, Arkins S, Li YM, Kelley KW (1993). Macrophages suppress lectin-induced proliferation of lymphocytes from aged rats. *Mech Ageing Dev.* 67: 33–46.

7.   Zhou D, Chrest FJ, Adler W, Munster A, Winchurch RA (1993). Increased production of TGF-beta and Il-6 by aged spleen cells. *Immunol Lett.* 36: 7–11.

8.   Rosenstein MM, Strausser HR (1980). Macrophage-induced T cell mitogen suppression with age. *J Reticuloendothel Soc.* 27: 159–66.

9.   Goodwin JS (1982). Changes in lymphocyte sensitivity to prostaglandin E, histamine, hydrocortisone, and X irradiation with age: studies in a healthy elderly population. *Clin Immunol Immunopathol.* 25: 243–51.

10. Vercammen C, Ceuppens JL (1987). Prostaglandin E2 inhibits human T-cell proliferation after crosslinking of the CD3-Ti complex by directly affecting T cells at an early step of the activation process. *Cell Immunol.* 104: 24–36.

11. Grewe M (2001). Chronological ageing and photoageing of dendritic cells. *Clin Exp Dermatol.* 26: 608–12.

12. Sato H, Dobashi M (1998). The distribution, immune complex trapping ability and morphology of follicular dendritic cells in popliteal lymph nodes of aged rats. *Histol Histopathol.* 13: 99–108.

13. Maletto BA, Gruppi A, Moron G, Pistoresi-Palencia MC (1996). Age-associated changes in lymphoid and antigen-presenting cell functions in mice immunized with *Trypanosoma cruzi* antigens. *Mech Ageing Dev.* 88: 39–47.

14. Daynes RA, Araneo BA, Ershler WB, Maloney C, Li GZ, Ryu SY (1993). Altered regulation of IL-6 production with normal aging. Possible linkage to the age-associated decline in dehydroepiandrosterone and its sulfated derivative. *J Immunol.* 150: 5219–30.

15. Rich EA, Mincek MA, Armitage KB, *et al.* (1993). Accessory function and properties of monocytes from healthy elderly humans for T lymphocyte responses to mitogen and antigen. *Gerontology* 39: 93–108.

16. Miller RA, Garcia G, Kirk CJ, Witkowski JM (1997). Early activation defects in T lymphocytes from aged mice. *Immunol Rev.* 160: 79–90.

17. Linton PJ, Haynes L, Tsui L, Zhang X, Swain S (1997). From naive to effector–alterations with aging. *Immunol Rev.* 160: 9–18.

18. Jenkinson EJ, Owen JJ, Aspinall R (1980). Lymphocyte differentiation and major histocompatibility complex antigen expression in the embryonic thymus. *Nature* 284: 177–9.

19. Prindull G (1974). Maturation of cellular and humoral immunity during human embryonic development. *Acta Paediatr Scand.* 63: 607–15.

20. Ortman CL, Dittmar KA, Witte PL, Le PT (2002). Molecular characterization of the mouse involuted thymus: aberrations in expression of transcription regulators in thymocyte and epithelial compartments. *Int Immunol.* 14: 813–22.

21. Stutman O (1986). Postthymic T-cell development. *Immunol Rev.* 91: 159–94.

22. Hollander GA, Widmer B, Burakoff SJ (1994). Loss of normal thymic repertoire selection and persistence of autoreactive T cells in graft vs host disease. *J Immunol.* 152: 1609–17.

23. Mackall CL, Punt JA, Morgan P, Farr AG, Gress RE (1998). Thymic function in young/old chimeras: substantial thymic T cell regenerative capacity despite irreversible age-associated thymic involution. *Eur J Immunol.* 28: 1886–93.

24. Steinmann GG (1986). Changes in the human thymus during aging. *Curr Top Pathol.* 75: 43–88.

25. Steinmann GG, Klaus B, Muller-Hermelink HK (1985). The involution of the ageing human thymic epithelium is independent of puberty. A morphometric study. *Scand J Immunol.* 22: 563–75.

26. Wu L, Scollay R, Egerton M, Pearse M, Spangrude GJ, Shortman K (1991). CD4 expressed on earliest T-lineage precursor cells in the adult murine thymus. *Nature* 349: 71–4.

27. Rodewald HR, Kretzschmar K, Swat W, Takeda S (1995). Intrathymically expressed c-kit ligand (stem cell factor) is a major factor driving expansion of very immature thymocytes *in vivo*. *Immunity* 3: 313–19.

28. Godfrey DI, Kennedy J, Suda T, Zlotnik A (1993). A developmental pathway involving four phenotypically and functionally distinct subsets of CD3-CD4-CD8- triple-negative adult mouse thymocytes defined by CD44 and CD25 expression. *J Immunol.* 150: 4244–52.

29. Muegge K, Vila MP, Durum SK (1993). Interleukin-7: a cofactor for V(D)J rearrangement of the T cell receptor beta gene. *Science* 261: 93–5.

30. von Boehmer H, Fehling HJ (1997). Structure and function of the pre-T cell receptor. *Annu Rev Immunol.* 15: 433–52.

31. Jameson SC, Hogquist KA, Bevan MJ (1995). Positive selection of thymocytes. *Annu Rev Immunol.* 13: 93–126.

32. Douek DC, McFarland RD, Keiser PH, *et al.* (1998). Changes in thymic function with age and during the treatment of HIV infection. *Nature* 396: 690–5.

33. Tyan ML (1976). Impaired thymic regeneration in lethally irradiated mice given bone marrow from aged donors. *Proc Soc Exp Biol Med.* 152: 33–5.

34. Tyan ML (1977). Age-related decrease in mouse T cell progenitors. *J Immunol.* 118: 846–51.

35. Eren R, Globerson A, Abel L, Zharhary D (1990). Quantitative analysis of bone marrow thymic progenitors in young and aged mice. *Cell Immunol.* 127: 238–46.

36. Thoman ML (1995). The pattern of T lymphocyte differentiation is altered during thymic involution. *Mech Ageing Dev.* 82: 155–70.

37. Aspinall R (1997). Age-associated thymic atrophy in the mouse is due to a deficiency affecting rearrangement of the TCR during intrathymic T cell development. *J Immunol.* 158: 3037–45.

38. Lacorazza HD, Guevara Patino JA, Weksler ME, Radu D, Nikolic-Zugic J (1999). Failure of rearranged TCR transgenes to prevent age-associated thymic involution. *J Immunol.* 163: 4262–8.

39. Eren R, Zharhary D, Abel L, Globerson A (1988). Age-related changes in the capacity of bone marrow cells to differentiate in thymic organ cultures. *Cell Immunol.* 112: 449–55.

40. Aspinall R, Andrew D (2001). Age-associated thymic atrophy is not associated with a deficiency in the CD44($^{+}$)CD25(–)CD3(–)CD4(–)CD8(–) thymocyte population. *Cell Immunol.* 212: 150–7.

41. Mocchegiani E, Muzzioli M, Cipriano C, Giacconi R (1998). Zinc, T-cell pathways, aging: role of metallothioneins. *Mech Ageing Dev.* 106: 183–204.

42. Kelley KW, Brief S, Westly HJ, *et al.* (1986). GH3 pituitary adenoma cells can reverse thymic aging in rats. *Proc Natl Acad Sci USA* 83: 5663–7.

43. Murphy WJ, Rui H, Longo DL (1995). Effects of growth hormone and prolactin immune development and function. *Life Sci.* 57: 1–14.

44. Taub DD, Tsarfaty G, Lloyd AR, Durum SK, Longo DL, Murphy WJ (1994). Growth hormone promotes human T cell adhesion and migration to both human and murine matrix proteins *in vitro* and directly promotes xenogeneic engraftment. *J Clin Invest.* 94: 293–300.

45. Pierpaoli W, Regelson W (1994). Pineal control of aging: effect of melatonin and pineal grafting on aging mice. *Proc Natl Acad Sci USA* 91: 787–91.

46. Mocchegiani E, Bulian D, Santarelli L, *et al.* (1994). The immuno-reconstituting effect of melatonin or pineal grafting and its relation to zinc pool in aging mice. *J Neuroimmunol.* 53: 189–201.

47. Goya RG, Gagnerault MC, Sosa YE, Bevilacqua JA, Dardenne M (1993). Effects of growth hormone and thyroxine on thymulin secretion in aging rats. *Neuroendocrinology* 58: 338–43.

48. Mackall CL, Punt JA, Morgan P, Farr AG, Gress RE (1998). Thymic function in young/old chimeras: substantial thymic T cell regenerative capacity despite irreversible age-associated thymic involution. *Eur J Immunol.* 28: 1886–93.

49. Godfrey DI, Kennedy J, Mombaerts P, Tonegawa S, Zlotnik A (1994). Onset of TCR-beta gene rearrangement and role of TCR-beta expression during CD3-CD4-CD8-thymocyte differentiation. *J Immunol.* 152: 4783–92.

50. Oosterwegel MA, Haks MC, Jeffry U, Murray R, Kruisbeek AM (1997). Induction of TCR gene rearrangements in uncommitted stem cells by a subset of IL-7 producing, MHC class-II-expressing thymic stromal cells. *Immunity* 6: 351–60.

51. Sempowski GD, Gooding ME, Liao HX, Le PT, Haynes BF (2002). T cell receptor excision circle assessment of thymopoiesis in aging mice. *Mol Immunol.* 38: 841–8.

52. Pido-Lopez J, Imami N, Andrew D, Aspinall R (2002). Molecular quantitation of thymic output in mice and the effect of IL-7. *Eur J Immunol.* 32: 2827–36.

53. Sempowski GD, Hale LP, Sundy JS, *et al.* (2000). Leukemia inhibitory factor, oncostatin M, IL-6, and stem cell factor mRNA expression in human thymus increases with age and is associated with thymic atrophy. *J Immunol.* 164: 2180–7.

54. Godfrey DI, Zlotnik A, Suda T (1992). Phenotypic and functional characterization of c-kit expression during intrathymic T cell development. *J Immunol.* 149: 2281–5.

55. Betz UA, Muller W (1998). Regulated expression of gp130 and IL-6 receptor alpha chain in T cell maturation and activation. *Int Immunol.* 10: 1175–84.

56. Vella AT, Mitchell T, Groth B, *et al.* (1997). CD28 engagement and proinflammatory cytokines contribute to T cell expansion and long-term survival *in vivo*. *J Immunol.* 158: 4714–20.

57. Negoro S, Hara H, Miyata S, *et al.* (1986). Mechanisms of age-related decline in antigen-specific T cell proliferative response: IL-2 receptor expression and recombinant IL-2 induced proliferative response of purified Tac-positive T cells. *Mech Ageing Dev.* 36: 223–41.

58. Haynes L, Linton PJ, Eaton SM, Tonkonogy SL, Swain SL (1999). Interleukin 2, but not other common gamma chain-binding cytokines, can reverse the defect in generation of CD4 effector T cells from naive T cells of aged mice. *J Exp Med.* 190: 1013–24.

59. Mbawuike IN, Lange AR, Couch RB (1993). Diminished influenza A virus-specific MHC class I-restricted cytotoxic T lymphocyte activity among elderly persons. *Viral Immunol.* 6: 55–64.

60.   Keren G, Segev S, Morag A, Zakay-Rones Z, Barzilai A, Rubinstein E (1988). Failure of influenza vaccination in the aged. *J Med Virol.* 25: 85–9.

61.   Powers DC (1994). Effect of age on serum immunoglobulin G subclass antibody responses to inactivated influenza virus vaccine. *J Med Virol.* 43: 57–61.

62.   Song H, Price PW, Cerny J (1997). Age-related changes in antibody repertoire: contribution from T cells. *Immunol Rev.* 160: 55–62.

63.   Swain SL, Hu H, Huston G (1999). Class II-independent generation of CD4 memory T cells from effectors. *Science* 286: 1381–3.

64.   Po JL, Gardner EM, Anaraki F, Katsikis PD, Murasko DM (2002). Age-associated decrease in virus-specific CD8$^+$ T lymphocytes during primary influenza infection. *Mech Ageing Dev.* 123: 1167–81.

65.   Swain SL, Croft M, Dubey C, *et al.* (1996). From naive to memory T cells. *Immunol Rev.* 150: 143–67.

66.   Paganelli R, Scala E, Rosso R, *et al.* (1996). A shift to Th0 cytokine production by CD4$^+$ cells in human longevity: studies on two healthy centenarians. *Eur J Immunol.* 26: 2030–4.

67.   Cakman I, Rohwer J, Schutz RM, Kirchner H, Rink L (1996). Dysregulation between TH1 and TH2 T cell subpopulations in the elderly. *Mech Ageing Dev.* 87: 197–209.

68.   Sandmand M, Bruunsgaard H, Kemp K, *et al.* (2002). Is ageing associated with a shift in the balance between Type 1 and Type 2 cytokines in humans? *Clin Exp Immunol.* 127: 107–14.

69.   Miller RA (1991). Accumulation of hyporesponsive, calcium extruding memory T cells as a key feature of age-dependent immune dysfunction. *Clin Immunol Immunopathol.* 58: 305–17.

70.   McEachern MJ, Krauskopf A, Blackburn EH (2000). Telomeres and their control. *Annu Rev Genet.* 34: 331–58.

71.   Blackburn EH (1992). Telomerases. *Annu Rev Biochem.* 61: 113–29.

72.   Nakamura TM, Cech TR (1998). Reversing time: origin of telomerase. *Cell* 92: 587–90.

73.   Frenck RW Jr, Blackburn EH, Shannon KM (1998). The rate of telomere sequence loss in human leukocytes varies with age. *Proc Natl Acad Sci USA* 95: 5607–10.

74.   Iwama H, Ohyashiki K, Ohyashiki JH, *et al.* (1998). Telomeric length and telomerase activity vary with age in peripheral blood cells obtained from normal individuals. *Hum Genet.* 102: 397–402.

75.   Weng NP, Levine BL, June CH, Hodes RJ (1995). Human naive and memory T lymphocytes differ in telomeric length and replicative potential. *Proc Natl Acad Sci USA* 92: 11091–4.

76.   Zhang X, Fujii H, Kishimoto H, LeRoy E, Surh CD, Sprent J (2002). Aging leads to disturbed homeostasis of memory phenotype CD8(+). cells. *J Exp Med.* 195: 283–93.

77.   Zhang X, Sun S, Hwang I, Tough DF, Sprent J (1998). Potent and selective stimulation of memory-phenotype CD8$^+$ T cells *in vivo* by IL-15. *Immunity* 8: 591–9.

78.   Ku CC, Murakami M, Sakamoto A, Kappler J, Marrack P (2000). Control of homeostasis of CD8$^+$ memory T cells by opposing cytokines. *Science* 288: 675–8.

79.   Kennedy MK, Glaccum M, Brown SN, *et al.* (2000). Reversible defects in natural killer and memory CD8 T cell lineages in interleukin 15-deficient mice. *J Exp Med.* 191: 771–80.

80.   Akbar AN, Salmon M, Savill J, Jαnossy G (1993). A possible role for bcl-2 in regulating T-cell memory – a 'balancing act' between cell death and survival. *Immunol Today* 14: 526–32.

81.   Pilling D, Akbar AN, Girdlestone J, *et al.* (1999). Interferon-beta mediates stromal cell rescue of T cells from apoptosis. *Eur J Immunol.* 29: 1041–50.

82. Grayson JM, Zajac AJ, Altman JD, Ahmed R (2000). Cutting edge: increased expression of Bcl-2 in antigen-specific memory CD8$^+$ T cells. *J Immunol.* 164: 3950–4.

83. Garcia S, DiSanto J, Stockinger B (1999). Following the development of a CD4 T cell response *in vivo*: from activation to memory formation. *Immunity* 11: 163–71.

84. Suzuki H, Zhou YW, Kato M, Mak TW, Nakashima I (1999). Normal regulatory alpha/beta T cells effectively eliminate abnormally activated T cells lacking the interleukin 2 receptor beta *in vivo*. *J Exp Med.* 190: 1561–72.

85. Thornton AM, Shevach EM (1998). CD4$^+$CD25$^+$ immunoregulatory T cells suppress polyclonal T cell activation *in vitro* by inhibiting interleukin 2 production. *J Exp Med.* 188: 287–96.

86. Takahashi T, Kuniyasu Y, Toda M, *et al.* (1998). Immunologic self-tolerance maintained by CD25$^+$CD4$^+$ naturally anergic and suppressive T cells: induction of autoimmune disease by breaking their anergic/suppressive state. *Int Immunol.* 10: 1969–80.

87. Shevach EM (2000). Regulatory T cells in autoimmmunity. *Annu Rev Immunol.* 18: 423–49.

88. Piccirillo CA, Shevach EM (2001). Cutting edge: control of CD8$^+$ T cell activation by CD4$^+$CD25$^+$ immunoregulatory cells. *J Immunol.* 167: 1137–40.

89. Thornton AM, Shevach EM (2000). Suppressor effector function of CD4$^+$CD25$^+$ immunoregulatory T cells is antigen nonspecific. *J Immunol.* 164: 183–90.

90. Cederbom L, Hall H, Ivars F (2000). CD4$^+$CD25$^+$ regulatory T cells down-regulate co-stimulatory molecules on antigen-presenting cells. *Eur J Immunol.* 30: 1538–43.

91. Sinclair NR (2000). Immunoreceptor tyrosine-based inhibitory motifs on activating molecules. *Crit Rev Immunol.* 20: 89–102.

92. Beyersdorf NB, Ding X, Karp K, Hanke T (2001). Expression of inhibitory "killer cell lectin-like receptor G1" identifies unique subpopulations of effector and memory CD8 T cells. *Eur J Immunol.* 31: 3443–52.

93. Voehringer D, Blaser C, Brawand P, Raulet DH, Hanke T, Pircher H (2001). Viral infections induce abundant numbers of senescent CD8 T cells. *J Immunol.* 167: 4838–43.

94. Callahan JE, Kappler JW, Marrack P (1993). Unexpected expansions of CD8-bearing cells in old mice. *J Immunol.* 151: 6657–69.

95. Posnett DN, Sinha R, Kabak S, Russo C (1994). Clonal populations of T cells in normal elderly humans: the T cell equivalent to "benign monoclonal gammapathy". *J Exp Med.* 179: 609–18.

96. Mosley RL, Koker MM, Miller RA (1998). Idiosyncratic alterations of TCR size distributions affecting both CD4 and CD8 T cell subsets in aging mice. *Cell Immunol.* 189: 10–18.

97. Utsuyama M, Kasai M, Kurashima C, Hirokawa K (1991). Age influence on the thymic capacity to promote differentiation of T cells: induction of different composition of T cell subsets by aging thymus. *Mech Ageing Dev.* 58: 267–77.

98. Schwab R, Szabo P, Manavalan JS, *et al.* (1997). Expanded CD4$^+$ and CD8$^+$ T cell clones in elderly humans. *J Immunol.* 158: 4493–9.

99. Lehner PJ, Wang EC, Moss PA, *et al.* (1995). Human HLA-A0201-restricted cytotoxic T lymphocyte recognition of influenza A is dominated by T cells bearing the V beta 17 gene segment. *J Exp Med.* 181: 79–91.

100. Argaet VP, Schmidt CW, Burrows SR, *et al.* (1994). Dominant selection of an invariant T cell antigen receptor in response to persistent infection by Epstein-Barr virus. *J Exp Med.* 180: 2335–40.

101. Weekes MP, Wills MR, Mynard K, Carmichael AJ, Sissons JG (1999). The memory cytotoxic T-lymphocyte (CTL) response to human cytomegalovirus infection contains

individual peptide-specific CTL clones that have undergone extensive expansion *in vivo*. *J Virol.* 73: 2099–108.

102. Wang EC, Taylor-Wiedeman J, Perera P, Fisher J, Borysiewicz LK (1993). Subsets of CD8⁺, CD57⁺ cells in normal, healthy individuals: correlations with human cytomegalovirus (HCMV) carrier status, phenotypic and functional analyses. *Clin Exp Immunol.* 94: 297–305.

103. Khan N, Shariff N, Cobbold M, *et al.* (2002). Cytomegalovirus seropositivity drives the CD8 T cell repertoire toward greater clonality in healthy elderly individuals. *J Immunol.* 169: 1984–92.

104. Shi J, Miller RA (1992). Tyrosine-specific protein phosphorylation in response to anti-CD3 antibody is diminished in old mice. *J Gerontol.* 47: B147–53.

105. Grossmann A, Maggio-Price L, Jinneman JC, Rabinovitch PS (1991). Influence of aging on intracellular free calcium and proliferation of mouse T-cell subsets from various lymphoid organs. *Cell Immunol.* 135: 118–31.

106. Ghosh J, Miller RA (1995). Rapid tyrosine phosphorylation of Grb2 and Shc in T cells exposed to anti-CD3, anti-CD4, and anti-CD45 stimuli: differential effects of aging. *Mech Ageing Dev.* 80: 171–87.

107. Gorgas G, Butch ER, Guan KL, Miller RA (1997). Diminished activation of the MAP kinase pathway in CD3-stimulated T lymphocytes from old mice. *Mech Ageing Dev.* 94: 71–83.

108. Garcia GG, Miller RA (1998). Increased Zap-70 association with CD3zeta in CD4 T cells from old mice. *Cell Immunol.* 190: 91–100.

109. Clements JL, Boerth NJ, Lee JR, Koretzky GA (1999). Integration of T cell receptor-dependent signaling pathways by adapter proteins. *Annu Rev Immunol.* 17: 89–108.

110. Grakoui A, Bromley SK, Sumen C, *et al.* (1999). The immunological synapse: a molecular machine controlling T cell activation. *Science* 285: 221–7.

111. Tamir A, Eisenbraun MD, Garcia GG, Miller RA (2000). Age-dependent alterations in the assembly of signal transduction complexes at the site of T cell/APC interaction. *J Immunol.* 165: 1243–51.

112. Muszkat M, Friedman G, Schein MH, *et al.* (2000). Local SIgA response following administration of a novel intranasal inactivated influenza virus vaccine in community residing elderly. *Vaccine* 18: 1696–9.

113. Bridges CB, Thompson WW, Meltzer MI, *et al.* (2000). Effectiveness and cost-benefit of influenza vaccination of healthy working adults: a randomized controlled trial. *J Am Med Assoc.* 284: 1655–63.

114. Candore G, Di Lorenzo G, Mansueto P, *et al.* (1997). Prevalence of organ-specific and non organ-specific autoantibodies in healthy centenarians. *Mech Ageing Dev.* 94: 183–90.

115. Weksler ME, Goodhardt M, Szabo P (2002). The effect of age on B cell development and humoral immunity. *Springer Semin Immunopathol.* 24: 35–52.

116. Banerjee M, Mehr R, Belelovsky A, Spencer J, Dunn-Walters DK (2002). Age- and tissue-specific differences in human germinal center B cell selection revealed by analysis of IgVH gene hypermutation and lineage trees. *Eur J Immunol.* 32: 1947–57.

117. Dunn-Walters DK, Banerjee M, Mehr R (2003). Effects of age on antibody affinity maturation. *Biochem Soc Trans.* 31: 447–8.

Aging of the Nervous System

M.J. Saffrey

Dept. of Biological Sciences, Open University, Walton Hall, Milton Keynes, MK7 6AA, UK

Introduction

To most of us, the possible decline in memory, cognition and motor control is probably the most alarming prospect of impending old age. The changes that occur in the nervous system during aging can have severe consequences. The variation between individuals is great; for a few fortunate individuals, cognitive and motor function may be relatively unaffected even into very old age, while most suffer only mild decline. Many however, are affected by one of the neurodegenerative diseases, particularly Alzheimer's disease and Parkinson's disease. The distressing effects of these disorders and the increase in the number of sufferers as the aging population grows has made study of these diseases a major focus for gerontologists, but despite extensive study, their causes still remain to be firmly established. For many years, a full understanding of the pathology of these diseases has been hampered by the lack of knowledge about what constitutes "normal" or "healthy" aging of the nervous system.

The nervous system is the most complex system of the body, and analysis of changes that occur in it during aging is fraught with complications. The cellular architecture of different parts of the nervous system, such as the different brain areas, the spinal cord and the peripheral ganglia are distinct, and the neurones, and to a lesser extent, the glial cells, in these different regions, although sharing many common features also exhibit differing morphological, molecular and functional properties. In many cases, how this organization subserves complex neural functions such as consciousness and memory is still being unravelled. Together with the obvious problems of obtaining nervous system tissue from humans of different ages, and the known variation between individuals, these properties make elucidation of the changes that are common to "normal" nervous system aging a particular challenge to gerontologists.

R. Aspinall (ed.), Aging of Organs and Systems, 243–270.
© 2003 *Kluwer Academic Publishers. Printed in Great Britain.*

With increasingly sophisticated diagnosis, new methods of morphometric and molecular analysis and carefully controlled studies, the range of "normal" changes that occurs in the nervous system during aging is now beginning to be understood.

Evidence suggests that different neuronal populations and brain regions are differentially affected during normal aging. Particular types of cells in some regions are selectively vulnerable and these are often those affected in the neurodegenerative diseases, leading to the well-known symptoms of memory loss and dementia of Alzheimer's disease (AD) and the movement disorders of Parkinson's disease (PD), and motoneurone disease or amyotrophic lateral sclerosis (ALS). What is less well known however is that neural changes are also implicated in age-associated dysfunction of other systems, such as the cardiovascular, digestive and urinary systems. Some of the autonomic neurones that regulate the function of the visceral organs have been found to exhibit changes during "normal" aging. At a time when, in the developed world, average human lifespan has increased dramatically, the neurodegenerative diseases have become a major cause of morbidity in the elderly. The less publicised, but nonetheless common and distressing symptoms of autonomic dysfunction are also a major cause of morbidity.

The aim of this chapter is to review current knowledge of the changes that take place in some of the different regions of the nervous system during normal aging, considering not only the brain and spinal cord, but also the peripheral nervous system. There is not space to include an exhaustive description of the aging of all regions of the nervous system, so other reviews are referred to where appropriate. The neurodegenerative diseases will be touched upon, but not discussed in detail, again some recent reviews are referred to in the text. For detailed texts on aging of the nervous system, the reader is referred to Dani *et al.* [1] and Hof and Mobbs [2].

It is assumed that most readers will be familiar with the organization of the nervous system. For readers who are not, and as an "aide memoir" for those who are, the chapter will begin with a brief summary of the organization of the nervous system and its cellular components. Detailed information can be found in Purves *et al.* [3].

Overview of nervous system structure and function

The organization of the nervous system

The mammalian nervous system is composed of the central nervous system (CNS), that is, the brain and spinal cord, and the peripheral nervous system (PNS), which is usually defined as all neural cells and parts of neural cells that are present in the rest of the body. Each of these two major divisions has component subdivisions with specific organization and functions. Simplified illustrations of these divisions of the nervous system are shown in Figures 1 and 2.

The major functions of the different regions of the nervous system are also given in Table 1. There is not space to describe all regions and their functions here. Rather, a brief description of the areas focused on in this chapter is provided. For more detailed information, the reader is referred to Purves *et al.* [3].

The cells of the nervous system
The nervous system is comprised of two major cell types, neurones and glial cells. There are (of course), many types of neurone in terms of position, shape, projections, neurochemical properties and functions.

Neurones have a characteristic morphology, with a single long process known as an axon, and many long branched processes known as dendrites (see Figure 3). The number, length and degree of branching of these processes vary according to the type of neurone. Electrical impulses, the action potentials, pass along the axon until they reach the nerve terminals, which form specialized connections, or *synapses*, with target cells. At the nerve terminals, action potentials cause depolarization of the neuronal membrane which results in release of signaling molecules, *neurotransmitters,* into the synaptic cleft (a very narrow space between the pre and postsynaptic cells). Neurotransmitters then bind to specific receptor molecules on the postsynaptic target cells. Target cells may be other neurones, as is the case for most brain and some peripheral neurones, or effector cells such as muscle or glands, in the PNS. Different signals may impinge on a target cell, and these are integrated such that it may be stimulated to generate an action potential (or in the case of a muscle cell, to contract).

The main types of glial cells are the *astrocytes* and *oligodendrocytes* of the CNS, and the glia of the PNS, which include *satellite cells* and astrocyte-like *enteric glial cells* that reside in peripheral and enteric ganglia respectively, and both myelin and non-myelin forming peripheral glia, such as the *Schwann cells* that are intimately involved with the processes of motoneurones and sensory neurones within the peripheral nerves. Another kind of cell, the *microglia*, derived from the immune system, are also present in the CNS and play an essential role in inflammatory responses.

Astrocytes can have differing morphologies but tend to have long processes that together form a matrix in which neurones lie. Oligodendrocytes have extensive cytoplasmic sheets that wrap around nerve processes and contain a fatty substance known as myelin that forms an insulating layer around nerve fibres (see Figure 3). Schwann cells in the PNS perform a similar function.

A final group of cells that cannot be excluded from consideration in study of CNS aging are the endothelial cells of the blood vessels that supply the brain, which together with specialized astrocytes form the blood–brain barrier, the integrity of which is of major clinical importance.

Although, for obvious reasons, research has tended to focus on neuronal aging, the cells described above are, of course functionally interdependent. Hence, in order to fully understand aging of even a single part of the nervous system, study of the changes that occur in non-neuronal cells of the nervous system is of crucial importance.

Changes in the CNS during aging

Memory, cognition and motor control are among the key neural functions that are affected during aging. The centres and circuits that are primarily involved in these CNS functions are the cerebral cortex, the hippocampus, the basal ganglia (notably

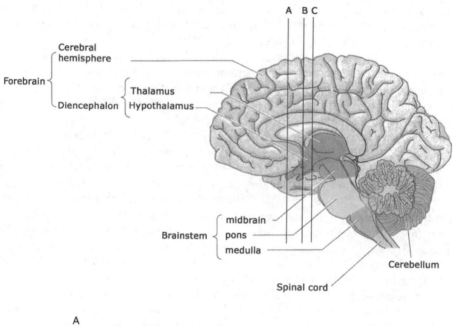

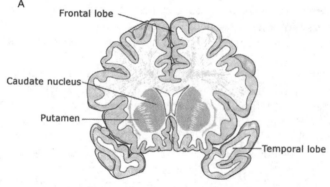

Figure 1. *The human brain. Simplified diagram illustrating some of the main brain regions. Transverse sections taken at positions A, B and C are shown. Functions of the regions shown, together with those of some other brain regions referred to in this chapter are given in Table 1.*

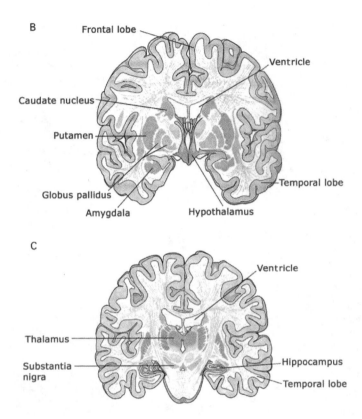

Figure 1 *(continued)*

the nigro-striatal system) and the spinal cord. These areas are also affected in the neurodegenerative diseases, and it is thus not surprising that they have been the focus of most attention in studies of normal aging. For many years the debate has been whether the dramatic pathological changes that underlie AD and PD occur, but to a lesser extent, during normal aging and if these changes are the cause of the mild cognitive and motor decline seen in many normal aged individuals. It is only very recently, with the development of more sophisticated methods of analysis that the answers to these questions are beginning to emerge. The consensus now is that AD and PD are pathologies distinct from the normal changes that take place during aging, although the neurones that they affect may be selectively vulnerable to other changes during aging and in injury [4].

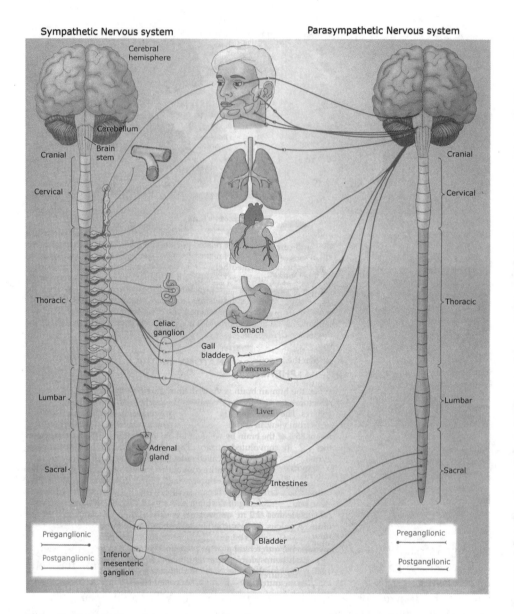

Sympathetic Nervous system

Parasympathetic Nervous system

Figure 2. *Organization of the human peripheral nervous system, focusing on the autonomic system, which regulates visceral functions. The motor nerves that innervate skeletal muscle are not shown; these leave the spinal cord along with the preganglionic sympathetic fibres shown on the left-hand side of the figure. (Redrawn with permission from Ref 3).*

Table 1. *Functions of major regions of the nervous system (note: this table is intended to provide an overview and does not include all brain nuclei)*

Region	Function
Central nervous system	
Forebrain	
Cerebral hemispheres	
– Cortex	Co-ordination of mental activity, conscious activities, controls voluntary movement
– Basal ganglia	Co-ordination of movements of skeletal muscle,
Striatum	relays information from cortical regions via thalamus
Substantia nigra	to motor cortex
– Limbic system	
Hippocampus	Short term memory
Amygdala	Emotion
Diencephalon	Links midbrain and cortex
– Thalamus	Relays and processes sensory information to cortex
– Hypothalamus	Integrates autonomic and endocrine systems
Cerebellum	Processing center involved in control of movements
Brainstem	
Midbrain	Visual reflexes, eye movements
Pons	Respiratory functions and relays information to higher centers
Medulla oblongata	Connects spinal cord with pons, and centers that regulate autonomic functions such as heart rate
Spinal cord	Conveys sensory information to brain, and signals from brain to peripheral organs e.g., skeletal muscle
Peripheral nervous system	
Primary sensory neurons	Detect and convey sensory information to the CNS
Peripheral nerves	Groups of nerve fibres consisting of the processes of motor and sensory neurones outside the spinal cord
Autonomic nervous system (ANS)	
Parasympathetic	Regulates functions of the body organs
Sympathetic	
Enteric	Regulates intestinal functions

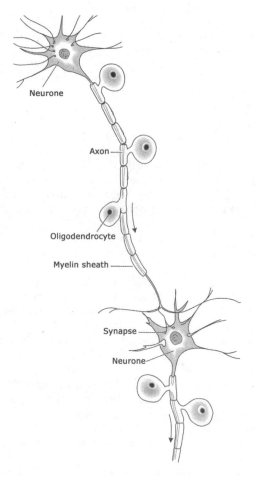

Figure 3. *Diagrammatic illustration of neurones and oligodendrocytes. (Redrawn with permission from Arking, Biology of Aging, Sinauer Associates, Sunderland, MA, USA).*

Macroscopic changes in the aging CNS
Analysis of changes in the overall volume and weight of the brain with age are hampered by a number of factors. For example, comparison of very old and young brains is affected by *secular acceleration*, that is, that there is an increase in the height of individuals in succeeding generations, and, as body size increases, so does body (and brain) weight [5, 6]. Moreover, there is great individual variation in these parameters, as was clearly demonstrated by Haug and colleagues, who also found no significant reduction in brain weight in individuals between 19 and 65 years of age, and that there was only a small decrease thereafter [see ref. 5]. With respect to volume, comparison of measurements of subjects of both sexes below and above 65 years of age showed that there is about a 10% decrease in the volume of the entire brain (excluding the ventricles, which show a large increase in size [8]).

Different brain regions were found to be differentially affected in this study; the grey matter of the cerebral cortex diminishing by about 5%, different cortical regions showing reductions of between 2 and 9% in volume with the exception of the parietal cortex, which showed an increase in volume. In general the motor cortex showed a larger diminution than the sensory cortex. The white matter showed the greatest reduction in volume, of 17% [5].

Studies of primate brain have revealed similar changes. For example, magnetic resonance imaging (MRI) reveals an increase in the size of the ventricles, a reduction in the area of the white matter (nerve fibre tracts containing much myelin), but no change in the grey matter (Figure 4).

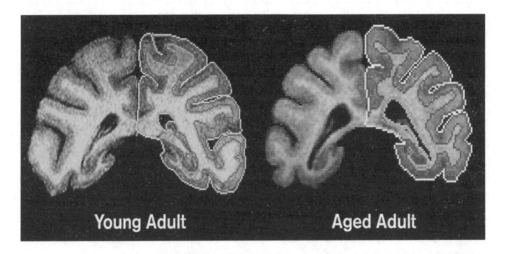

Figure 4. *MRI images comparing the brains of young (5 year old) and aged (32 year old) rhesus monkeys. The increase in the area of the ventricles, and decrease in the area of the white matter is clearly visible. From: I. Wickelgren, Science (1996) 273:48–50*

Neuronal changes in the CNS during aging
Neuronal numbers

For many years, it was believed that neuronal loss was an inevitable consequence of aging, and the cause of non-pathological as well as pathological cognitive decline. Recently however, there has been a major change in opinion and it is now widely accepted that there is no major reduction in neuronal numbers in the CNS during normal aging [see refs. 4–7]. This stability of neuronal numbers during aging has been established in many brain areas, of several species [for detailed review, see ref. 7]. Most recent studies have shown no or only small neuronal losses, the magnitude of which are often less than the normal variation in neuronal numbers between individuals.

The earliest report that neuronal numbers in the human brain diminish during aging was that of Ellis in 1919 and most, but not all studies over the ensuing 60–65 years supported this observation [reviewed in refs 6 and 7]. Losses of between 20–50% of neurones in various areas of the human brain were described. Data obtained since the mid-1980s, however, although not entirely consistent, have indicated that any losses that may occur are small, compared to the losses seen in AD, for example. Most studies that do show a loss reveal that it is less than about 10% of neurones in any area, but since inter-individual variation may be about 100% in some brain areas, this loss is not significant [see ref. 6]. There were three main reasons why the early data proved to be such dramatic overestimates of neuronal loss during normal aging. First, most early studies measured neuronal density rather than absolute numbers of neurones. Haug and colleagues established that, during the processing of brain samples for histological analysis, the brains of young individuals shrink more than those of older individuals (most probably due to the higher myelin content in the latter). This difference led to an apparent increased density of neurones in samples from young brains. Second, more accurate methods of counting, using a technique known as unbiased stereology, were introduced [8–10for reviews]. Finally, the early work on human samples is likely to have been confounded by inclusion of samples from individuals with undiagnosed early stage AD, since it is likely that neuronal losses precede the overt symptoms of the disease [11–13].

Despite the introduction of unbiased stereological methods, there is still considerable variation in the estimates of total neuronal numbers in different brain regions, and also in any changes that may occur during normal aging. Analysis of changes in neuronal number in aging humans are hampered by the variation not only between individuals, but also in the condition of the samples obtained for analysis. Studies using animals, although enabling greater standardization and control for epigenetic influences, may not relate closely to the human situation, especially for rodents. Study of non-human primates, which display cognitive changes, similar to some of those exhibited by humans [14], has many advantages. Even in the study of animals, however, differences in sampling of the areas chosen for counting may result in conflicting results.

With these caveats in mind, analysis of the available data from studies that have utilized unbiased stereological methods indicates that small neuronal losses occur during normal aging in only few areas of the human CNS. These are the substantia nigra, where dopaminergic neurones are reduced in normal aging [15–17], the spinal cord, where a loss of motoneurones has been described by several authors [see ref. 7] and possibly also the Purkinje cells of the cerebellum, although the data for this area are conflicting [see ref. 7]. Data on the hippocampus are also conflicting, but any non-pathological neuronal losses that may occur in this brain region during aging are insignificant when compared to the major losses seen in AD.

In support of the evidence that neuronal numbers do not normally decrease markedly with age, electron microscopic studies have failed to reveal evidence of degenerating neurones [see ref. 6]. These studies, however, have revealed that a number of other, more subtle changes occur in the brain during normal aging; these are very likely the causes of cognitive decline that occurs in non-pathological aging, and are described below (see sections on "Neuronal size, morphology and synaptic density" and "Histological changes"). First, however, two other issues with respect to neuronal numbers and the aging brain must be considered, briefly.

Neurogenesis in the adult brain: In recent years much attention has been given to the discovery that a population of undifferentiated precursor cells, known as *neural stem cells* exists in the adult brain. The idea that new neurones might be generated in the adult brain was not new, first being suggested in the late nineteenth century [see ref. 7] but definitive evidence for the existence of these cells has only recently been forthcoming. Neural stem cells are relatively small in number, and probably only play a significant role in the replenishing of neurones in the adult mammalian brain in a few areas, notably the dentate gyrus of the hippocampus, which plays a crucial role in memory formation, and the olfactory bulb. There are species differences in the extent and pattern of neurogenesis in the dentate gyrus, and the process is affected by a number of epigenetic factors such as exercise and stress. It has been estimated that some 9000 new cells are generated in the adult rat dentate gyrus each day, of which half develop into neurones [18]. This continued neurogenesis, even in the dentate gyrus and olfactory bulb, however, does not appear to result in an increase in neuronal numbers in these regions, although some studies have reported a small increase in the dentate gyrus. This relative stability of neuronal numbers in regions where significant numbers of new neurones are generated indicates that neuronal losses also occur and that there is a turnover of neurones in these areas, which are hence in a state of dynamic equilibrium.

The formation of new neurones in other regions of the adult CNS including the cortex [19] has been shown and the existence of neural precursors in several regions of the CNS and PNS has also been demonstrated. The significance of these findings remains to be determined, for example, it is possible that neurogenesis plays an important role in repair after injury. Recent development of techniques for the isolation of neural stem cells from several regions of the nervous system, and the ability to grow them in cell culture to provide a potential source of cells for cell replacement in neurodegenerative diseases or after injury, has raised much hope and excitement [20]. A key issue, however, of how neural stem cells and the process of adult neurogenesis are affected during normal aging have not yet been addressed in detail [21].

What is the link between neuronal loss and functional impairment? The clear functional consequences of the major neuronal losses that occur in AD, PD and motoneurone disease are indisputable. As described above, significant neuronal losses in normal aging are now thought to be an unlikely general occurrence. That is not to say, however, that some individuals exhibit some losses, which may be of a local nature, and these could lead to cognitive decline. In this respect, detailed studies on animals, particularly non-human primates, in which cognitive status as well as neuronal numbers can be measured are of importance. Such studies have shown that cognitive decline can occur without measurable changes in neuronal number [22]. What then are the causes of such decline?

Neuronal size, morphology and synaptic density

The size of neurones has long been known to be smaller in the aging brain, due to atrophy of large neurones [see refs 5, 23]. A number of detailed analyses of Golgi impregnated neurones have shown that there is a reduction in the extent of the dendritic tree of aging cortical neurones [e.g., ref. 24]. Importantly, a significant reduction in the number of dendritic spines, the site of most synaptic inputs, has been consistently reported in the cortex, in both light and electron microscopic studies [25–28]. Losses of 9–10% of total dendritic length and 50% of spines have been reported in human cortex [26], and similar losses of spines in monkey cortex have been described [27]. Most recently, in an elegant study of dye-filled cortical neurones, Page *et al.* [28] have confirmed these observations. Data on synaptic changes in the dentate gyrus, however, have failed to demonstrate a loss of either spines or synapses in aging monkeys [29], indicating that age-associated changes may be projection specific.

Age-associated morphological changes in the neurones of the basal ganglia have also been described. Striatal neurones have been most studied, the medium spiny GABAergic neurones in this region of the cat brain have been found to exhibit reduction in spine densities of some 40–49% [29–31]. Dendritic length of medium spiny neurones in the mouse striatum was also found to be reduced in motor-impaired aging animals [32]. Neurones of the substantia nigra, which project to the medium spiny neurones, have been less studied but are also reported to exhibit age-associated morphological changes such as stunted neurites [33].

In support of the morphological evidence that synaptic density decreases in normal aging, molecular markers of synapses, such as the presynaptic vesicle protein *synaptophysin,* have been shown to be altered in aging rats [34]. Clearly, a loss of synapses in normally aging brain may have profound functional consequences and is likely to at least contribute to age-associated cognitive decline.

Histological changes

General age-associated changes: lipofuscin: A characteristic general change that occurs within neurones as well as other cell types during aging is accumulation of the pigment, known as lipofuscin. Lipofuscin is in fact the name given to a heterogeneous group of conjugates formed by interaction between products of lipid peroxidation, protein degradation and non-enzymatic glycosylation of long-lived molecules such as some proteins and nucleic acids [23]. Lipofuscin accumulation within cells appears to be a function of metabolic activity, and free radicals have been implicated in its formation. Lipofuscin can sometimes be autofluorescent and forms "granules" that appear dark under the electron microscope. The effects of lipofuscin accumulation are not established, although it has been suggested that it may impair cellular function, and that antioxidants may reduce its formation.

Plaques and tangles occur in normally aging brain: The accumulation of neuritic or senile plaques (which consist of dystrophic neurites and glial processes sometimes with a central amyloid core) and neurofibrillary tangles (which consist of accumulations of abnormal neuronal cytoskeletal components) are a hallmark of AD. These structures are, however, also found in normally aging brain and the current view is that their occurrence is not a prelude to the severe changes that occur in AD, which are accompanied by neuronal loss and disruption of key neuronal circuits [see refs 4, 35]. There is not space to describe the properties of these structures, or the pathology of AD in detail here, but evidence suggests that particular groups of neurones are vulnerable and that these neurones share common properties with neurones that are affected in other neurodegenerative disorders including PD and ALS [see ref. 4]. One such property is a high level of neurofilament protein, particularly non-phosphory-lated neurofilament protein, and there is now considerable evidence that several different neuronal populations that express high levels of this type of neurofilament protein are vulnerable to degeneration [reviewed in ref. 35].

Neurotransmitter systems
Many studies to investigate age-associated changes in the different neurotransmitter systems of the brain have been performed, and a detailed review of the resulting literature is beyond the scope of this chapter. Rather, an overview of changes in some selected areas will be described, and the reader is referred to reviews cited in the sections below.

The glutaminergic system: Glutamate plays an essential role as an excitatory transmitter in the CNS, and is involved in the control of cognition, memory and motor function, among others. Evidence suggests that glutaminergic transmission may be affected in cortico-cortical and cortical-striatal pathways during normal aging, as well as in neurodegenerative diseases such as AD and PD [see ref. 36].

The most marked and consistent change in the glutaminergic system during normal aging is a change in expression of glutamate receptors. For example, Hof *et al.* [37] have recently demonstrated changes in both GluR2 (a subunit of the AMPA receptor) and NR1 (an NMDA receptor subunit) immunoreactivity in the monkey cortex. A decrease in expression of both subunits with aging was seen, which was different in different parts of the cortico-cortical circuitry [see ref. 35]. This decrease is consistent with the previous demonstration of an age-associated reduction in spines, where signaling via these receptors plays an important role. No change in neurofilament immunoreactivity was seen, either in level or distribution, indicating that neurodegeneration was not occurring in the tissues analysed [35, 37] Since the GluR2 subunit plays a role in maintaining impermeability of the AMPA receptor to calcium, the reduction in GluR2 expression by these aging neurones would also be expected to make them more prone to calcium-mediated toxicity [37].

The glutaminergic system is essential for hippocampal-dependent learning, the NMDA receptor having a key role in this process. The expression of NR1 in the dentate gyrus has been found to change during normal aging [38]. A problem with most studies of changes in expression, however, is to establish if any observed changes actually affect function. This question has recently been addressed in an important study in which spatial learning ability of young and old rats has been correlated with NMDA receptor levels [39]. A finding of fundamental importance was made, that functional impairment was correlated with reduced NR1 subunit expression only in the CA3 region, and that the correlation was across all subjects, that is, it was independent of age [39]. No consistent changes in GluR2 expression with age or learning ability were detected in the same study.

The dopaminergic system of the basal ganglia: The major loss of dopaminergic neurones that occurs in the substantia nigra of PD patients is well established. In normal aging some loss of these neurones may occur [see refs 16, 17], but this loss is likely to be relatively small, and certainly less than that seen in PD [see refs 40, 41]. However, motor deterioration does occur in normal old age, indicating that some impairment of the function of this neuronal system may occur during normal aging. In addition to the morphological changes described above, neurochemical changes in this system are also implicated in normal aging. The nigral dopaminergic neurones project to the striatum, where reductions in the levels of dopamine, dopamine receptors (both D1 and D2 receptors) and dopamine transporters have been described in normal aging animals [see ref. 41 for a detailed review]. mRNA for dopamine receptors has been found to be reduced in the striatum of aged rats and mRNA for dopamine transporters has also been shown to be reduced in the substantia nigra of normal aging animals [42]. A significant body of evidence suggests that these changes in expression are associated with functional changes in the system, notably in reuptake and in postsynaptic responses. With respect to dopamine synthesis, storage and release, however, data are ambiguous [see ref. 41]. In summary, evidence indicates that neurochemically, the nigrostriatal dopaminergic system is impaired in normal aging, even when nigral dopaminergic, and striatal medium spiny neurones are not reduced in number. This change is likely to lead to impaired function in normal aging that may well underlie the deterioration in motor function seen in many elderly individuals.

Global changes in gene expression: This section has focused on changes in the neurotransmitter system in aging. There is however, evidence from many studies that changes in the expression of a number of other molecules may occur during aging. Recently, microarray technology has been applied to study of the aging brain [e.g., ref. 43; see refs 11, 41, 44]. Despite the problems inherent in the analysis of such data, these studies indicate that molecules involved in cell structure and signaling are among those affected during aging.

Glial changes

As in many aspects of neurobiology, study of glial cells in aging has tended to lag behind that of neurones. There is evidence, however, that important changes occur in these cells during normal aging, and these changes may have significant effects on neuronal function. In fact, somewhat paradoxically, it is reported that upon microscopic examination of cerebral cortex, changes are more apparent in the glia than in neurones [see refs 45, 46].

Astrocytes

A characteristic feature of astrocytes is their response to insults of varying kinds, after which they undergo the process known as *reactive gliosis*. This can often occur as a secondary event after neuronal damage, and involves astrocyte proliferation, cytoplasmic hypertrophy and increased expression of glial fibrillary acidic protein (GFAP). Establishing the nature and incidence of primary age-associated changes of astrocytes is thus problematic.

In the cortex, there is no evidence for a change in the number of astrocytes with aging [47]. Astrocytes in non-pathologically aging individuals and animals do, however, exhibit several characteristic changes in morphology. These are the presence of cytoplasmic inclusions, probably of phagocytosed material, and hypertrophy, resulting in an increase in the number and thickness of their processes [see ref. 46]. Increased gliofilaments are present, and increased levels of GFAP expression have been reported in several studies [e.g., ref. 48]. Interestingly, caloric restriction, which has beneficial effects on stress responses and a positive effect on longevity of several species, attenuates the age-associated changes in glial cells [48, 49].

Oligodendrocytes

Decreases in the white matter of the normally aging brain have already been mentioned. Electron microscopy reveals that, in addition to cytoplasmic inclusions and swellings, changes in the organization of myelin sheaths also occur in aging [see ref. 46]. Changes indicative of proliferation of oligodendrocyte precursors also occur, and several changes are seen in the myelin sheaths. Peters and colleagues made detailed studies of myelinated nerve fibres and found changes in the myelin in several cortical areas [15, 45, 50] These changes include splitting of the sheath, ballooning out of the myelin, and formation of sheaths with increased thickness of myelin, caused by an increase in the number of lamellae [51; see also ref. 46]. Estimates indicate that about 5% of fibres may be affected in monkey primary visual cortex. These changes have been found to correlate with age and importantly, with impairment of cognition in individual monkeys [15]. Although changes in visual function are not usually seen in aged monkeys, this finding has led to the suggestion that the changes in myelin seen in the visual cortex are representative of changes elsewhere in the cortex, which are therefore likely to underlie the changes in cognition [see refs 15, 46]. Evidence for this hypothesis is that age-associated changes in myelin have been reported in other cortical areas, and also the corpus callosum [see refs 51, 52]. The mechanism by which abnormalities in myelin sheaths might affect neuronal function is hypothesized to be due to decreases in conduction velocity reported in other species [see ref. 46]. For example, a reduction in conduction velocity of 43% has been reported in the pyramidal tracts of old cats [53]. Such a reduction would slow signals along affected axons, and lead to impaired timing of neural response, that may affect function in some neural circuits.

Changes in other cells: microglia, and the effects of the aging cardiovascular system
Microglia in aged animals have, like both astrocytes and oligodendrocytes, been found to have increased numbers of cytoplasmic inclusions, in this case typical of

phagocytosed material [46]. The number of microglia also increases in the aging brain.

Although not an example of primary aging of the nervous system, it is important to consider that age-associated change in the nervous system may be the result of changes in the cardiovascular system. Stroke and anoxia can cause mild effects as well as the well-known major tissue damage. Minor cerebrovascular changes during aging may well contribute to the general decline in CNS function in the elderly.

Changes in the peripheral nervous system during aging

Peripheral nerves supplying skeletal muscle

As described above, there is some evidence for a loss of motoneurones from the spinal cord during aging. The axons of these neurones are myelinated and leave the spinal cord alongside the preganglionic autonomic nerves of the sympathetic system (see Figure 2) via the ventral roots. These fibres then join with sensory fibres (which enter the spinal cord via the dorsal roots) to form the peripheral nerves, some of which innervate the skeletal muscle. A variety of age-associated changes in the motor nerves that innervate the skeletal muscles have been described [see ref. 54 for review]. Changes in the morphology of the motor end plates are amongst these. End plates undergo remodeling throughout life, but there is some evidence for an overall shift in the balance towards denervation in old age. A reduction in the number of end plates has been reported in some studies, but was accompanied by an increase in the area of terminals, which may compensate functionally for the numerical loss [see ref. 54]. Many other morphological and neurochemical age-associated changes to the neuromuscular junction have been described, some of which are contradictory, and the changes also appear to be depended to the particular muscle being studied.

Sensory neurones

Many morphological changes have been described in peripheral nerves during aging. A loss of sensory nerve endings in the skin, receptor organs and visceral organs has also been described. Data on age-associated changes in the number of dorsal root sensory ganglion neurones however, are conflicting. Increases in cell number of rat L4 and L5 ganglia have been reported in some studies [e.g., refs 55, 56], while other studies showed no changes in neuronal number [e.g., ref. 57]. Recently, however, using serial section reconstruction and total profile counting, Mohammed and Santer [58] have shown that there is no age-associated change in total neuronal numbers in lumbosacral primary afferent neurones. Bergman and colleagues however have reported a small neuronal loss in sensory ganglia of aged rats [59]. Whether this discrepancy is a reflection of strain differences remains to be determined. Nevertheless, analysis of neuronal subpopulations in aged sensory ganglia indicates that their relative numbers did not change, although changes in expression of peptides were detected [60]. These changes included an upregulation of neuropeptide Y (NPY) expression and a decrease in calcitonin gene-related peptide (CGRP) expression [60]. In another study, the proportion of NOS immunopositive neurones in the lumbosacral DRG was also reduced in aged rats [61].

The sensory neurones that innervate the pelvic viscera also exhibit some changes in aged rats. The sensory innervation of the bladder and ureters, studied by immuno-histochemistry, show a reduction in the density of innervation [62]. It has been hypothesized that changes in signaling by target derived neurotrophic factors may be involved in these changes, and also in the changes in neuropeptide expression, described above [see ref. 63].

The autonomic nervous system

Autonomic dysfunction increases in incidence in the elderly, and may result in disease or an increased risk of failure in autonomic reflexes, and also complicates management of other diseases [see ref. 64]. The neurones of the autonomic nervous system, which play a vital role in autonomic control, are classified into the sympathetic, parasympathetic and enteric divisions. Preganglionic neurones, with cell bodies located either in the brainstem or the spinal cord innervate the effector neurones, which are located in small ganglia that are distributed throughout the body. Despite their importance, these neurones have, in comparison to the neurones of the CNS, been relatively little studied in aging humans or animals. The possibility that age-associated changes of these neurones underlies some of the autonomic problems known to occur in the elderly, such as dysfunction of the cardiovascular, gastrointestinal and urinogenital systems are now, however, receiving increasing attention.

Age-associated changes have been reported in some, but not all regions of the autonomic nervous system. The majority of studies have been performed on rats, although there are some data from other animals and humans. As in the brain, only some populations of neurones appear to be affected.

The parasympathetic nervous system

Parasympathetic preganglionic neurones: Parasympathetic neurones and intramural neurones of the heart, airways and some enteric neurones are innervated by preganglionic neurones that originate in the dorsal motor nucleus of the vagus (DMN), the nucleus ambiguous (NA) and the nucleus tractus solitarius (NTS) and their fibres leave the brain in the vagus nerve. Some sensory neurones that innervate the viscera are also located in these brainstem nuclei and their fibres also join the vagus nerve, which thus contains a mixture of visceral efferent, afferent and also some somatic nerves.

Although some impairment of vagal function has been reported in aging, in both humans [65] and, less consistently, in rats [66], other evidence has shown only a small decrease in conduction velocity of vagal nerve fibres in aged rats [67]. Morphological studies have shown only small age-associated changes in the vagal nuclei and vagus nerve. Apart from some changes in size, no changes have been detected in the preganglionic neural cell bodies in aged rats [68]. With respect to the vagus nerve itself, again, only few changes have been measured in the numbers or size of unmyelinated fibres in either the cervical vagus of rats or abdominal vagus in humans [69]. Some changes in the myelinated fibres, which are a minor population in this nerve trunk, have been described in the aged rat. These changes include an increase in the thickness of the myelin sheath and a decrease in axon diameter [69]. Thus the evidence indicates that, unlike somatic nerves, the vagus nerve is relatively unaffected by aging.

Parasympathetic neurones that supply the pelvic viscera such as the terminal bowel and bladder are innervated by preganglionic neurones that originate in the sacral spinal cord. These neurones also appear to be relatively unaffected in aging [69–71].

Neurones of the parasympathetic ganglia: The effects of aging on postganglionic parasympathetic neurones, such as those of the ciliary ganglion, have been relatively little studied. In a study that combined morphological and electrophysiological analysis of aging of neurones in the rat ciliary ganglion, no change in either neuronal numbers or the density of synaptic boutons on the ciliary ganglion neurones was detected, and electrophysiological properties were also unaffected by aging [72].

The sympathetic nervous system
The neurones of the sympathetic nervous system are probably the best-studied of the autonomic neurones with regard to aging [see refs 73, 74].

Sympathetic preganglionic neurones: In contrast to the parasympathetic system, sympathetic preganglionic neurones in the lumbosacral spinal cord show marked age-related changes, indicative of degeneration [61, 70, 71]. For example, reductions in the length and complexity of the dendrites of sympathetic preganglionic neurones that project to the major pelvic ganglion have been reported [75]. A significant reduction in the area of synaptic contact made by glutaminergic immunoreactive boutons on to the dendrites of these neurones has also been described, indicating that the spinal inputs to these cells may also be affected [75]. Although absolute quantitation was not performed, the authors also described an apparent decrease in the number of sympathetic preganglionic neurones. No changes in the number or morphology of preganglionic neurones that innervate parasympathetic pelvic ganglion cells were seen, in the same study.

Age-associated changes in innervation of sympathetic ganglia: The sympathetic neurones innervated by the preganglionic neurones described above are located in the small paracervical ganglia of the sympathetic chain, and also in the larger superior cervical ganglion (SCG), the superior and inferior mesenteric ganglia and several other ganglia including the pelvic and hypogastric ganglia, which also contain parasympathetic neurones. In addition to the preganglionic inputs to the neurones in these ganglia, some fibres from sensory, parasympathetic and other autonomic neurones, such as afferents from the enteric nervous system, either pass through or innervate neurones in some sympathetic ganglia.

Degeneration and a reduction in the density of nerve terminals of the preganglionic sympathetic neurones has been described in some sympathetic ganglia of both human [76] and rats [77]. The morphological evidence has been supported by immunohistochemical demonstration of a decrease in synapsin I, a marker of nerve terminals, in the hypogastric ganglion of aged rats [78]. A decrease of about 50% of innervation to individual sympathetic was measured; importantly, the loss was confined to terminals supplying the sympathetic postganglionic neurones, the parasympathetic neurones also present in this ganglion were not affected. No such changes were observed in the density of synapsin immunolabeling in other sympathetic ganglia, including the celiac/SMG, thoracic, stellate or SCG [78].

Changes in sympathetic neurones: The different sympathetic neurones innervate very different target cells. For example different neurones of the SCG innervate cerebral blood vessels and the smooth muscle of the iris. SMG neurones innervate the bladder, ureters, and terminal gut. Thus it is perhaps not surprising that different populations of sympathetic neurones are differentially affected in aging. While some of the described changes in aging sympathetic neurones have not been consistently observed, a number of distinctive morphological and neurochemical age-associated changes are consistently seen, and are likely to have significant effects of autonomic reflexes [see ref. 64].

It is clear from the studies published to date that there is no widespread loss of sympathetic neurones in aged animals [79–81], or indeed in aged humans [77, 82–85]. Although stereological methods are only now being applied to confirm these data, it is unlikely that a major reduction in neuronal number in these small ganglia would have gone unmissed in the detailed investigations that have been performed by several different groups, on a number of different sympathetic ganglia. Other changes in the cellular architecture of sympathetcic ganglia, however, may occur during aging. For example, an increase in intraganglionic connective tissue has been described in the aged rat hypogastric and thoracic ganglia, leading to an apparent reduction in neuronal density along with an increase in the overall size of the ganglia [79]. Immunocytochemical studies also indicate that changes in the *composition* of the extracellular matrix may change; changes in the balance of the different extracellular matrix molecules occur in these ganglia during aging [79].

A change in the size of some sympathetic neurones has been described in aged animals. Using ionto phoretic injection of Lucifer Yellow to label cells an increase in the size of mouse SCG neurones was measured in aged neurones [86]. In rat SCG, however, using the same technique, results showed a significant decrease in somal size of aged SCG neurones that was independent of the target to which the neurones projected [87]. In yet another study using standard morpholmentric methods, no change in somal size of neurones in the rat hypogastric or thirteenth thoracic ganglion was measured [79].

Significant decreases in dendritic branching and length in aged sympathetic neurones have been observed more consistently. This atrophy of dendritic processes has been described in neurones of aged rat SCG [79, 86, 87] but was not observed in the neurones of the celiac/SMG [86]. Evidence suggests that only some SCG neurones are affected; dendritic atrophy was seen in the population of neurones that innervate the middle cerebral artery or submandibular gland, whilst those innervating the iris were not affected [87]. Interestingly, local application of the neurotrophic factor NGF reversed the age associated changes seen in the former population [86–89].

A marked age-associated change that has been reported consistently in sympathetic ganglia of all species examined is a neuroaxonal dystrophy that involves intraganglionic axon terminals and synapses [see ref. 64]. This dystrophy resembles that seen in some neuropathies, such as diabetic neuropathy. In human ganglia, the characteristic features of this axonal dystrophy include swollen axons that arise from preterminal axons. These swollen axons are most often argyrophilic and contain cytoskeletal elements such as neurofilaments that contain extensively phosphorylated 200 kDa NF-H epitopes [90] or, less frequently the swelling is caused by tubulovesicular elements [84, 85]. The changes increased with age and were more common in the superior mesenteric ganglion than the superior cervical ganglion, which was only slightly affected. Similar neuroaxonal dystrophic changes have been described in rats [91], Chinese hamsters [92] and mice [86].

Immunohistochemical studies indicate that only some populations of neurones are affected by these dystrophic changes. In aged human SMG, only tyrosine hydroxylase and neuropeptide Y -immunoreactive fibres were found to be dystrophic; substance P, GRP/bombesin, CGRP and enkephalin-immunoreactive fibres appeared normal [83], indicating that the axons of the sympathetic neurones themselves were dystrophic. Dystrophic axons in the SMG were also found to be DBH immunopositive and to express the high-affinity NGF receptor, TrkA and the low affinity neurotrophin receptor, p75 [64], further evidence that the dystrophic axons arise from sympathetic neurones, possibly by collateral sprouting.

A selective and marked reduction in expression of the calcium-binding protein, calbindin D28K by sympathetic neurones in the pelvic ganglia has been described [93]. Expression of calcineurin, however, is reduced in both sympathetic and parasympathetic neurones [94]. Such changes would be expected to have effects on calcium buffering and calcium signaling by these cells.

Sympathetic terminals in target organs: The terminals of sympathetic neurones in the target tissues have also been reported to be affected in aging animals. Decreases in the density of noradrenergic nerve terminals have been reported in cerebral blood vessels [73, 81, 89, 95], rat heart [96], ileum [97], bladder [98], kidney [99], and pineal [100].

The enteric nervous system
The ENS is the largest division of the peripheral nervous system; its ganglia are embedded within the gut wall and are distributed along the entire length of the digestive tract. There are thus many enteric neurones; more than there are sympathetic neurones, for example, and they are a heterogenous population that coordinate gastrointestinal function. The possibility that degenerative changes in the ENS underlie the gastrointestinal disorders that are common in the elderly, such as constipation, have received increasing attention [see refs 101–103].

Neuronal losses have been described in the aged ENS of several species, including man [104–106], guinea-pig [107], rats [108, 109] and mice [110]. Accurate quantitation of neuronal numbers in the ENS however, is hampered by the irregularity in size of enteric ganglia, and the varied size of the gut at different ages. Strict measurements of the area of the tissue containing the ganglia, and corrections for changes that occur in stretch and during fixation are essential. Recent evidence from work on rat ileum suggests that the neuronal losses that occur in the myenteric plexus are diet-dependent, and only occur to any significant extent in *ad libitum* fed animals; the loss being prevented by a calorically-restricted diet. Analysis of neuronal subpopulations indicates that the cholinergic population are vulnerable, but that nitrergic neurones are unaffected [111]. Studies on the colon, however, indicate that nitrergic neurones may be lost in this region of the gastrointestinal tract. Thus there may be regional differences in the age-associated changes in the ENS.

Lifestyle and brain aging

Most of the above sections have, perhaps somewhat naturally, focused on negative aspects of normal nervous system aging. However, all is not "doom and gloom"; many individuals age very successfully and there is also some recent evidence that relatively simple lifestyle changes may have a positive impact on nervous system aging. These changes include dietary supplementation, and exercise.

Dietary supplementation
The possible influence of nutrition on cognitive function and its decline during aging have been increasingly studied. The special metabolic needs of the brain, which requires a constant supply of glucose, are well established, and recently the possibility that even mild deficiencies in micronutrients may have an impact on brain metabolism and therefore brain function have been explored [see ref. 112].

B vitamins and cognitive status
The possible roles of the B vitamins; vitamin B12 (or cobalamin), B6 (pyridoxine) and folate (folic acid or folacin) have been extensively investigated, since minor deficiencies in these vitamins are common, particularly in the elderly [113]. Almost all data in this area have come from cross-sectional studies in which the nutritional status of individuals with respect to these vitamins is related to performance of a variety of cognitive tasks, and also with the later onset of Alzheimer's disease. Many of these studies have shown that there is a positive correlation between the levels of these nutrients and cognitive status. Data on dietary supplementation with the B vitamins are still only preliminary, but also suggest that the B vitamins have a positive effect on cognition [see ref. 113].

How do the B vitamins exert a beneficial effect? There are two possible mechanisms. First, the B vitamins are known to be essential for the "one carbon" cycle, in which homocysteine combines with a folate derivative to form L-methionine, a reaction in which vitamin B12 acts as an essential cofactor. L-Methionine in turn is an essential precursor needed for many methylation reactions that are crucial for normal brain function. Deficiencies in vitamin B12 and folate therefore can lead to a state known as hypomethylation which has been linked to neuropathological changes, as well as cognitive impairment and mood disturbances [see ref. 113]. Second, high levels of homocysteine have been associated with vascular disease, and may affect the cerebral vasculature; they have also been associated with Alzheimer's disease [see ref. 112]. Homocysteine levels are raised not only by deficiencies in vitamin B12 and folate, but also by deficiencies in vitamin B6, which is needed for the conversion of homocysteine into cystathione. Lack of any of the B vitamins, but particularly folate can therefore result in an elevation of homocysteine levels.

So, although details of the mechanisms by which B vitamins and homocysteine produce positive effects on cognitive function remain to be firmly determined, the case for a beneficial effect of dietary supplementation with the B vitamins is strong.

Exercise
The beneficial effects of regular physical exercise on stress, depression and anxiety are well known. Over the last 10–15 years, evidence indicating that exercise, particularly aerobic exercise, may also improve cognitive function in the elderly has emerged [see refs 114, 115].

A number of prospective and retrospective cross-sectional studies have shown that physical activity is associated with a lower risk of cognitive impairment and AD. Intervention studies, in which the effects of aerobic and non-aerobic (e.g., strength and flexibility, or toning and stretching) exercise and "no exercise" programs are compared have also been performed. While some of the results of such studies are clearly dependent upon the particular tests used to analyse cognitive skills, they do allow clear comparisons to be made. The results of these studies indicate that performance in a range of tasks is improved in groups exercising aerobically, but that continued exercise is needed for the effect to be maintained.

The beneficial effects of exercise on brain function have also been demonstrated in animal studies, which have also provided some evidence about the possible cellular and molecular mechanisms that might mediate these effects [see refs 114, 115]. For example, voluntary wheel-running by mice and rats, training in obstacle courses or provision of an enriched environment can have beneficial effects, both on performance in some tasks designed to measure learning, and also on morphological correlates, such as synapse density, or neurogenesis. The molecular changes underlying the behavioral and morphological changes are beginning to be unravelled [see ref. 115].

Finally, the well-known phrase "use it or lose it" which appears to apply to several body systems, is also likely to apply to the nervous system. Mental exercises, such as the solving of crossword puzzles, and the repeated practise, or training, of certain mental tasks, have been shown to improve "performance" in those tasks, and may therefore help retain cognitive abilities during aging.

Conclusion

The long-held belief that neuronal loss is widespread during aging of the nervous system, and is the cause of cognitive and motor decline has now been refuted. Recent work, however, has shown that many subtle changes take place in the nervous system during aging and that different neuronal populations are differentially affected. Elucidation of the nature of these changes will greatly inform treatment of the functional impairment common in the elderly. The plasticity of the nervous system has also been found to be much greater than previously thought; the existence of neural stem cells in adult animals, and the attenuation of age-associated neuronal losses and glial changes by simple interventions such as exercise and caloric restriction and dietary supplementation offer much promise for the future.

References

1. Dani SU, Hori A, Walter GF (1997). *Principles of Neural Aging*. Amsterdam: Elsevier.
2. Hof PR, Mobbs CV (2001). *Functional Neurobiology of Aging*. San Diego: Academic Press.
3. Purves D, Augustine GJ, Fitzpatrick D, Katz L, LaMantia A-S, McNamara JO. *Neuroscience*. Sinauer: Sunderland, MA, USA.
4. Morrison JH, Hof PR (1997). Life and death of neurons in the aging brain. *Science* 278: 412–19.
5. Haug H (1997). The aging human cerebral cortex: morphometry of areal differences and their functional meaning. In: Dani SU, Walter GF, Hori A, eds. *Principles of Neural Aging*. Amsterdam: Elsevier.
6. Peters A, Morrison JH, Rosene DL, Hyman BT (1998). Feature article: are neurons lost from the primate cerebral cortex during normal aging? *Cereb Cortex* 8: 295–300.
7. Turlejski K, Djavadian R (2002). Life-long stability of neurons: a century of research on neurogenesis, neuronal death and neuron quantification in adult CNS. *Prog Brain Res.* 136: 39–65.
8. Gundersen HJ, Bagger P, Bendtsen TF, *et al.* (1988). The new stereological tools: disector, fractionator, nucleator and point sampled intercepts and their use in pathological research and diagnosis. *Apmis* 96: 857–81.
9. Sterio DC (1984). The unbiased estimation of number and sizes of arbitrary particles using the disector. *J Microsc.* 134(Pt 2): 127–36.
10. West MJ (1999). Stereological methods for estimating the total number of neurons and synapses: issues of precision and bias. *Trends Neurosci.* 22: 51–61.
11. Weindruch R, Kayo T, Lee CK, Prolla TA (2002). Gene expression profiling of aging using DNA microarrays. *Mech Ageing Dev.* 123: 177–93.
12. Gomez-Isla T, Price JL, McKeel DW Jr, Morris JC, Growdon JH, Hyman BT (1997). Neuronal loss correlates with but exceeds neurofibrillary tangles in Alzheimer's disease. *Ann Neurol.* 41: 17–24.
13. Gomez-Isla T, Price JL, McKeel DW Jr, Morris JC, Growdon JH, Hyman BT (1996). Profound loss of layer II entorhinal cortex neurons occurs in very mild Alzheimer's disease. *J Neurosci.* 16: 4491–500.
14. Peters A, Rosene DL, Moss MB, *et al.* (1996). Neurobiological bases of age-related cognitive decline in the rhesus monkey. *J Neuropathol Exp Neurol.* 55: 861–74.
15. Peters A, Moss MB, Sethares C (2000). Effects of aging on myelinated nerve fibers in monkey primary visual cortex. *J Comp Neurol.* 419: 364–76.
16. Ma SY, Roytt M, Collan Y, Rinne JO (1999). Unbiased morphometrical measurements show loss of pigmented nigral neurones with ageing. *Neuropathol Appl Neurobiol.* 25: 394–99.
17. Ma SY, Roytt M, Collan Y, Rinne JO (1999). Dopamine transporter-immunoreactive neurons decrease with age in the human substantia nigra. *J Comp Neurol.* 409: 25–37.

18. Cameron HA, McKay RD (2001). Adult neurogenesis produces a large pool of new granule cells in the dentate gyrus. *J Comp Neurol.* 435: 406–17.

19. Gould E, Reeves AJ, Graziano MS, Gross CG (1999). Neurogenesis in the neocortex of adult primates. *Science* 286: 548–52.

20. Svendsen CN, Caldwell MA, Ostenfeld T (1999). Human neural stem cells: isolation, expansion and transplantation. *Brain Pathol.* 9: 499–513.

21. Rao MS, Mattson MP (2001). Stem cells and aging: expanding the possibilities. *Mech Ageing Dev.* 122: 713–34.

22. Rapp PR, Gallagher M (1996). Preserved neuron number in the hippocampus of aged rats with spatial learning deficits. *Proc Natl Acad Sci USA* 93: 9926–30.

23. Dani S (1997). Histological markers of neuronal aging and their meaning. In: Dani SU, Walter GF, Hori A (eds). *Principles of Neural Aging.* Amsterdam: Elsevier.

24. Cupp CJ, Uemura E (1980). Age-related changes in prefrontal cortex of Macaca mulatta: quantitative analysis of dendritic branching patterns. *Exp Neurol.* 69: 143–63.

25. de Brabander JM, Kramers RJ, Uylings HB (1998). Layer-specific dendritic regression of pyramidal cells with ageing in the human prefrontal cortex. *Eur J Neurosci.* 10: 1261–9.

26. Jacobs B, Driscoll L, Schall M (1997). Life-span dendritic and spine changes in areas 10 and 18 of human cortex: a quantitative Golgi study. *J Comp Neurol.* 386: 661–80.

27. Peters A, Sethares C, Moss MB (1998). The effects of aging on layer 1 in area 46 of prefrontal cortex in the rhesus monkey. *Cereb Cortex* 8: 671–84.

28. Page TL, Einstein M, Duan H, *et al.* (2002). Morphological alterations in neurons forming corticocortical projections in the neocortex of aged Patas monkeys. *Neurosci Lett.* 317: 37-41.

29. Tigges J, Herndon JG, Rosene DL (1995). Mild age-related changes in the dentate gyrus of adult rhesus monkeys. *Acta Anat (Basel)* 153: 39–48.

30. Levine MS (1988). Neurophysiological and morphological alterations in caudate neurons in aged cats. *Ann NY Acad Sci.* 515: 314–28.

31. Levine MS, Adinolfi AM, Fisher RS, Hull CD, Guthrie D, Buchwald NA (1988). Ultrastructural alterations in caudate nucleus in aged cats. *Brain Res.* 440: 267–79.

32. McNeill TH, Koek LL (1990). Differential effects of advancing age on neurotransmitter cell loss in the substantia nigra and striatum of C57BL/6N mice. *Brain Res.* 521: 107–17.

33. Emborg ME, Ma SY, Mufson EJ, *et al.* (1998). Age-related declines in nigral neuronal function correlate with motor impairments in rhesus monkeys. *J Comp Neurol.* 401: 253–65.

34. Smith TD, Adams MM, Gallagher M, Morrison JH, Rapp PR (2000). Circuit-specific alterations in hippocampal synaptophysin immunoreactivity predict spatial learning impairment in aged rats. *J Neurosci.* 20: 6587–93.

35. Morrison JH, Hof PR (2002). Selective vulnerability of corticocortical and hippocampal circuits in aging and Alzheimer's disease. *Prog Brain Res.* 136: 467–86.

36. Segovia G, Porras A, Del Arco A, Mora F (2001). Glutamatergic neurotransmission in aging: a critical perspective. *Mech Ageing Dev.* 122: 1–29.

37. Hof PR, Duan H, Page TL, *et al.* (2002). Age-related changes in GluR2 and NMDAR1 glutamate receptor subunit protein immunoreactivity in corticocortically projecting neurons in macaque and patas monkeys. *Brain Res.* 928: 175–86.

38. Gazzaley AH, Siegel SJ, Kordower JH, Mufson EJ, Morrison JH (1996). Circuit-specific alterations of N-methyl-D-aspartate receptor subunit 1 in the dentate gyrus of aged monkeys. *Proc Natl Acad Sci USA* 93: 3121–5.

39. Adams MM, Smith TD, Moga D, *et al.* (2001). Hippocampal dependent learning ability correlates with N-methyl-D-aspartate (NMDA) receptor levels in CA3 neurons of young and aged rats. *J Comp Neurol.* 432: 230–43.

40. Joyce JN (2001). The basal ganglia dopaminergic systems in normal aging and Parkinson's disease. In: Hof PR, Mobbs CV, eds. *Functional Neurobiology of Aging.* San Diego: Academic Press.

41. Stanford JA, Herbert MA, Gerhardt G (2001). Biochemical and anatomical changes in basal ganglia of aging animals. In: Hof PR, Mobbs CV, eds. *Functional Neurobiology of Aging.* San Diego: Academic Press.

42. Himi T, Cao M, Mori N (1995). Reduced expression of the molecular markers of dopaminergic neuronal atrophy in the aging rat brain. *J Gerontol A Biol Sci Med Sci.* 50: B193–200.

43. Jiang CH, Tsien JZ, Schultz PG, Hu Y (2001). The effects of aging on gene expression in the hypothalamus and cortex of mice. *Proc Natl Acad Sci USA* 98: 1930–4.

44. Weindruch R, Prolla TA (2002). Gene expression profile of the aging brain. *Arch Neurol.* 59: 1712–14.

45. Peters A, Sethares C (2002). Aging and the myelinated fibers in prefrontal cortex and corpus callosum of the monkey. *J Comp Neurol.* 442: 277–91.

46. Peters A (2002). Structural changes in the normally aging cerebral cortex of primates. Prog *Brain Res.* 136: 455–65.

47. Peters A, Josephson K, Vincent SL (1991). Effects of aging on the neuroglial cells and pericytes within area 17 of the rhesus monkey cerebral cortex. *Anat Rec.* 229: 384–98.

48. Major DE, Kesslak JP, Cotman CW, Finch CE, Day JR (1997). Life-long dietary restriction attenuates age-related increases in hippocampal glial fibrillary acidic protein mRNA. *Neurobiol Aging* 18: 523–6.

49. Morgan TE, Xie Z, Goldsmith S, *et al.* (1999). The mosaic of brain glial hyperactivity during normal ageing and its attenuation by food restriction. *Neuroscience* 89: 687–99.

50. Sandell JH, Peters A (2001). Effects of age on nerve fibers in the rhesus monkey optic nerve. *J Comp Neurol.* 429: 541–53.

51. Peters A, Sethares C, Killiany RJ (2001). Effects of age on the thickness of myelin sheaths in monkey primary visual cortex. *J Comp Neurol.* 435: 241–8.

52. Peters A, Moss MB, Sethares C (2001). The effects of aging on layer 1 of primary visual cortex in the rhesus monkey. *Cereb Cortex* 11: 93–103.

53. Xi MC, Liu RH, Engelhardt JK, Morales FR, Chase MH (1999). Changes in the axonal conduction velocity of pyramidal tract neurons in the aged cat. *Neuroscience* 92: 219–25.

54. Larsson L, Ansved T (1995). Effects of ageing on the motor unit. *Prog Neurobiol.* 45: 397–458.

55. Popken GJ, Farel PB (1997). Sensory neuron number in neonatal and adult rats estimated by means of stereologic and profile-based methods. *J Comp Neurol.* 386: 8–15.

56. Devor M, Govrin-Lippmann R (1991). Neurogenesis in adult rat dorsal root ganglia: on counting and the count. *Somatosens Mot Res.* 8: 9–12.

57. La Forte RA, Melville S, Chung K, Coggeshall RE (1991). Absence of neurogenesis of adult rat dorsal root ganglion cells. *Somatosens Mot Res.* 8: 3–7.

58. Mohammed HA, Santer RM (2001). Total neuronal numbers of rat lumbosacral primary afferent neurons do not change with age. *Neurosci Lett.* 304: 149–52.

59. Bergman E, Ulfhake B (1998). Loss of primary sensory neurons in the very old rat: neuron number estimates using the disector method and confocal optical sectioning. *J Comp Neurol.* 396: 211–22.

60. Bergman E, Carlsson K, Liljeborg A, Manders E, Hokfelt T, Ulfhake B (1999). Neuropeptides, nitric oxide synthase and GAP-43 in B4-binding and RT97 immunoreactive primary sensory neurons: normal distribution pattern and changes after peripheral nerve transection and aging. *Brain Res.* 832: 63–83.

61. Mohammed H, Santer RM (2001). Distribution and changes with age of nitric oxide synthase-immunoreactive nerves of the rat urinary bladder, ureter and in lumbosacral sensory neurons. *Eur J Morphol.* 39: 137–44.

62. Mohammed H, Hannibal J, Fahrenkrug J, Santer R (2002). Distribution and regional variation of pituitary adenylate cyclase activating polypeptide and other neuropeptides in the rat urinary bladder and ureter: effects of age. *Urol Res.* 30: 248–55.

63. Ulfhak B, Bergman E, Fundin BT (2002). Impairment of peripheral sensory innervation in senescence. *Auton Neurosci.* 96: 43–9.

64. Schmidt RE (2002). Age-related sympathetic ganglionic neuropathology: human pathology and animal models. *Auton Neurosci.* 96: 63–72.

65. Low PA, Opfer-Gehrking TL, Proper CJ, Zimmerman I (1990). The effect of aging on cardiac autonomic and postganglionic sudomotor function. *Muscle Nerve* 13: 152–7.

66. Ferrari AU, Daffonchio A, Gerosa S, Mancia G (1991). Alterations in cardiac parasympathetic function in aged rats. *Am J Physiol.* 260: H647–9.

67. Sato A, Sato Y, Suzuki H (1985). Aging effects on conduction velocities of myelinated and unmyelinated fibers of peripheral nerves. *Neurosci Lett.* 53: 15–20.

68. Soltanpour N, Santer RM (1997). Vagal nuclei in the medulla oblongata: structure and activity are maintained in aged rats. *J Auton Nerv Syst.* 67: 114–17.

69. Soltanpour N, Santer RM (1996). Preservation of the cervical vagus nerve in aged rats: morphometric and enzyme histochemical evidence. *J Auton Nerv Syst.* 60: 93–101.

70. Dering MA, Santer RM, Watson AH (1996). Age-related changes in the morphology of preganglionic neurons projecting to the rat hypogastric ganglion. *J Neurocytol.* 25: 555–63.

71. Dering MA, Santer RM, Watson AH (1998). Age-related changes in the morphology of preganglionic neurons projecting to the paracervical ganglion of nulliparous and multiparous rats. *Brain Res.* 780: 245–52.

72. Wigston DJ (1983). Maintenance of cholinergic neurones and synapses in the ciliary ganglion of aged rats. *J Physiol.* 344: 223–31.

73. Cowen T, Gavazzi I (1998). Plasticity in adult and ageing sympathetic neurons. *Prog Neurobiol.* 54: 249–88.

74. Kuchel G, Cowen T (2001). The aged sympathetic nervous system. In: Hof PR, Mobbs CV, eds. *Functional Neurobiology of Aging.* San Diego: Academic Press, pp. 929–40.

75. Santer RM, Dering MA, Ranson RN, Waboso HN, Watson AH (2002). Differential susceptibility to ageing of rat preganglionic neurones projecting to the major pelvic ganglion and of their afferent inputs. *Auton Neurosci.* 96: 73–81.

76. Schroer JA, Plurad SB, Schmidt RE (1992). Fine structure of presynaptic axonal terminals in sympathetic autonomic ganglia of aging and diabetic human subjects. *Synapse* 12: 1–13.

77. Schmidt RE, Dorsey DA, McDaniel ML, Corbett JA (1993). Characterization of NADPH diaphorase activity in rat sympathetic autonomic ganglia – effect of diabetes and aging. *Brain Res.* 617: 343–8.

78. Warburton AL, Santer RM (1995). Decrease in synapsin I staining in the hypogastric ganglion of aged rats. *Neurosci Lett.* 194: 157–60.

79. Warburton AL, Santer RM (1997). The hypogastric and thirteenth thoracic ganglia of the rat: effects of age on the neurons and their extracellular environment. *J Anat*. 190(Pt 1): 115–24.

80. Santer RM (1991). Morphological evidence for the maintenance of the cervical sympathetic system in aged rats. *Neurosci Lett*. 130: 248–50.

81. Cowen T (1993). Ageing in the autonomic nervous system: a result of nerve-target interactions? A review. *Mech Ageing Dev*. 68: 163–73.

82. Schmidt RE, Plurad SB, Parvin CA, Roth KA (1993). Effect of diabetes and aging on human sympathetic autonomic ganglia. *Am J Pathol*. 143: 143–53.

83. Schmidt RE, Chae HY, Parvin CA, Roth KA (1990). Neuroaxonal dystrophy in aging human sympathetic ganglia. *Am J Pathol*. 136: 1327–38.

84. Schmidt RE (1996). Neuropathology of human sympathetic autonomic ganglia. *Microsc Res Tech*. 35: 107–21.

85. Schmidt RE (1996). Synaptic dysplasia in sympathetic autonomic ganglia. *J Neurocytol*. 25: 777–91.

86. Schmidt RE, Beaudet L, Plurad SB, Snider WD, Ruit KG (1995). Pathologic alterations in pre- and postsynaptic elements in aged mouse sympathetic ganglia. *J Neurocytol*. 24: 189–206.

87. Andrews TJ, Thrasivoulou C, Nesbit W, Cowen T (1996). Target-specific differences in the dendritic morphology and neuropeptide content of neurons in the rat SCG during development and aging. *J Comp Neurol*. 368: 33–44.

88. Andrews TJ, Cowen T (1994). *In vivo* infusion of NGF induces the organotypic regrowth of perivascular nerves following their atrophy in aged rats. *J Neurosci*. 14: 3048–58.

89. Andrews TJ, Cowen T (1994). Nerve growth factor enhances the dendritic arborization of sympathetic ganglion cells undergoing atrophy in aged rats. *J Neurocytol*. 23: 234–41.

90. Schmidt RE, Beaudet LN, Plurad SB, Dorsey DA (1997). Axonal cytoskeletal pathology in aged and diabetic human sympathetic autonomic ganglia. *Brain Res*. 769: 375–83.

91. Schmidt RE, Plurad SB, Modert CW (1983). Neuroaxonal dystrophy in the autonomic ganglia of aged rats. *J Neuropathol Exp Neurol*. 42: 376–90.

92. Schmidt RE, Plurad DA, Plurad SB, Cogswell BE, Diani AR, Roth KA (1989). Ultrastructural and immunohistochemical characterization of autonomic neuropathy in genetically diabetic Chinese hamsters. *Lab Invest*. 61: 77–92.

93. Corns RA, Boolaky UV, Santer RM (2000). Decreased calbindin-D28k immunoreactivity in aged rat sympathetic pelvic ganglionic neurons. *Neurosci Lett*. 292: 91–4.

94. Corns RA, Hidaka H, Santer RM (2001). Decreased neurocalcin immunoreactivity in sympathetic and parasympathetic neurons of the major pelvic ganglion in aged rats. *Neurosci Lett*. 297: 81–4.

95. Gavazzi I, Cowen T (1996). Can the neurotrophic hypothesis explain degeneration and loss of plasticity in mature and ageing autonomic nerves? *J Auton Nerv Syst*. 58: 1–10.

96. Goldberg PB, Kreider MS, McLean MR, Roberts J (1986). Effects of aging at the adrenergic cardiac neuroeffector junction. *Fed Proc*. 45: 45–7.

97. Baker DM, Watson SP, Santer RM (1991). Evidence for a decrease in sympathetic control of intestinal function in the aged rat. *Neurobiol Aging* 12: 363–5.

98. Warburton AL, Santer RM (1994). Sympathetic and sensory innervation of the urinary tract in young adult and aged rats: a semi-quantitative histochemical and immunohistochemical study. *Histochem J* 26: 127–33.

99. Vega JA, Ricci A, Amenta F (1990). Age-dependent changes of the sympathetic innervation of the rat kidney. *Mech Ageing Dev*. 54: 185–96.

100. Kuchel GA, Crutcher KA, Naheed U, Thrasivoulou C, Cowen T (1999). NGF expression in the aged rat pineal gland does not correlate with loss of sympathetic axonal branches and varicosities. *Neurobiol Aging* 20: 685–93.

101. Wiley JW (2002). Aging and neural control of the GI tract: III. Senescent enteric nervous system: lessons from extraintestinal sites and nonmammalian species. *Am J Physiol Gastrointest Liver Physiol.* 283: G1020–6.

102. Wade PR (2002). Aging and neural control of the GI tract. I. Age-related changes in the enteric nervous system. *Am J Physiol Gastrointest Liver Physiol.* 283: G489–95.

103. Hall KE (2002). Aging and neural control of the GI tract. II. Neural control of the aging gut: can an old dog learn new tricks? *Am J Physiol Gastrointest Liver Physiol.* 283: G827–32.

104. de Souza RR, Moratelli HB, Borges N, Liberti EA (1993). Age-induced nerve cell loss in the myenteric plexus of the small intestine in man. *Gerontology* 39: 183–8.

105. Gomes OA, de Souza RR, Liberti EA (1997). A preliminary investigation of the effects of aging on the nerve cell number in the myenteric ganglia of the human colon. *Gerontology* 43: 210–17.

106. Meciano Filho J, Carvalho VC, de Souza RR (1995). Nerve cell loss in the myenteric plexus of the human esophagus in relation to age: a preliminary investigation. *Gerontology* 41: 18–21.

107. Gabella G (1989). Fall in the number of myenteric neurons in aging guinea pigs. *Gastroenterology* 96: 1487–93.

108. Johnson RJ, Schemann M, Santer RM, Cowen T (1998). The effects of age on the overall population and on sub-populations of myenteric neurons in the rat small intestine. *J Anat.* 192(Pt 4): 479–88.

109. Santer RM, Baker DM (1988). Enteric neuron numbers and sizes in Auerbach's plexus in the small and large intestine of adult and aged rats. *J Auton Nerv Syst.* 25: 59–67.

110. El-Salhy M, Sandstrom O, Holmlund F (1999). Age-induced changes in the enteric nervous system in the mouse. *Mech Ageing Dev.* 107: 93–103.

111. Cowen T, Johnson RJ, Soubeyre V, Santer RM (2000). Restricted diet rescues rat enteric motor neurones from age related cell death. *Gut* 47: 653–60.

112. Smith AD (2002). Homocysteine, B vitamins, and cognitive deficit in the elderly. *Am J Clin Nutr.* 75: 785–6.

113. Calvaresi E, Bryan J (2001). B vitamins, cognition, and aging: a review. *J Gerontol B Psychol Sci Soc Sci.* 56: P327–39.

114. Churchill JD, Galvez R, Colcombe S, Swain RA, Kramer AF, Greenough WT (2002). Exercise, experience and the aging brain. *Neurobiol Aging* 23: 941–55.

115. Cotman CW, Berchtold NC (2002). Exercise: a behavioral intervention to enhance brain health and plasticity. *Trends Neurosci.* 25: 295–301.

Aging of the Liver

Joseph M. Dhahbi[1,2] and Stephen R. Spindler[2]

[1] *Biomarker Pharmaceuticals, 5055 Canyon Crest Drive, Riverside, CA 92507;* [2] *Department of Biochemistry, University of California, Riverside, Riverside, CA 92521, USA*

Introduction

Most organs are altered morphologically and functionally in old animals. To a varying extent, these age-related changes lead to a progressive loss of differentiated functions and physiologic capacities [1]. For the liver, the data are inconsistent and conflicting regarding the effects of aging on structure and function [2]. A number of reviews dealing with the physiology, structure and function of the aging liver have been published recently [3–5]. Reported morphological and structural changes do not generally correlate with the functional alterations found in the liver with age. These disparities raise the question of whether or not the observed changes with age actually compromise liver function, which led to the idea that the liver ages well when compared to other organs.

However, as described in a number of the reviews cited above, the liver does decline functionally with age. Our recent microarray studies of mice found that aging was accompanied by changes in gene expression linked to the development of the characteristic age-related liver pathologies [6]. These include hepatocellular carcinoma, fibrosis, cirrhosis, and unhealthful apolipoprotein and fatty acid biosynthesis. Aging increased gene expression associated with inflammation, cellular stress, and fibrosis, and reduced capacity for apoptosis, negative cell-growth control, and phase I and II xenobiotic metabolism. In this study, caloric restriction (CR) from weaning (long-term CR (LT-CR)), reversed the majority of these changes. Surprisingly, in very old mice just 2 to 4 weeks of CR (short-term CR (ST-CR)) reversed approximately 70% of these age-related changes in gene expression.

LT-CR and ST-CR also produced changes in the expression of genes which did not change with age. These *CR-specific* changes involved increased gluconeogenesis and disposal of the byproducts of extrahepatic protein catabolism, reduced glycolysis, and healthful changes in apolipoprotein and fatty acid biosynthesis. In addition, LT-

R. Aspinall (ed.), Aging of Organs and Systems, 271–291.
© 2003 *Kluwer Academic Publishers. Printed in Great Britain.*

CR and ST-CR produced changes in gene expression associated with enhanced anti-proliferative growth control, increased apoptosis and reduced chemical carcinogenesis. A number of other alterations in gene expression are associated with enhanced longevity in mice.

Microarrays

Quantitative change in the activity of specific genes can control the rate of aging in invertebrates and mammals [7, 8]. Although there have been many studies of the relationship between aging, CR and hepatic gene expression, there are serious shortcomings in this literature. There are numerous cross-sectional studies of gene expression in animals of various ages which are interpreted as showing that the major effect of CR is to *prevent* age-related changes in gene expression [e.g., ref 9]. This interpretation has become pervasive in the literature, despite the cross-sectional nature of the studies. Funding and publication bias has reinforced this notion, producing a literature replete with reports of age-related changes in gene expression which appear to be *prevented* by CR.

Genome-wide DNA-microarrays are capable of quantifying the expression of all known genes in a single experiment. A significant strength of this approach is the absence of hypothesis-based bias in the choice of genes which are studied. Instead, a comprehensive profile of the relationship between a physiological state and gene expression is generated. Application of this technology has revealed the gene expression signatures underlying the physiological effects of aging, CR, and the dwarf mutations [6, 10–14]. In this way, microarrays are providing insights into aging, the development of age-related diseases, and the ameliorative actions of CR.

Our studies using this technology suggest that rather than simply preventing age-related changes in gene expression, CR instead acts rapidly to establish a new profile of gene expression which may better resist aging. Overall, ST-CR reproduced nearly 70% of the effects of LT-CR on genes that changed expression with age [6]. Thus, CR rapidly reversed, rather than prevented, many age-related changes in gene expression.

Another important effect of LT- and ST-CR was to establish the CR-specific patterns of gene expression. These CR-specific changes were in the same functional categories as the age-related changes. Further, CR in young mice produced gene expression changes which were a subset of those produced in old CR mice. These similarities between CR in young and old mice, and between ST- and LT-CR led us to conclude that CR rapidly produces a new pattern of gene expression which better resists aging.

Intra- and intercellular signaling

Recently, a small family of single gene mutations which interfere with growth hormone (GH)/insulin-like growth factor-1 (IGF-1) signaling, resulting in dwarfism, have been shown to increase mean and maximal lifespans of mice by 40% to 70% beyond those of their heterozygous siblings [8, 15, 16]. The dwarf mice are homozygous for loss-of-function mutations in the Pit1 (Snell dwarf mice), Prop1 (Ames dwarf mice), or GH receptor (GHR KO mice) loci. The Pit1 and Prop1 mutations prevent differentiation of the anterior pituitary, decreasing levels of thyroid hormone, growth hormone, IGF-1 and prolactin. The GHR KO is a more focused mutation, which prevents receptor-mediated GH responsiveness. However, the extension of lifespan in GHR KO mice may be somewhat less robust than that of the Snell and Ames dwarf mice. The mutations appear to slow the intrinsic rate of aging. Snell dwarf mice show delays in age-dependent collagen cross-linking and in six age-sensitive indices of immune system status. These findings demonstrate that a single gene can control maximum lifespan and the timing of both cellular and extracellular senescence in mammals. The already enhanced lifespan of Ames dwarf mice can be further extended $\sim 25\%$ by CR [17].

Our microarray studies found that the hepatic expression of IFG-I binding protein 1 decreased with age. This protein plays an important role in the negative regulation of the IGF-1 system, a stimulator of mitogenesis [18]. Given what is now known about the apparent role of IGF-1 signaling in aging, this change may have both pro-cancer and anti-aging effects.

CR repressed expression of GH receptor and iodothyronine deiodinase type I mRNA in the liver of both young and old mice, and induced overexpression of IGF binding protein mRNA, which inhibits IGF-1 signaling. Reduction in iodothyronine deiodinase type I expression should reduce hepatic conversion of the pro-hormone form of thyroid hormone (T_4) to the active form (T_3). Down-regulation of this enzyme is likely responsible for the reduced levels of circulating T_3 found in CR rodents [19]. Short-term treatment with low-calorie diets rapidly reduces circulating T_3 levels in morbidly obese men, apparently by reducing type I deiodinase activity [20]. Thus, CR appears to reduce thyroid hormone action in CR mice and humans.

Thus, some of the changes in gene expression induced by CR in mouse liver are associated with decreased GH receptor, IGF-1, and thyroid hormone signaling. This is highly suggestive of the lifespan extending effects of the Prop-1 and Pit-1 mutations [21]. It suggests that the dwarf mutations and CR may work in part through the same signal transduction pathways.

Age-related inflammation

Published microarray studies of mammalian aging found that aging was associated with changes in gene expression linked to the development of the characteristic age-related pathologies of tissues such as liver, muscle and brain [6, 13, 22, 23]. Our microarray studies of mouse liver revealed that aging was associated with other gene expression changes consistent with liver pathogenesis. We found age-associated induction in the expression of several genes important in inflammation including lysozyme and complement component 1, q, β [6]. Lysozyme is a myeloid cell-specific marker. Induction of this gene is normally associated with macrophage activation [24]. Complement component 1, q, β, a macrophage expressed protein, is a part of the recognition set of the complement C system, the primary humoral mediator of antigen-antibody reactions [25]. Activated macrophages, along with other inflammatory cells, are involved in a large number of liver diseases including cirrhosis, hepatitis, and sepsis- and endotoxin-induced liver injury [26].

Old mice also overexpressed the mRNA for biglycan, a proteoglycan of the hepatic extracellular matrix, serum amyloid P-component, a glycoprotein present in all amyloid deposits, and cystatin B, an inhibitor of cysteine proteinases. In areas of inflammation, fibrogenic myofibroblasts express biglycan and other proteoglycans, leading to hepatic fibrosis [27]. Serum amyloid P-component is one of the major acute phase reactants induced by inflammation in hepatocytes [28]. Cystatins and their target enzymes play a role in many pathological events, including inflammatory disease [29]. In the liver, an imbalance between cystatins and their targets can disregulate matrix degradation and accumulation, leading to hepatic fibrosis [30].

CR suppressed the age-associated increase in inflammatory and stress response genes. Consistent with decreased inflammatory response gene expression, CR delays the onset and diminishes the severity of autoimmune and inflammatory diseases in mice [31]. Decreased chaperone and stress response gene expression suggests that CR reduces the age-related physiological stress on the liver. Further, as discussed below, reduced chaperone expression is proapoptotic and anti-neoplastic. Thus, these effects may explain the delayed onset of hepatoma in CR mice [32].

Apoptosis and tumorigenesis

Hepatocytes are mitotically competent, although they have long, mostly intermitotic lifespans. These lifespans appear to lead to the incremental accumulation of damage, and the gradual impairment of physiological functions. Thus, there is an important role for apoptosis in the maintenance of hepatic function. Apoptosis was initially viewed as potentially injurious to tissues, because it destroys cells. The current view recognizes that the role of apoptosis in aging is most likely tissue-specific. In every tissue, a balance must be struck between the need to maintain cell number and function, and the need to eliminate damaged, potentially toxic or neoplastic cells. This decision is crucial in largely postmitotic tissues such as brain. Brain apoptosis can contribute to neurological disorders of aging, including Alzheimer's disease, Parkinson's disease and stroke [33].

Hepatocytes are exposed to genotoxins from the diet and from free radicals generated by xenobiotic metabolism and beta-oxidation. These can produce elevated levels of DNA damage, a potential source of neoplasia. Apoptosis acts to eliminate the damaged and preneoplastic cells, which are then replaced by cell proliferation, thus maintaining homeostatic liver function.

The predominant morphologic change in aging human liver is termed *brown atrophy* [34]. A brown color in aged liver cells results from the accumulation of lipofuscin in lysosomes. Liver atrophy results from an age-related decline in liver mass, resulting in fewer, larger hepatocytes. These observations suggest that aging is accompanied by a disregulation of apoptosis and cell division which fails to maintain youthful hepatocyte number and function during aging. Consistent with this idea, aging is associated with a decline in the rate of liver regeneration and apoptosis [35, 36]. Likewise, aging is accompanied by an increase in liver tumors [37, 38].

Aging, CR and hepatic cell division

Our microarray studies found that 23% of the genes which decreased expression with age are involved in DNA replication and regulation of the cell cycle [6]. Most of these genes have a negative effect on cell growth and division. Thus, hepatic aging may be accompanied by a general loss of negative control of cell division. Among these genes, the product of phosphatase and tensin homolog gene is a tumor suppressor which induces cell-cycle arrest through inhibition of the phosphoinositide 3-kinase pathway [39]. B-cell translocation gene 2 is a tumor suppressor which increases expression in response to DNA damage [40]. The murine gene product of the amino-terminal enhancer of split is a potent co-repressor of gene expression and cellular proliferation [41]. Calcium binding protein A11 binds to and regulates the activity of annexin II, which is involved in the transduction of calcium-related mitogenic signals [42]. As discussed above, IGF binding protein 1 negatively regulates the IGF-1 signaling [18]. Therefore, this change may be mitogenic.

Seventy-eight percent of the mice of this strain and sex fed the control diet used here die of some form of neoplasia, and the death rate from neoplasia accelerates dramatically with age [32]. Approximately 21% of these mice die of hepatoma, mostly late in life. Decreased expression of the negative growth regulators and overexpression of the chaperone genes with age are consistent with this higher incidence of hepatoma in aged mice.

LT- and ST-CR induced the expression of cyclin-dependent kinase 2-associated protein 1, a putative tumor suppressor gene [43]. Overexpression of this gene suggests that LT- and ST-CR enhance anti-proliferative growth control. Consistent with this idea, IGF binding protein 7 gene expression was induced by LT-CR. The product of this gene functions both as an IGF binding protein and independently of IGF as a growth-suppressing factor [44]. The expression of IGF binding protein 1, which has anti-growth activity through its inhibition of IGF-1 signaling, was reduced by age and restored by ST-CR. Thus, LT- and ST-CR may produce additional anti-proliferative effects on preneoplastic cells of the liver through their effects on the expression of these IGF binding protein family members.

Aging and apoptosis

Our microarray studies revealed that aging in mice was accompanied by elevated chaperone levels and the overexpression of the other anti-apoptotic genes, myeloid cell leukemia sequence 1 and apoptosis inhibitory protein 6. These observations suggested that aging should be accompanied by a decrease in apoptotic potential of the liver. In contrast to this expectation, a number of studies reported that aging is accompanied by an increase in the intrinsic rate of apoptosis in rodent liver [35, 45, 46].

However, a recent study has clarified this conundrum. Suh et al. showed that the intrinsic rate of apoptosis in liver does increase slightly with age. But the increase was not significant in their study. However, they found a large and significant decrease in the apoptotic potential of the liver with age [36]. Brief exposure to a direct-acting genotoxic alkylating agent produced high rates of apoptosis in the liver of young rats, but little apoptosis in the livers of old rats. These results suggest that the apoptotic capacity of the liver declines with age, while the basal rate of apoptosis may increase slightly. These data suggest that damaged and preneoplastic cells likely accumulate with age in the liver. This interpretation is consistent with the increase in brown atrophy and hepatocellular neoplasms with age in mouse and man [37, 47].

CR and apoptosis

Our genome-wide microarray studies found that 21% of the genes which changed expression in response to LT- and ST-CR are associated with apoptosis, cell growth, or cell survival [6]. LT-CR induced the expression of the Bcl2 homologous antagonist/killer and voltage-dependent anion channel 1 (porin) genes. Bcl2 homologous antagonist/killer is a pro-apoptotic member of the Bcl2 family of apoptosis regulators. It directly interacts with porin to release the pro-apoptotic factor cytochrome c from mitochondria, initiating apoptosis [48]. The overexpression of porin found in ST-CR mice is consistent with the increase in apoptosis and reduction in chemical carcinogenesis found in fasting rodents [49, 50]. LT-CR decreased the expression of the anti-apoptotic genes interferon inducible ds RNA dependent inhibitor, X-box binding protein, and lymphocyte antigen 6 complex, locus E [51–53].

ST-CR reproduced the effects of LT-CR on the expression of 50% of the cell-cycle/DNA replication and apoptosis genes. The combination of these effects on gene expression suggests that ST-CR may be capable of rapidly reproducing the anti-neoplastic effects of LT-CR in very old animals. This conclusion is consistent with studies showing that short-term fasting increases apoptosis in preneoplastic lesions, and reduces rates of chemical carcinogenesis in the liver [49].

There is compelling evidence that CR increases the rate of apoptosis in preneoplastic and normal cells. The rate of apoptosis in the liver of mice, as measured using terminal dUTP nick end labeling (TUNEL) of apoptotic bodies, was 3 times higher in CR mice [54]. Increased hepatocyte apoptosis was associated with a significantly lower incidence of spontaneous hepatomas throughout the lifespan of the CR mice. Using glutathione S-transferase-II (GST-II) as an immunohistochemical marker of preneoplastic liver cells, a progressive rise in GST-II labeling was seen with age in control mice [38]. This increase was associated with a high incidence of GST-II positive liver tumors. GST-II expression was negligible in CR mice, which had a significant decrease in tumor incidence. One week of CR induced apoptosis in the GST-II-positive hepatocytes. In another study, CR eliminated 20–30% of liver cells by apoptosis, decreasing the number of preneoplastic liver foci by 85% [55]. Apoptosis is significantly higher at all ages in hepatocytes from CR mice [35]. CR enhances apoptosis in other organs as well, including gastrointestinal tract, bladder, spleen and lymph nodes [56–58].

Chaperones, aging, and CR

A consistent finding of our genome-wide microarray and conventional studies was that the mRNA and protein levels of essentially every endoplasmic-reticulum chaperone increased with age and decreased with CR in the liver [6, 59–61]. Similar results were obtained in several other tissues. The induction of chaperone gene expression in the livers of aged mice may be a physiological adaptation to increased oxidative or possibly other stress during aging. For a number of years the meaning of these changes was unclear. In the past few years, the relationship between chaperones and health is beginning to be understood.

Stress-inducible chaperones respond to a diverse group of stimuli including heat, oxidative and ischemic stress, inflammation, hemodynamics, and exposure to toxic chemicals [62–64]. Under such conditions, these inducible chaperones associate with abnormally folded proteins to promote their renaturation, prevent their aggregation, or promote their degradation if they cannot be properly refolded. A number of years ago, Richardson and his colleagues found that the heat inducibility of the stress-responsive chaperone hsp70 was significantly reduced in hepatocytes isolated from old rats [65]. Similar results were found in fibroblasts from donors of various ages [66]. Richardson and colleagues also found that in old rats maintained on LT-CR, there was no decrease in the response of hsp70 to hyperthermia [65].

However, inducible chaperones like hsp70 cannot be detected in the absence of physiological stress. They play a different role than the chaperones which are present continuously in cells in the absence of physiological stress, which is by far the most common physiological state. Most proteins require interactions with constitutively expressed molecular chaperones for their biosynthesis, maturation, processing, intracellular transport, and secretion [67]. Chaperones also perform cytoprotective functions, including prevention of protein denaturation and aggregation, the repair of structurally damaged proteins [68], and promotion of the ubiquination and proteasomal degradation of malfolded, damaged proteins [69, 70]. In this context, it might appear that constitutive overexpression of chaperones would be healthful, perhaps by preventing the accumulation of lipofuscin. However, another function of chaperones appears to mitigate this possible benefit.

Chaperone levels are a part of molecular decision making following genotoxic stress. Elevated chaperone levels tip the balance away from apoptosis and toward cell survival [71, 72]. As described above, aging increases chaperone expression and decreases the apoptotic response to genotoxic stress [36]. The increase in chaperone expression with age may explain why hepatocellular neoplasms are the most common lesions in older mice [37, 47]. In contrast, CR, which reduces endoplasmic reticulum chaperone levels in the liver and other tissues, enhances apoptosis [72–76]. Enhanced apoptosis by CR may account for its well-documented anti-cancer benefits [77].

The linkage between chaperone levels and apoptosis also extends to fasting and feeding. While feeding increases chaperone levels, fasting reduces the levels of nearly every endoplasmic reticulum and cytoplasmic chaperone we investigated [61, 78]. Fasting also increases apoptosis of preneoplastic lesions and reduces the rate of chemical carcinogenesis [49, 50]. This connection between caloric intake and chaperone levels may link food intake to the capacity for protein folding, assembly, and processing within cells. The level of chaperone expression in response to feeding does not depend on endoplasmic reticulum protein trafficking [78]. It appears to be regulated by the blood insulin-to-glucagon ratio.

Molecular mechanisms linking chaperones, protein secretion and carcinogenesis

Chaperone induction has emerged as a new anti-apoptotic mechanism [79, 80]. Elevated chaperone levels during tumorigenesis allow cells to survive carcinogenesis and tumor formation [81]. Induced GRP78, GRP94 and GRP170 are essential for the survival, growth and immuno-resistance of transformed cells [82–84]. Tumorigenesis-associated chaperone induction confers drug resistance to the tumors [74, 85–89]. Chaperone induction allows precancer cells to survive the DNA damage and mutations which result in transformation, proliferation and onset of carcinogenesis [73–76, 90].

Chaperone induction might reduce the production or secretion of apoptogenic signals, or increase the production or secretion of apoptosis inhibitory proteins. Several studies indicate that the abundance of endoplasmic reticulum chaperones influences the secretion efficiency of many liver proteins [91–93]. The interaction between chaperones and other proteins can enhance either protein folding, maturation and processing, or enhance the degradation of proteins [94, 95]. It appears that the longer a protein spends in association with chaperones, the greater the chance it will undergo degradation [95–98]. We found that CR, which decreased the level of most endoplasmic reticulum chaperones, increased the rate, efficiency and level of hepatic protein secretion [61]. It is thus possible, that the effect of CR on endoplasmic reticulum chaperone levels and secretion efficiency may change the activity of receptor mediated apoptotic pathways. It may change the display or secretion of pro- or anti-apoptotic receptors or ligands.

The increase in secretory protein output in response to CR may also enhance the turnover of serum proteins. This may decrease circulating levels of glycated serum proteins, which are associated with micro- and macrovascular damage, nephropathy, neuropathy, retinopathy and atherosclerotic disease [99, 100]. Modified plasma proteins appear to be significant contributors to the development of age- and diabetes-related renal, vascular, ocular and neurological pathologies, and to aging itself [101–103]. CR reduces the age-related accumulation of glycoxidation products in blood and tissue proteins [104–106].

Xenobiotic metabolism

The effect of aging on hepatic drug metabolism is extremely important due to its effects on both carcinogenesis and its practical implications in determining the drug doses that are safe for older individuals. A decline in hepatic drug metabolism and clearance, and an increase in adverse drug reactions are common hallmarks of human and rodent aging. The liver's capacity to metabolize xenobiotics declines with age [4]. Pharmacokinetic evidence in humans indicates that aging is accompanied by reduced liver phase I drug metabolism. For example, cytochrome P450 content in human subjects decreases 30% after 70 years of age [107]. Altered drug metabolism has been attributed to a decline in liver volume and blood flow in humans, although these changes may only partly account for the decline in the metabolism and clearance of drugs with aging in man [107]. In rodent studies, there is compelling evidence for a decline in phase I and phase II enzyme activities and expression, although the specific enzymes which are altered may vary with strain and species [6, 108–110].

In our microarray and conventional studies, aging decreased expression of genes for the xenobiotic metabolism genes [6]. This is an additional class of pro-neoplastic changes in gene expression encountered in our microarray studies. The genes for the phase I enzymes amine N-sulfotransferase and three cytochrome P450 isozymes, as well as the gene for the phase II enzyme glutathione S-transferase-like gene were negatively regulated by age. Decreased expression of Phase I enzyme genes in the liver of aged rodents has been reported in many studies [108, 111, 112]. Decreased expression of such genes is likely responsible in part for the age-related decline in the xenobiotic metabolizing capacity of the liver. This decline is a recognized source of adverse drug reactions in aged mammals [2]. It may contribute to the increase in neoplasms with age in mice.

LT- and ST-CR reversed the age-related decrease in the expression of genes such as the B-cell translocation gene 2, amino-terminal enhancer of split, glutathione-S-transferase like, amine N-sulfotransferase, and cytochrome P450, 2f2 mRNAs. This CR effect is consistent with the delayed onset of hepatoma in CR mice. Partial restoration of the hepatic drug metabolizing and detoxifying functions of the liver may be a source of the anti-aging and anti-cancer effects of CR. These results suggest that ST-CR may rapidly restore some differentiated functions in tissues of older animals.

Intermediary metabolism

Energy metabolism in the liver is altered by aging. For example, at least two studies have shown decreased mitochondrial respiratory rates in the liver with age [113, 114]. Perhaps the major effects of age are on homeostatic glucose regulation. The liver plays a critical role in maintaining glucose homeostasis. This homeostasis is controlled by hormones such as insulin, glucagon, growth hormone, and IGF-1. High levels of glucose and insulin are implicated in many age-associated pathologies [115]. Likewise, loss of homeostatic glucose regulation is a hallmark of mammalian aging [116]. CR reduces blood glucose and insulin concentrations in rodents, primates and humans [117–119]. Disorders associated with elevated glucose levels are reduced or mitigated entirely by CR. These facts indicate that the anti-aging effects of CR may be mediated by alteration of the normal sequence of age-related metabolic changes in the liver.

Aging and hepatic energy metabolism

In general terms, our studies of the effects of aging on key hepatic and muscle enzymes of glucose homeostasis indicated that aging is accompanied by a decline in the enzymatic capacity for the turnover and utilization of peripheral protein for the production of glucose by the liver (Figures 1 and 2). We found an age-related decrease in the expression of phosphoenolpyruvate carboxykinase (PEPCK) and glucose-6-phosphatase mRNA in the liver of mice (Figure 1) [120, 121]. An age-related decrease in PEPCK mRNA also was reported in hepatocytes isolated from rats of various ages [122]. This enzyme catalyzes the committing step in gluconeogenesis, the conversion of oxaloacetate to phosphoenolpyruvate (Figure 1). Once carbon is converted to phosphoenolpyruvate it will be converted to glucose in the liver. PEPCK controls the flow of carbon for hepatic glucose production. This carbon is derived from amino acid intermediates (principally glutamine) derived from the turnover of protein in the periphery for energy generation. There are no known allosteric modifiers of the activity of any PEPCK isoform [123]. PEPCK mRNA and activity are excellent indicators of the enzymatic capacity for gluconeogenesis in the liver. Thus, aging appears to reduce the gluconeogenic capacity of the liver (Figure 1).

Liver gluconeogenesis derives its substrates mainly from protein turnover in the peripheral tissues, suggesting that aging is accompanied by a decrease in the turnover of peripheral protein. During the postabsorptive state, muscle and other tissues utilize amino acids derived from protein turnover to generate energy via the TCA cycle. This amino acid catabolism is initiated in the muscle by two enzymatic steps, collectively called the transdeamination reaction (Figure 2). Transdeamination leads to the liberation of the amino nitrogen as ammonia. Because of its extreme toxicity, this ammonia is transferred to glutamate by glutamine synthetase, producing glutamine. Glutamine serves to transfer both carbon and nitrogen to the liver. Aging leads to a decrease in the activity of muscle glutamine synthetase. This is consistent with an age-related decrease in the turnover of peripheral protein for energy production. It is also consistent with decreased expression of hepatic carbamylphosphate synthase-1, glutamine synthetase, and tyrosine aminotransferase (TAT; Figure 2).

Glutamine produced in the muscle is metabolized in the liver by glutaminase into glutamate and ammonia. The ammonia derived from this reaction can be returned to the glutamine pool by liver glutamine synthetase (Figure 2). An age-related decrease in glutamine synthetase activity would channel glutamine into gluconeogenesis. The nitrogen from this glutamine would be channeled by carbamylphosphate synthase-1 into the urea cycle for detoxification and disposal. These effects are likely responsible for a part of the decrease in muscle protein synthesis and turnover known to occur with age.

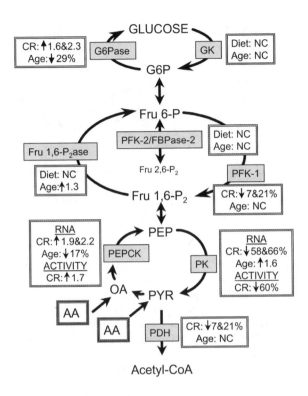

Figure 1. Summary of the effects of age and CR on the glycolytic and gluconeogenic pathways of the liver. Glycolytic metabolism in the liver involves three irreversible, regulated steps. Glucokinase (GK) initiates glucose metabolism by phosphorylation of C6 yielding glucose 6-phosphate (G6P). The committed step in glycolysis, and the second irreversible and regulated step, is the phosphorylation of Fru 6-P by phosphofructokinase (PFK-1) to produce fructose 1,6-bisphosphate (Fru 1,6-P$_2$). The third irreversible step controls the outflow of the pathway. Phosphoenolpyruvate (PEP) and ADP are utilized by pyruvate kinase (PK) to produce pyruvate (PYR) and ATP. Pyruvate dehydrogenase (PDH) oxidatively decarboxylates pyruvate to form acetyl-CoA, which is a bridge between glycolysis and the tricarboxylic acid cycle. Phosphoenolpyruvate carboxykinase (PEPCK) catalyzes the first committed step in gluconeogenesis. The main non-carbohydrate precursors for gluconeogenesis are amino acids from the diet, and from muscle protein breakdown. Other organs also contribute amino acids, but muscle is the major source. Most of these amino acids are converted to oxaloacetate (OA), which is metabolized to PEP by PEPCK. In the second regulated and essentially irreversible step in gluconeogenesis, fructose 1,6-bisphosphatase (Fru 1,6-P$_2$ase) catalyzes the formation of fructose 6-phosphate (Fru 6-P) from fructose 1,6-bisphosphate (Fru 1,6-P$_2$). Finally, in the third essentially irreversible reaction of gluconeogenesis, glucose is formed by the hydrolysis of G6P in a reaction catalyzed by glucose 6-phosphatase (G6Pase). Substrates are not boxed, enzyme names are in shaded boxes, summaries of experimental results are in double bordered boxes, and amino acids are indicated by "AA" in triple bordered boxes. When two values are given following "CR", they represent the fold change in the young and old mice, respectively. The value after "Age" is the main effect of age. A down arrow indicates the percent decrease, an up arrow indicates the fold increase. The value given for age is a combination of both dietary groups. NC is no change.

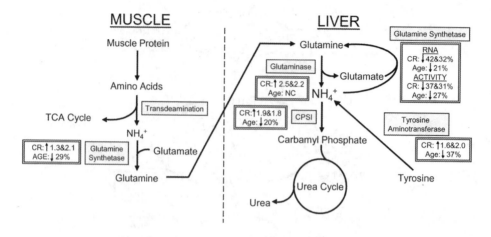

Figure 2. *Summary of the effects of age and diet on muscle and liver nitrogen metabolism. In muscle and other extrahepatic tissues, the degradation of proteins to amino acids is utilized for generating metabolic energy. Transdeamination of amino acids produces tricarboxylic acid cycle intermediates and ammonia. Glutamine synthetase synthesizes glutamine from glutamate and ammonia. Glutamine is transported to the liver where glutaminase releases the ammonia, regenerating glutamate. CPSI converts this ammonia to carbamyl phosphate, which is converted to urea in the urea cycle. The amino group of excess tyrosine is released by TAT as ammonia, which is also detoxified beginning with the action of CPSI. In the figure, substrates are not boxed, enzyme names are in shaded boxes, and summaries of experimental results are in double bordered boxes. When two values are given following "CR", they represent the fold change in the young and old mice, respectively. The value after "Age" is the main effect of age. A down arrow indicates the percent decrease, an up arrow indicates the fold increase. The value given for age is a combination of both dietary groups. NC is no change.*

CR and hepatic energy metabolism

In our microarray studies, CR modified the expression of a significant number of genes coding for key metabolic enzymes [6]. CR increased expression of glutamate oxaloacetate transaminase 1 and decreased expression of pyruvate dehydrogenase E1α subunit. These changes are consistent with our conventional molecular-biological and biochemical studies showing that CR increases shuttling of nitrogen and carbon to the liver from the peripheral tissues. It increases the enzymatic capacity of the liver for gluconeogenesis and the disposal of the byproducts of extrahepatic protein catabolism for energy production, while reducing the enzymatic capacity for glycolysis [121, 125]. These CR effects are consistent with theories of aging, such as the oxidative stress theory, which postulate that the accumulation of damaged proteins contributes to the rate of aging [126].

CR increased fasting levels of the mRNA and activity of PEPCK and mRNA of glucose-6-phosphatase (G6Pase) [121]. The abundance of PEPCK mRNA was greater in the liver of young and old CR mice than it was in control mice of the same ages. PEPCK activity also was higher in the liver of CR mice. As discussed above, aging decreased the mRNA for PEPCK and G6Pase. In addition, when CR and control mice were fasted overnight and fed their normal daily ration of food, PEPCK mRNA and activity decreased within 1.5 hours of feeding in both control and CR mice. However, its mRNA abundance and activity increased rapidly thereafter, especially in CR mice. By 5 hours after feeding, PEPCK activity in CR mice was approximately twice that of controls. Similarly, G6Pase mRNA abundance was higher in CR mice for the 5 hours following feeding. G6Pase catalyzes the terminal step in hepatic glucose production, the hydrolysis of glucose 6-phosphate to glucose and inorganic phosphate (Figure 1). This step leads to the release of glucose from the liver into the circulation when blood glucose levels would otherwise fall.

Together, these results suggest that the enzymatic capacity for gluconeogenesis returns rapidly after feeding. Thus, higher levels of peripheral tissue turnover persist in CR mice, even after feeding. These mice are at approximate weight equilibrium. Therefore, in CR mice feeding is accompanied by intensified protein biosynthetic activity followed immediately by peripheral protein turnover. CR mice are approximately 4 times more insulin sensitive than control mice [120].

Consistent with the interpretation offered above, CR and age decreased the expression of glutamine synthetase activity and mRNA in the liver, while age decreased and CR increased its expression in muscle (Figure 2). These differential effects should lead to a transfer of carbon and nitrogen in the form of glutamine from the periphery to the liver, where it would increase the hepatic pool of glutamine. The increase in glutaminase expression would increase hepatic catabolism of glutamine, producing glutamate and ammonia. mRNA levels closely reflect the levels of glutaminase activity [127, 128]. Ammonia production by glutaminase is closely coupled to urea synthesis by CPSI. CPSI mRNA levels in young and old CR mice were twice that of control mice [121, 125]. CR leads to coordinate induction of carbamylphosphate synthase-1 transcription, mRNA, protein, and activity [129]. The resulting glutamate accumulation would fuel CR enhanced gluconeogenesis.

These data support the interpretation that CR leads to enhanced carbon flux from amino acid degradation in the peripheral tissues to the liver. This amino acid degradation extends to tyrosine, an amino acid that requires a liver specific enzyme, TAT, for catabolism [121]. TAT degradation of tyrosine is well known to provide ketogenic and gluconeogenic substrates to the liver when glucose is limiting and amino acids are utilized as a major source of energy. Aging decreased TAT mRNA in the liver by an average of 37%. TAT mRNA in CR mice was approximately double the level in control mice. The age-related changes in nitrogen metabolizing enzymes are consistent with a decrease in catabolism of extrahepatic protein for energy. CR appears to enhance the capacity for mobilizing and transporting carbon and nitrogen products of muscle protein catabolism to the liver. CR mice also have enhanced hepatic capacity for the biosynthesis of glucose from this carbon, and for the detoxification of this nitrogen.

Conclusions

While the physiological and structural studies of the liver suggest that it ages well, the molecular biology and biochemistry of the liver indicate that it undergoes changes with age that have serious systemic effects. Genome-wide microarray and conventional molecular and biochemical studies indicate that there is an age-related shift in liver toward a state associated with oncogenesis, fibrosis, cirrhosis, and unhealthful apolipoprotein and fatty acid biosynthesis. Evidence was found for age-related increases in inflammation, cellular stress, and fibrosis; and for reduced capacity for apoptosis, negative cell-growth control, and phase I and II xenobiotic metabolism. LT- and ST-CR reversed the majority of these changes. LT-CR also produced CR-specific changes in signal transduction-associated gene expression known to lead to enhanced longevity. Evidence for a CR-related increase in the turnover and renewal of peripheral protein also was found. In addition, healthful changes in apolipoprotein and fatty acid biosynthesis-related gene expression were found. LT- and ST-CR produced changes in gene expression associated with enhanced anti-proliferative growth control, increased apoptosis and reduced chemical carcinogenesis. Together these studies make it clear that aging and its mitigation by CR are multifaceted processes which affect many aspects of liver function at the molecular level. It also appears that unbiased, exploratory approaches such as the genome-wide microarray studies described here are providing new and valuable insights into these processes.

References

1. Dice JF (1993). Cellular and molecular mechanisms of aging. *Physiol Rev.* 73: 149–59
2. Schmucker DL (1998). Aging and the liver: An update. *J Gerontol Biol Sci.* 53A: B315–20.
3. Joh T, Itoh M, Katsumi K, *et al.* (1986). Physiological concentrations of human epidermal growth factor in biological fluids: use of a sensitive enzyme immunoassay. *Clin Chim Acta* 158: 81–90.
4. Durnas C, Loi CM, Cusack BJ (1990). Hepatic drug metabolism and aging. *Clin Pharmacokinet.* 19: 359–89.
5. Sotaniemi EA, Arranto AJ, Pelkonen O, Pasanen M (1997). Age and cytochrome P450-linked drug metabolism in humans: an analysis of 226 subjects with equal histopathologic conditions. *Clin Pharmacol Ther.* 61: 331–9.
6. Cao SX, Dhahbi JM, Mote PL, Spindler SR (2001). Genomic profiling of short- and long-term caloric restriction in the liver of aging mice. *Proc Natl Acad Sci USA* 98: 10630–5.
7. Guarente L, Kenyon C (2000). Genetic pathways that regulate ageing in model organisms [In Process Citation]. *Nature* 408: 255–62.
8. Brown-Borg HM, Borg KE, Meliska CJ, Bartke A (1996). Dwarf mice and the ageing process [letter]. *Nature* 384: 33.
9. Ward W, Richardson A (1991). Effect of age on liver protein synthesis and degradation. *Hepatology* 14: 935–48.
10. Kaminski N, Allard JD, Pittet JF, *et al.* (2000). Global analysis of gene expression in pulmonary fibrosis reveals distinct programs regulating lung inflammation and fibrosis. *Proc Natl Acad Sci USA* 97: 1778–83.

11. Welsh JB, Zarrinkar PP, Sapinoso LM, *et al.*(2001). Analysis of gene expression profiles in normal and neoplastic ovarian tissue samples identifies candidate molecular markers of epithelial ovarian cancer. *Proc Natl Acad Sci USA* 98: 1176–81

12. Golub TR, Slonim DK, Tamayo P, *et al.* (1999). Molecular classification of cancer: class discovery and class prediction by gene expression monitoring. Science 286: 531–7.

13. Lee CK, Weindruch R, Prolla TA (2000). Gene-expression profile of the ageing brain in mice. *Nat Genet.* 25: 294–7.

14. Kayo T, Allison DB, Weindruch R, Prolla TA (2001). Influences of aging and caloric restriction on the transcriptional profile of skeletal muscle from rhesus monkeys. *Proc Natl Acad Sci USA* 98: 5093–8.

15. Coschigano KT, Clemmons D, Bellush LL, Kopchick JJ (2000). Assessment of growth parameters and life span of GHR/BP gene-disrupted mice. *Endocrinology* 141: 2608–13.

16. Flurkey K, Papaconstantinou J, Miller RA, Harrison DE (2001). Lifespan extension and delayed immune and collagen aging in mutant mice with defects in growth hormone production. *Proc Natl Acad Sci USA* 98: 6736–41.

17. Bartke A, Wright JC, Mattison JA, Ingram DK, Miller RA, Roth GS (2001). Extending the lifespan of long-lived mice. *Nature* 414: 412.

18. Frystyk J, Delhanty PJ, Skjaerbaek C, Baxter RC (1999). Changes in the circulating IGF system during short-term fasting and refeeding in rats. *Am J Physiol.* 277: E245–52.

19. Herlihy JT, Stacy C, Bertrand HA (1990). Long-term food restriction depresses serum thyroid hormone concentrations in the rat. *Mech Ageing Dev.* 53: 9–16.

20. Katzeff HL, Yang M-U, Presta E, Leibel RL, Hirsch J, Van Itallie TB (1990). Calorie restriction and iopanoic acid effects on thyroid hormone metabolism. *Am J Clin Nutr.* 52: 263–6.

21. Kealy RD, Lawler DE, Ballam JM, *et al.* (2002). Effects of diet restriction on life span and age-related changes in dogs. *J Am Vet Med Assoc.* 220: 1315–20.

22. Jiang CH, Tsien JZ, Schultz PG, Hu Y (2001). The effects of aging on gene expression in the hypothalamus and cortex of mice. *Proc Natl Acad Sci USA* 98: 1930–4.

23. Lee CK, Klopp RG, Weindruch R, Prolla TA (1999). Gene expression profile of aging and its retardation by caloric restriction. *Science* 285: 1390–3.

24. Keshav S, Chung P, Milon G, Gordon S (1991). Lysozyme is an inducible marker of macrophage activation in murine tissues as demonstrated by in situ hybridization. *J Exp Med.* 174: 1049–58.

25. Petry F, Reid KB, Loos M (1989). Molecular cloning and characterization of the complementary DNA coding for the B-chain of murine Clq. *FEBS Lett.* 258: 89–93.

26. Jaeschke H (1997). Cellular adhesion molecules: regulation and functional significance in the pathogenesis of liver diseases. *Am J Physiol.* 273: G602–11.

27. Friedman SL (1993). Seminars in medicine of the Beth Israel Hospital, Boston. The cellular basis of hepatic fibrosis. Mechanisms and treatment strategies. *N Engl J Med.* 328: 1828–35.

28. Pepys MB, Baltz M, Gomer K, Davies AJ, Doenhoff M (1979). Serum amyloid P-component is an acute-phase reactant in the mouse. *Nature* 278: 259–61.

29. Buttle DJ, Abrahamson M, Burnett D, *et al.* (1991). Human sputum cathepsin B degrades proteoglycan, is inhibited by alpha 2-macroglobulin and is modulated by neutrophil elastase cleavage of cathepsin B precursor and cystatin C. *Biochem J.* 276(Pt 2): 325–31.

30. Kos J, Lah TT (1998). Cysteine proteinases and their endogenous inhibitors: target proteins for prognosis, diagnosis and therapy in cancer (review). *Oncol Rep.* 5: 1349–61.

31. Cherry, Engelman RW, Wang BY, Kinjoh K, El Badri NS, Good RA (1998). Calorie restriction delays the crescentic glomerulonephritis of SCG/Kj mice. *Proc Soc Exp Biol Med.* 218: 218–22.

32. Weindruch R, Walford RL, Fligiel S, Guthrie D (1986). The retardation of aging in mice by dietary restriction: longevity, cancer, immunity and lifetime energy intake. *J Nutr.* 116: 641–54.

33. Warner HR (1997). Aging and regulation of apoptosis. *Curr Top Cell Regul.* 35: 107–21.

34. Pieri C, ZS-Nagy I, Mazzufferi G, Giuli C (1975). The aging of rat liver as revealed by electron microscopic morphometry – I. Basic parameters. *Exp Gerontol.* 10: 291–304.

35. Muskhelishvili L, Hart RW, Turturro A, James SJ (1995). Age-related changes in the intrinsic rate of apoptosis in livers of diet-restricted and ad libitum-fed B6C3F1 mice. *Am J Pathol.* 147: 20–4.

36. Suh Y, Lee KA, Kim WH, Han BG, Vijg J, Park SC (2002). Aging alters the apoptotic response to genotoxic stress. *Nat Med.* 8: 3–4.

37. Weindruch R, Walford RL (1982). Dietary restriction in mice beginning at 1 year of age: effect on life-span and spontaneous cancer incidence. *Science* 215: 1415–18.

38. Muskhelishvili L, Turturro A, Hart RW, James SJ (1996). Pi-class glutathione-S-transferase-positive hepatocytes in aging B6C3F1 mice undergo apoptosis induced by dietary restriction. *Am J Pathol.* 149: 1585–91.

39. Cantley LC, Neel BG (1999). New insights into tumor suppression: PTEN suppresses tumor formation by restraining the phosphoinositide 3-kinase/AKT pathway. *Proc Natl Acad Sci USA* 96: 4240–5.

40. Cortes U, Moyret-Lalle C, Falette N, *et al.* (2000). BTG gene expression in the p53-dependent and -independent cellular response to DNA damage. *Mol Carcinog.* 27: 57–64.

41. Pinto M, Lobe CG (1996). Products of the grg (Groucho-related gene) family can dimerize through the amino-terminal Q domain. *J Biol Chem.* 271: 33026–31.

42. Puisieux A, Ji J, Ozturk M (1996). Annexin II up-regulates cellular levels of p11 protein by a post-translational mechanisms. *Biochem J.* 313(Pt 1): 51–5.

43. Todd R, McBride J, Tsuji T, *et al.* (1995). Deleted in oral cancer-1 (doc-1), a novel oral tumor suppressor gene. *FASEB J.* 9: 1362–70.

44. Oh Y, Nagalla SR, Yamanaka Y, Kim HS, Wilson E, Rosenfeld RG (1996). Synthesis and characterization of insulin-like growth factor-binding protein (IGFBP)-7. Recombinant human mac25 protein specifically binds IGF-I and -II. *J Biol Chem.* 271: 30322–5.

45. Higami Y, Shimokawa I, Okimoto T, Tomita M, Yuo T, Ikeda T (1997). Effect of aging and dietary restriction on hepatocyte proliferation and death in male F344 rats. *Cell Tissue Res.* 288: 69–77.

46. Ando K, Higami Y, Tsuchiya T, Kanematsu T, Shimokawa I (2002). Impact of aging and life-long calorie restriction on expression of apoptosis-related genes in male F344 rat liver. *Microsc Res Tech.* 59: 293–300.

47. Muskhelishvili L, Hart RW, Turturro A, James SJ (1995). Age-related changes in the intrinsic rate of apoptosis in livers of diet-restricted and ad libitum-fed B6C3F1 mice. *Am J Pathol.* 147: 20–4.

48. Shimizu S, Tsujimoto Y (2000). Proapoptotic BH3-only Bcl-2 family members induce cytochrome c release, but not mitochondrial membrane potential loss, and do not directly modulate voltage-dependent anion channel activity. *Proc Natl Acad Sci USA* 97: 577–82.

49. Pitot HC, Hikita H, Dragan Y, Sargent L, Haas M (2000). Review article: the stages of gastrointestinal carcinogenesis – application of rodent models to human disease. *Aliment Pharmacol Ther.* 14(Suppl 1): 153–60.

50. Hikita H, Vaughan J, Babcock K, Pitot HC (1999). Short-term fasting and the reversal of the stage of promotion in rat hepatocarcinogenesis: role of cell replication, apoptosis, and gene expression. *Toxicol Sci.* 52: 17–23.

51. Tang NM, Korth MJ, Gale M, *et al.* (1999). Inhibition of double-stranded RNA- and tumor necrosis factor alpha-mediated apoptosis by tetratricopeptide repeat protein and cochaperone P58(IPK). *Mol Cell Biol.* 19: 4757–65.

52. Reimold AM, Etkin A, Clauss I, *et al.* (2000). An essential role in liver development for transcription factor XBP-1. *Genes Dev.* 14: 152–7.

53. Treister A, Sagi-Assif O, Meer M, *et al.* (1998). Expression of Ly-6, a marker for highly malignant murine tumor cells, is regulated by growth conditions and stress. *Int J Cancer* 77: 306–13.

54. James SJ, Muskhelishvili L (1994). Rates of apoptosis and proliferation vary with caloric intake and may influence incidence of spontaneous hepatoma in C57BL/ 6 × C3H F1 mice. *Cancer Res.* 54: 5508–10.

55. Grasl-Kraupp B, Bursch W, Ruttkay-Nedecky B, Wagner A, Lauer B, Schulte-Hermann R (1994). Food restriction eliminates preneoplastic cells through apoptosis and antagonizes carcinogenesis in rat liver. *Proc Natl Acad Sci USA* 91: 9995–9.

56. Holt PR, Moss SF, Heydari AR, Richardson A (1998). Diet restriction increases apoptosis in the gut of aging rats. *J Gerontol: Biol Sci.* 53A: B168–72.

57. Dunn SE, Kari FW, French J, *et al.* (1997). Dietary restriction reduces insulin-like growth factor I levels, which modulates apoptosis, cell proliferation, and tumor progression in p53-deficient mice. *Cancer Res.* 57: 4667–72.

58. Luan X, Zhao W, Chandrasekar B, Fernandes G (1995). Calorie restriction modulates lymphocyte subset phenotype and increases apoptosis in MRL/lpr mice. *Immunol Lett.* 47: 181–6.

59. Spindler SR, Crew MD, Mote PL, Grizzle JM, Walford RL (1990). Dietary energy restriction in mice reduces hepatic expression of glucose-regulated protein 78 (BiP) and 94 mRNA. *J Nutr.* 120: 1412–17.

60. Dhahbi JM, Mote PL, Tillman JB, Walford RL, Spindler SR (1997). Dietary energy tissue-specifically regulates endoplasmic reticulum chaperone gene expression in the liver of mice. *J Nutr.* 127: 1758–64.

61. Dhahbi JM, Cao SX, Tillman JB, *et al.* (2001). Chaperone-mediated regulation of hepatic protein secretion by caloric restriction. *Biochem Biophys Res Commun.* 284: 335–9.

62. Marber MS, Mestril R, Chi SH, Sayen MR, Yellon DM, Dillmann WH (1995). Overexpression of the rat inducible 70-kD heat stress protein in a transgenic mouse increases the resistance of the heart to ischemic injury. *J Clin Invest.* 95: 1446–56.

63. Kyriakis JM, Avruch J (1996). Sounding the alarm: protein kinase cascades activated by stress and inflammation. *J Biol Chem.* 271: 24313–16.

64. Jacquier-Sarlin MR, Fuller K, Dinh-Xuan AT, Richard MJ, Polla BS (1994). Protective effects of hsp70 in inflammation. *Experientia* 50: 1031–8.

65. Heydari AR, Wu B, Takahashi R, Strong R, Richardson A (1993). Expression of heat shock protein 70 is altered by age and diet at the level of transcription. *Mol Cell Biol.* 13: 2909–18.

66. Gutsmann-Conrad A, Heydari AR, You SH, Richardson A (1998). The expression of heat shock protein 70 decreases with cellular senescence in vitro and in cells derived from young and old human subjects. *Exp Cell Res.* 241: 404–13.

67. Wickner S, Maurizi MR, Gottesman S (1999). Posttranslational quality control: folding, refolding, and degrading proteins. *Science* 286: 1888–93.

68. Hartl FU (1996). Molecular chaperones in cellular protein folding. *Nature* 381: 571–9.

69. Medina R, Wing SS, Goldberg AL (1995). Increase in levels of polyubiquitin and proteasome mRNA in skeletal muscle during starvation and denervation atrophy. *Biochem J.* 307(Pt 3): 631–7.

70. Sherman MY, Goldberg AL (1996). Involvement of molecular chaperones in intracellular protein breakdown. *EXS* 77: 57–78.

71. Ciocca DR, Fuqua SA, Lock-Lim S, Toft DO, Welch WJ, McGuire WL (1992). Response of human breast cancer cells to heat shock and chemotherapeutic drugs. *Cancer Res.* 52: 3648–54.

72. McMillan DR, Xiao X, Shao L, Graves K, Benjamin IJ (1998). Targeted disruption of heat shock transcription factor 1 abolishes thermotolerance and protection against heat-inducible apoptosis. *J Biol Chem.* 273: 7523–8.

73. McCormick TS, McColl KS, Distelhorst CW (1997). Mouse lymphoma cells destined to undergo apoptosis in response to thapsigargin treatment fail to generate a calcium-mediated grp78/grp94 stress response. *J Biol Chem.* 272: 6087–92.

74. Jamora C, Dennert G, Lee AS (1996). Inhibition of tumor progression by suppression of stress protein GRP78/BiP induction in fibrosarcoma B/C10ME. *Proc Natl Acad Sci USA* 93: 7690–4.

75. Migliorati G, Nicoletti I, Crocicchio F, Pagliacci C, D'Adamio F, Riccardi C (1992). Heat shock induces apoptosis in mouse thymocytes and protects them from glucocorticoid-induced cell death. *Cell Immunol.* 143: 348–56.

76. Wei YQ, Zhao X, Kariya Y, Fukata H, Teshigawara K, Uchida A (1994). Induction of apoptosis by quercetin: involvement of heat shock protein. *Cancer Res.* 54: 4952–7.

77. Schulte-Hermann R, Bursch W, Grasl-Kraupp B, Mullauer L, Ruttkay-Nedecky B (1995). Apoptosis and multistage carcinogenesis in rat liver. *Mutat Res.* 333: 81–7.

78. Dhahbi JM, Cao SX, Mote PL, Rowley BC, Wingo JE, Spindler SR (2002). Postprandial induction of chaperone gene expression is rapid in mice. *J Nutr.* 132: 31–7.

79. Rao RV, Peel A, Logvinova A, *et al.* (2002). Coupling endoplasmic reticulum stress to the cell death program: role of the ER chaperone GRP78. *FEBS Lett.* 514: 122–8.

80. Rao RV, Hermel E, Castro-Obregon S, *et al.* (2001). Coupling endoplasmic reticulum stress to the cell death program. Mechanism of caspase activation. *J Biol Chem.* 276: 33869–74.

81. Koong AC, Chen EY, Lee AS, Brown JM, Giaccia AJ (1994). Increased cytotoxicity of chronic hypoxic cells by molecular inhibition of GRP78 induction. *Int J Radiat Oncol Biol Phys.* 28: 661–6.

82. Cai JW, Henderson BW, Shen JW, Subjeck JR (1993). Induction of glucose regulated proteins during growth of a murine tumor. *J Cell Physiol.* 154: 229–37.

83. Patierno SR, Tuscano JM, Kim KS, Landolph JR, Lee AS (1987). Increased expression of the glucose-regulated gene encoding the Mr 78,000 glucose-regulated protein in chemically and radiation-transformed C3H 10T1/2 mouse embryo cells. *Cancer Res.* 47: 6220–4.

84. Menoret A, Meflah K, Le Pendu J (1994). Expression of the 100-kda glucose-regulated protein (GRP100/endoplasmin). is associated with tumorigenicity in a model of rat colon adenocarcinoma. *Int J Cancer* 56: 400–5.

85.   Chatterjee S, Cheng MF, Berger SJ, Berger NA (1994). Induction of M(r) 78,000 glucose-regulated stress protein in poly(adenosine diphosphate-ribose) polymerase- and nicotinamide adenine dinucleotide-deficient V79 cell lines and its relation to resistance to the topoisomerase II inhibitor etoposide. *Cancer Res.* 54: 4405–11.

86.   Chatterjee S, Cheng MF, Berger RB, Berger SJ, Berger NA (1995). Effect of inhibitors of poly(ADP-ribose) polymerase on the induction of GRP78 and subsequent develop- ment of resistance to etoposide. *Cancer Res.* 55: 868–73.

87.   Shen J, Hughes C, Chao C, *et al.* (1987). Coinduction of glucose-regulated proteins and doxorubicin resistance in Chinese hamster cells. *Proc Natl Acad Sci USA* 84: 3278–82.

88.   Sugawara S, Nowicki M, Xie S, Song HJ, Dennert G (1990). Effects of stress on lysability of tumor targets by cytotoxic T cells and tumor necrosis factor [published erratum appears in *J Immunol.* 1991; 146(3): 1083]. *J Immunol.* 145: 1991–8.

89.   Sugawara S, Takeda K, Lee A, Dennert G (1993). Suppression of stress protein GRP78 induction in tumor B/C10ME eliminates resistance to cell mediated cytotoxicity. *Cancer Res.* 53: 6001–5.

90.   Johnson RJ, Liu N, Shanmugaratnam J, Fine RE (1998). Increased calreticulin stability in differentiated NG-108-15 cells correlates with resistance to apoptosis induced by antisense treatment. *Brain Res Mol Brain Res.* 53: 104–11.

91.   Knittler MR, Haas IG (1992). Interaction of BiP with newly synthesized immunoglo- bulin light chain molecules: cycles of sequential binding and release. *EMBO J.* 11: 1573– 81.

92.   Dorner AJ, Krane MG, Kaufman RJ (1988). Reduction of endogenous GRP78 levels improves secretion of a heterologous protein in CHO cells. *Mol Cell Biol.* 8: 4063–70.

93.   Dorner AJ, Wasley LC, Kaufman RJ (1992). Overexpression of GRP78 mitigates stress induction of glucose regulated proteins and blocks secretion of selective proteins in Chinese hamster ovary cells. *EMBO J.* 11: 1563–71.

94.   Beggah A, Mathews P, Beguin P, Geering K (1996). Degradation and endoplasmic reticulum retention of unassembled alpha- and beta-subunits of Na,K-ATPase correlate with interaction of BiP. *J Biol Chem.* 271: 20895–902.

95.   Sato R, Imanaka T, Takatsuki A, Takano T (1990). Degradation of newly synthesized apolipoprotein B-100 in a pre-Golgi compartment. *J Biol Chem.* 265: 11880–4.

96.   Bonifacino JS, Lippincott-Schwartz J (1991). Degradation of proteins within the endoplasmic reticulum. *Curr Opin Cell Biol.* 3: 592–600.

97.   Amara JF, Lederkremer G, Lodish HF (1989). Intracellular degradation of unas- sembled asialoglycoprotein receptor subunits: a pre-Golgi, nonlysosomal endoproteoly- tic cleavage. *J Cell Biol.* 109: 3315–24.

98.   Lippincott-Schwartz J, Bonifacino JS, Yuan LC, Klausner RD (1988). Degradation from the endoplasmic reticulum: disposing of newly synthesized proteins. *Cell* 54: 209– 20.

99.   Vlassara H (1997). Recent progress in advanced glycation end products and diabetic complications. *Diabetes* 46(Suppl 2): S19–25.

100. Yan SD, Stern D, Schmidt AM (1997). What's the RAGE? The receptor for advanced glycation end products (RAGE) and the dark side of glucose. *Eur J Clin Invest.* 27: 179– 81.

101. Masoro EJ, Katz MS, McMahan CA (1989). Evidence for the glycation hypothesis of aging from the food-restricted rodent model. *J Gerontol.* 44: B20–2.

102. Vlassara H, Bucala R (1996). Recent progress in advanced glycation and diabetic vascular disease: role of advanced glycation end product receptors. *Diabetes* 45(Suppl 3): S65–6.

103. Beisswenger PJ, Makita Z, Curphey TJ, *et al.* (1995). Formation of immunochemical advanced glycosylation end products precedes and correlates with early manifestations of renal and retinal disease in diabetes. *Diabetes* 44: 824–9.

104. Cefalu WT, Bell-Farrow AD, Wang ZQ, *et al.* (1995). Caloric restriction decreases age-dependent accumulation of the glycoxidation products, N epsilon-(carboxymethyl)lysine and pentosidine, in rat skin collagen. *J Gerontol A Biol Sci Med Sci.* 50: B337–41.

105. Dyer DG, Dunn JA, Thorpe SR, Lyons TJ, McCance DR, Baynes JW (1992). Accumulation of Maillard reaction products in skin collagen in diabetes and aging. *Ann NY Acad Sci.* 663: 421–2.

106. Sell DR, Lane MA, Johnson WA, *et al.* (1996). Longevity and the genetic determination of collagen glycoxidation kinetics in mammalian senescence. *Proc Natl Acad Sci USA* 93: 485–90.

107. Thompson PD, Hsieh JC, Whitfield GK, *et al.* (1999). Vitamin D receptor displays DNA binding and transactivation as a heterodimer with the retinoid X receptor, but not with the thyroid hormone receptor. *J Cell Biochem.* 75: 462–80.

108. Mote PL, Grizzle JM, Walford RL, Spindler SR (1990). Age-related down regulation of hepatic cytochrome P1-450, P3-450, catalase and CuZn-superoxide dismutase RNA. *Mech Ageing Dev.* 53: 101–10.

109. Sun JQ, Lau PP, Strobel HW (1986). Aging modifies the expression of hepatic microsomal cytochromes P-450 after pretreatment of rats with beta-naphthoflavone or phenobarbital. *Exp Gerontol.* 21: 65–73.

110. Richardson A, Rutherford MS, Birchenall-Sparks MC, Robert MS, Wu WT, Cheung HT (1985). Levels of specific messenger RNA species as a function of age. In: Sohal RS, Birnbaum LS, Cutler RG, eds. *Molecular Biology of Aging: Gene Stability and Gene Expression.* New York: Raven Press, pp. 229–41.

111. Ali B, Walford RL, Imamura T (1985). Influence of aging and poly IC treatment on xenobiotic metabolism in mice. *Life Sci.* 37: 1387–93.

112. Mote PL, Grizzle JM, Walford RL, Spindler SR (1991). Influence of age and caloric restriction on expression of hepatic genes for xenobiotic and oxygen metabolizing enzymes in the mouse. *J Gerontol.* 46: B95–100.

113. Yen TC, Chen YS, King KL, Yeh SH, Wei YH (1989). Liver mitochondrial respiratory functions decline with age. *Biochem Biophys Res Commun.* 165: 944–1003.

114. Muller-Hocker J, Aust D, Rohrbach H, *et al.* (1997). Defects of the respiratory chain in the normal human liver and in cirrhosis during aging. *Hepatology* 26: 709–19.

115. Rossetti L, Giaccari A, DeFronzo RA (1990). Glucose toxicity. *Diabetes Care* 13: 610–30.

116. Halter JB (1995). Carbohydrate metabolism. In: Masoro EJ, ed. *Handbook of Physiology. Section 11: Aging.* pp. 119–145. New York: Oxford University Press, New York,

117. Lane MA, Ball SS, Ingram DK, Cutler RG, Engel J, Read V, Roth GS (1995). Diet restriction in rhesus monkeys lowers fasting and glucose-stimulated glucoregulatory end points. *Am J Physiol.* 268: E941–8.

118. Harris SB, Gunion MW, Rosenthal MJ, Walford RL (1994). Serum glucose, glucose tolerance, corticosterone and free fatty acids during aging in energy restricted mice. *Mech Ageing Dev.* 73: 209–21.

119. Walford RL, Harris SB, Gunion MW (1992). The calorically restricted low-fat nutrient-dense diet in Biosphere 2 significantly lowers blood glucose, total leukocyte count, cholesterol, and blood pressure in humans. *Proc Natl Acad Sci USA* 89: 11533–7.

120. Spindler SR (2001). Caloric restriction enhances the expression of key metabolic enzymes associated with protein renewal during aging. *Ann NY Acad Sci.* 928: 296–304.

121.   Dhahbi JM, Mote PL, Wingo J, Tillman JB, Walford RL, Spindler SR (1999). Calories and aging alter gene expression for gluconeogenic, glycolytic, and nitrogen-metabolizing enzymes. *Am J Physiol.* 277: E352–60.

122.   Wimonwatwatee T, Heydari AR, Wu WT, Richardson A (1994). Effect of age on the expression of phosphoenolpyruvate carboxykinase in rat liver. *Am J Physiol.* 267: G201–4.

123.   Hanson RW, Reshef L (1997). Regulation of phosphoenolpyruvate carboxykinase (GTP) gene. *Annu Rev Biochem.* 66: 581–611.

124.   Van Remmen H, Ward WF, Sabia RV, Richardson A (1995). Gene expression and protein degradation. In: Masoro EJ, ed. *Handbook of Physiology. Section 11: Aging.* New York: Oxford University Press, New York, pp. 171–234.

125.   Dhahbi JM, Mote PL, Wingo J, *et al.* (2001). Caloric restriction alters the feeding response of key metabolic enzyme genes. *Mech Ageing Dev.* 122: 35–50.

126.   Stadtman ER, Berlett BS (1998). Reactive oxygen-mediated protein oxidation in aging and disease. *Drug Metab Rev.* 30: 225–43.

127.   Watford M, Vincent N, Zhan Z, Fannelli J, Kowalski T, Kovacevic Z (1994). Transcriptional control of rat hepatic glutaminase expression by dietary protein level and starvation. *J Nutr.* 124: 493–99.

128.   Zhan Z, Vincent NC, Watford M (1994). Transcriptional regulation of the hepatic glutaminase gene in the streptozotocin-diabetic rat. *Int J Biochem.* 26: 263–68.

129.   Tillman JB, Dhahbi JM, Mote PL, Walford RL, Spindler SR (1996). Dietary calorie restriction in mice induces carbamyl phosphate synthetase I gene transcription tissue specifically. *J Biol Chem.* 271: 3500–6.

The Aging Male Reproductive System

Rafi T. Kevorkian and John E. Morley

Saint Louis University Health Sciences Center, Division of Geriatric Medicine and St. Louis VA Medical Center, GRECC, St. Louis, Missouri, USA

The reproductive glands are an essential part of the human body. Extensive changes occur as we develop in the womb, to the period of puberty, and then as we age. Through out this time, intricate and timely hormonal changes occur in a unique manner that allow these changes to occur. As the process unwinds our aging body reflects the decades of repetitive assimilation and protection of the germ lines, as well as hormonal decline that affects every male. The revolution of science has allowed us to better understand the aging process so that maintenance of well being and functionality can be preserved. Hopefully, better understanding of this process will allow us to promote successful aging.

The initial process of reproductive gland formation begins in the womb. The gonads of both sexes develop by 5 weeks of gestation, and are indistinguishable from each other. Primordial germ cells within the gonad generate oogonia and spermatogonia. They eventually mature into large numbers of ova and sperm. Two types of cells are present the first being the Sertoli cell which forms within the seminiferous tubules of the indifferent gonad. They help maintain the germ cells, foster their maturation, and guide their movement into the genital duct system. Interstitial cells are the second type of cell that gives rise to theca cells of the ovary and the Leydig cells in the testes, which secrete androgenic hormones.

As the fetus subsequently matures, genetic determination begins to take effect. The presence of the Y chromosome is the single most constant determinant of maleness. Neither testes nor a masculine genital pattern can develop in its absence. Other factors necessary for masculinization include the Sex determining region of the Y (SRY gene) which encodes the testis determining factor (TDF). A similar gene to the SRY gene encodes a histocompatibility antigen known as the H-Y. This glycoprotien antigen is one of two present on all male cells. Autosomal and X-chromosomal genes also play a role in virilization of the genital ducts and external genitalia. An androgen receptor hormone is encoded by genes on the X chromosome.

R. Aspinall (ed.), Aging of Organs and Systems, 293–308.

In conjunction with genetic determination, proper migration of embryonic tissue, followed by hormone production is necessary for continued propagation and development of the genital organs. Primordial germ cells have been noted to differentiate in the five-day-old blastocyst. At 3 weeks gestation, they are present in the yolk sac endoderm with subsequent migration to the genital ridge. They then associate with mesonephric tissue to form an indifferent gonad. The primitive gonad consists of the coelomic epithelium, which is the precursor of Sertoli cells in males, the mesenchymal stromal cells, which are precursors to the Leydig cells, and the germ cells. The seminiferous tubules begin to form by 6–7 weeks gestation as the Sertolli cells enclose the germ cells. For further male development, Leydig cells need to appear by 8–9 weeks gestation, because they secrete testosterone. Female genitals will develop if testosterone is not secreted. As a result the Wolffian, or Mesonephric duct begins to grow and eventually gives rise to the epidiymis, the vas deferens, the seminal vesicles, and the ejaculatory duct by 9–10 weeks of gestation. The Sertoli cells then produce an Anti-Mullerian hormone, which leads to the regression of the Mullerian duct, which is the precursor to the inner female genitalia. This hormone may participate in organizing the testes into the seminiferous tubules, stimulating the development of Leydig cells, and initiating descent of the testis into the inguinal area. The external genitalia of both sexes begin to differentiate at 9–10 weeks gestation. They are derived from the genital tubercle, which develops into the glans penis, the genital swelling which fold and fuse into the scrotum, the urethral folds, which enlarge and enclose the penile urethra and corpus spongiosa, and the urogenital sinus which gives rise to the prostate gland. The above will not occur without the presence of the androgen receptor on target tissue (Figure 1).

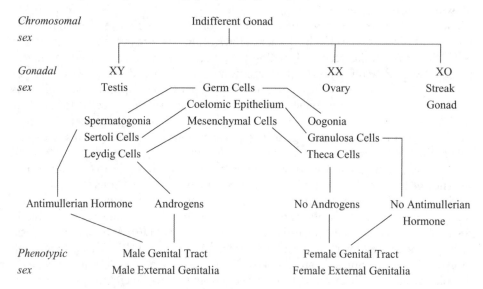

Figure 1. Development of the cells of the ovary and testis.

Sexual differentiation in males does not depend on fetal pituitary gonadotropins in early stages of development. A leutinizing like hormone is produced from the placenta called chorionic gonadotropin. It stimulates testosterone production by the fetal Leydig cells. As development continues, growth of the male external genitalia requires fetal pituitary leutinizing hormone to stimulate the necessary quantity of androgen during the last six months of gestation. The Sertoli and Lydia cells also synthesize and secrete numerous peptide and protein products that act systemically and locally to modulate gametogenesis. One of these products is Inhibin which is a glycoprotein that circulates in plasma and inhibits gonadotropin-releasing hormone stimulated follicle-stimulating hormone (FSH) secretion by the pituitary gland. Activin has the opposite action and stimulates FSH secretion. Follistatin is another FSH-suppressing protein, but its structure is different from Inhibin. It acts by binding and neutralizing activin. The main regulators of gonadal function are luteinizing hormone (LH), and the follicle-stimulating hormone (FSH). Through negative feedback, their synthesis and secretion are increased by decreases in gonadal steroids. LH stimulates the interstitial cell line of male and female gonads (Leydig and thecal cells) mainly to secrete androgens. FSH acts on ovarian granulosa cells and testicular Sertoli cells. It stimulates the aromatase gene leading to an increase in estrogen production. FSH also stimulates inhibin and numerous other protein products of the Sertoli and granulosa cells.

Levels of FSH and LH initially increase at birth followed by a second burst at 2 months of age. They are then produced at very low but detectable levels. As puberty approaches the low-level pulsatile pattern of LH and FSH secretion becomes more pronounced. This leads to an increase of the appropriate sex steroids in each gender, which augments growth hormone secretion, leading to accelerated linear growth and secondary maturation of the reproductive organs. Women develop cyclic monthly bursts with LH exceeding FSH. This does not occur in men. By the fifth decade, a decline in gonadal responsiveness occurs in both sexes. It is more gradual in men allowing reproductive capability into the eight decade. In both sexes, continued negative feedback leads to elevated plasma gonadotropin levels with FSH rising more than LH (more distinct in females).

The pulsatile release of gonadotropin releasing hormone (GnRH) and subsequent arrival of LH and FSH at their target cells is critical for reproduction (Figure 2). Men with congenital GnRH deficiency can be made fertile only if exogenous GnRH is given in appropriately sized pulses and timed intervals. Fertility does not occur if it is given continuously. Testosterone is then synthesized and released by the Leydig cells, which are regulated by LH. Therefore, plasma testosterone levels undergo small coordinate pulses throughout the day. In addition, plasma testosterone levels follow a superimposed diurnal trend. Plasma testosterone is about 25% lower at 20:00 than 08:00. Within the male testes, production of sperm occurs daily through an intricate system of cell division. Approximately 100–200 million sperm are produced daily. Primary spermatocytes divide to form the secondary spermatocytes. They divide to form spermatids. They lie near the lumen of the seminiferous tubule. They are attached to the Sertoli cells by specialized junctions. The spermatids then undergo a process called spermiogenesis. There is nuclear condensation, shrinkage of cyto-

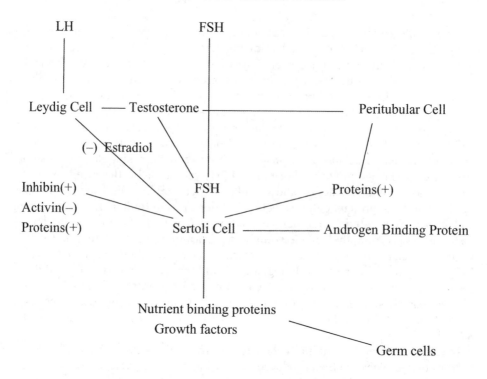

Figure 2. Hormonal regulation of the testis during spermatogensis. (+) stimulating, (–) inhibiting.

plasm, formation of an acrosome, and development of a tail to emerge as flagellated spermatozoa. The spermatozoa are then extruded into the lumen of the tubule by a process called spermiation, during which most of the cytoplasm of the spermatozoa is ejected as the residual body. This entire sequence from spermatogonia to spematozoa takes 60–70 days. They then reach the epididymis, where they undergo further maturation, gaining motility and losing all their cytoplasm. The growth and differentiation of the epididymis, as well as the motility and fertility of the sperm that migrate through it, depend on androgens. Proteins provided by epididymal and seminiferous tubular fluid bind to the membranes of sperm and enhance their motility and fertilizing ability. The sperm are protected by the testes-blood barrier, which prevents passage of unwanted noxious substances.

Delivery of sperm into the female genital tract for reproductive occurs by ejaculation from the vas deferens. Prostatic and seminal vesicle secretions help neutralize the acid pH of the vagina and help stimulate contraction of the uterus and fallopian tubes. Ejaculation requires penile erection, which is caused by blood filling the venous sinuses of the corpora cavernosa and spongiosa. This process can be initiated in the brain. It may also begin with afferent sensory impulses from the penis carried by the pudendal nerve. This is further mediated by impulses in the pelvic nerves, which contain both parasympathetic and sympathetic fibers. There are five phases to an erection, the first being the flaccid phase leading to the filling phase. It is followed by the tumescence phase. During tumescence, nitric oxide and prostanglandin E1 cause relaxation of the cavernosa smooth muscle and an increase in its compliance allowing easy entry of blood and engorgement. Full erection then occurs followed by the rigid phase where cavernosa pressure exceeds systolic pressure. Ejaculation occurs during this phase as a result of sympathetic stimuli that cause contraction of the ischiocavernosa and bulbocavernosa muscles.

Testosterone gives rise to two other potent androgens: dihydrotestotsterone (DHT) and 5α-androstanediol (Figure 3). Sixty- five percent of testosterone is bound to a liver-derived glycoprotein called sex hormone binding globulin. Only 1–2% of circulating testosterone is in the free form while the remainder is bound to albumin. Adult levels of total testosterone can reach as high as 600 ng/dl by the age of 17, which is then maintained through the third decade. Testosterone has many actions, which includes sebum formation, beard growth, balding, enlargement of the genital organs, thickening of the larynx which leads to a deeper voice, stimulation of pubertal growth, enlargement of muscle mass, alterations in lipid levels (increase in low-density lipoprotein) and cholesterol, accumulation of visceral fat which is associated with a greater risk of cardiovascular disease, stimulation of synthesis and maturation of erythroid precursors, stimulation of renal sodium reabsorption, initiation of sexual drive (libido) and potency, and possible stimulation of aggressive behavior (Figure 4).

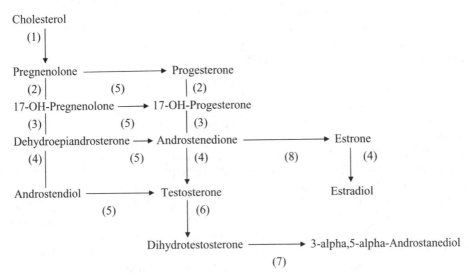

Figure 3. Pathway of synthesis of gonadal steroid hormones. (1) 20,22-desmolase; (2) 17-hydroxylase; (3) 17,20-desmolase; (4) 17-beta-OH-steroid dehydorogenase; (5) 3-beta-ol-dehydrogenase, and delta-4,5-isomerase; (6) 5-alpha-reductase; (7) 3-alpha-reductase; (8) aromatase.

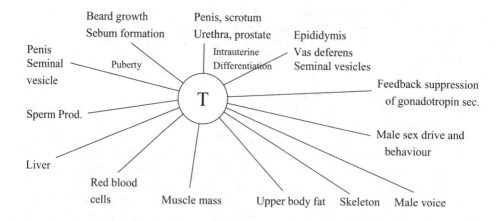

Figure 4. Effects of testosterone on body.

Once having reached full maturation, the reproductive organs and its hormonal and sexual functions, begin to age in a rather predictable way. Several long-term studies have been performed to look at the changes that occur with aging. Levels start to decline at age 30 and this rate of decline parallels the rate of decline of the physiologic changes associated with aging as has been shown in several large cross-sectional studies [1–5]. In an Australian study using the free androgen index, it was reported that a decline in testosterone began after age 31 and continued to decline each decade thereafter [6]. Community-based samples from a male population in Belgium showed that 7% of men aged 40 to 60 years, 21% of men aged 60 to 80 years, and 35% of men older than 80 years have subnormal testosterone levels [7]. In a Canadian study of 319 physicians, bioavailable testosterone was measured and 49% of men aged 40 to 49 years and 70% of men aged 70 years and older had subnormal levels [8]. This decline in testosterone occurs despite the fact that there is an increase in sex hormone binding globulin (SHBG) with aging [9]. The rate of decline appears to be between 1% and 2% per year. This fall in testosterone levels occurs in extremely healthy men, and the rate of decline in testosterone appears to be more rapid when illness intervenes [10–12]. Because of the increase in SHBG, it would be expected that free and bioavailable testosterone (testosterone bound to albumin as well as free testosterone) would demonstrate an even more dramatic fall with aging. A number of studies have now confirmed this expectation [13–18]. Although younger men have a circadian rhythm in both testosterone and bioavailable testosterone, this rhythm is markedly attenuated in older men [19–20]. With decreasing levels of testosterone, clinical symptoms in men have been observed. Werner described the presence of male "climacteric" symptoms in 273 men in 1946 [21]. He gave testosterone propionate to 181 of these men and reported symptomatic improvement in 176. Common symptoms that were reported in these "climacteric" include altered potency and libido, psychological symptoms such as nervousness, irritability, fatigue, depression, and decreased memory, hot flushes, headache, vague pains, numbness and tingling, and obesity. A questionnaire can be used called the ADAM questionnaire (Table 1) to look for these symptoms. In a study over 300 Canadian physicians, this questionnaire was demonstrated to have high sensitivity (88%), and a specificity of 60% in identifying men with low bioavailable testosterone [22].

Table 1. ADAM Questionnaire

1.	Do you have a decrease in libido (sex drive)?	Yes —	No —
2.	Do you have a lack of energy?	Yes —	No —
3.	Do you have a decrease in strength and/or endurance?	Yes —	No —
4.	Have you lost height?	Yes —	No —
5.	Have you noticed a decreased enjoyment of life?	Yes —	No —
6.	Are you sad and/or grumpy?	Yes —	No —
7.	Are your erections less strong?	Yes —	No —
8.	Have you noticed a recent deterioration in your ability to play sports?	Yes —	No —
9.	Are you falling asleep after dinner?	Yes —	No —
10.	Has there been a recent deterioration in your work performance?	Yes —	No —

If questions 1 or 7 are answered yes, or at least 3 other questions are answered yes, testosterone levels may be low

After utilizing the ADAM questionnaire as a tool to screen for hypogonadism, one can then look for physical exam findings which should include the evaluation of hair and fat distribution, gynecomastia, breast masses, and the examination of the penis, testicles, and prostate. Laboratory tests such as complete blood cell count, serum chemistry profile, and thyroid function should be ordered to confirm a disease process suggested by history and physical examination. Total testosterone, free testosterone, and bioavailable testosterone should also be ordered. Prolactin levels can be ordered if hypogonadism is found, since hyperprolactinemia can decrease testosterone levels by decreasing the secretion of gonadotropin releasing hormone (Figure 5). Medication should also be assessed since some have been associated with testicular dysfunction, gynecomastia or both (Table 2). Total and free testosterone levels are not adequate for diagnosing hypogonadism. In a study by Morely et al. [23], a study was performed in which 50 male subjects 28 to 90 years who were healthy and not hypogonadal, did not have disease or took medication that produces a decline in testosterone, had their serum analyzed at 08:00 and 10:00, and had their serum analyzed for (1) total serum testosterone, (2) serum SHBG, (3) bioavailable testosterone (BT) by ammonium sulfate precipitation using a commercially available radioiummunoassay kit (lower limit of normal for BT is 70 ng/dl or if divided by

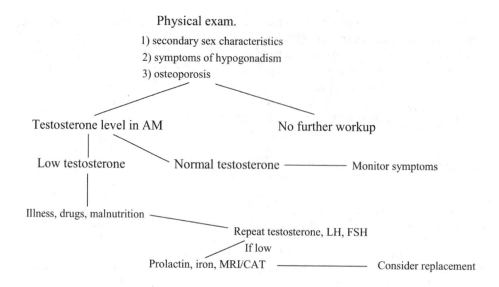

Figure 5. *Approach to patient with androgen deficiency.*

28.84 to convert to nmol/L the lower limit of normal is 2.1 nmol/L), (4) free testosterone by ultracentrifugation, (5) direct estimation of serum free testosterone by a ligand radioimmunossay, (6) free androgen index (FAI) which is the calculated total testosterone (T) and SHBG (FAI=100 T/SHBG), and (7) the free testosterone index (FTI) which is calculated from SHBG and T using the method of vermeulen [24]. All measures were statistically significant with the best correlation being with FTI (r = 0.807, $p<0.001$), and second best being with BT (r = 0.670, $p<0.001$). If total T <300 ng/dl is used as the cutoff for hypogonadism, BT correlated best with FTI (r = 0.871, $p<0.001$). Using total T, 42 % of patients would have been misclassified, in which a total of 26% with normal T were hypogonadal by BT, and 16% of those hypogonadal by T were normal by BT. Testosterone levels and age correleated best with BT (r = -0.744, $p<0.001$). All the correlations were significant except for total T (r = -0.126). In conclusion, some measure of free T (using FTI or free testosterone by dialysis), or BT should be used to identify hypogonadism in men across the life span. Total T misclassifies hypogonadism in a third of cases and seems an inappropriate measure to use. Also, if the initial BT level is normal with a male with symptoms of hypogonadism, a second level should be obtained since there seems to be variability form week to week in BT. In an 8 week period, BT was checked in 16 men and 10 of them men would have been hypogonad at one point and eugonadal at a second point. Salivary testosterone levels may be a good screening test [25].

Numerous anatomic changes also occur with aging. The Massachusetts male aging study showed that the prevalence of erectile dysfunction was 52% [26]. There is a

Table 2. Drugs that affect sexual function

Agent	Biologic effect	Endocrine effect	Type of sexual dysfunction
1. Clonidine	Alpha-blocker	None	Erection
2. Ethanol	CNS depressant	Decrease (T, E)	Libido, erection
3. Opiates	Endorphin agonist	Decrease (T, LH)	Libido, erection
		Decreased response to GNRH	Amenorrhea
4. Cimetidine	H-2 receptor antagonist	Anti-androgen, including prolactin including estrogen	Libido, erection
5. Digoxin		Decrease T, LH Decrease T and gonadotropins	Libido, erection Libido, erection
6. Estrogens			Libido, erection, ejaculation
7. Tricyclic antidepressants	Central anticholinergic		Libido, erection
8. Metoprolol	Beta blocker		Libido, erection
9. Spironolactone	Antialdosterone	Antiandrogen	Erection, ejaculation
10. Hydroclorothiazide	Diuretic		Erection, libido, ejaculation
11. Methyldpoa	Alpha-blocker		Libido, erection
12. Diphenhydramine	Sedative antihistaminic		Libido
13. Barbituates	Central depression		Libido
14. Benzodaizpenes	Gaba receptor		Libido, ejaculation, erection
15. Seretonin reuptake inhibitors	Central anticholinergic		

decrease in testicular size and weight, a reduction in the number of Leydig cells [27–29], and development of vacuolization and lipofuscin within the Leydig cells [30]. Function of the seminiferous tubule compartment of the testis also declines with aging. In older compared with younger men, spermatogenesis assessed histologically is reduced [31–33], but ejaculated sperm concentration is unchanged or increased as a result of diminished ejaculatory volume and frequency [34–35]. The number of sperm with normal motility and morphology also decreases but *in vitro* fertilizing capacity is relatively well preserved in older men [36–37]. Despite overall well-preserved fertility potential [38], and documented instances of paternity in men older than 90 years, overall fertility rates decline with age [38, 39]. With older paternal age, the risk of inherited autosomal dominant diseases increases in offspring [39]. It is noted that the circadian variation is blunted in older men especially with the morning peak [40]. It has been shown that there is a decline in testosterone production in older persons when the testes are stimulated by human chorionic gonadotropin [41–44]. The impact of reduced testosterone (T) production on circulating T levels is lessened by the decrease in metabolic clearance of T that also occurs with aging [45]. However, the major problem with aging appears to be within the hypothalamic-pituitary axis (Table 3). The major change with aging appears to be due to a failure of the hypothalamus to generate an appropriate amplitude of pulsatile secretion bursts of gonadotropin-releasing hormone [46]. There is evidence that the hypothalamic-pituitary unit is more sensitive to the suppressive effects of testosterone in old people (over 60 years of age) than in young people under the age of 40 [47, 48]. Beta-endorphin suppresses the LH surge [49]. In middle-aged persons with low testosterone levels there is evidence for increased opiodergic suppression of LH [50]. In older individuals, this is no longer the case [51], and may explain the rise in LH that occurs in the old-old (greater than 80 years of age). The increase in follicle-stimulating hormone levels that occur with aging in men appears related both to the fall in

Table 3. Prevalence of hypogonadism in older men

| Age (y) | Percent hypogonada | | |
	Baltimore Longitudinal Study	Mayo Clinic	Canadian Physicians
40–49	2	2	5
50–59	9	6	30
60–69	34	20	45
70–79	68	34	70
80 and above	91	–	–

inhibin levels [52], and an increase in dimeric activin A levels [53]. In the Massachusetts Male Aging Study, serum total and free T levels were found to be associated with the CAG repeat length in the androgen receptor gene [54]. It is has been shown that a shorter CAG repeat length on the androgen receptor is associated with greater androgen receptor activity and possible overall greater androgen action [55]. Thus, it is hypothesized that, in older men with shorter CAG repeat length, increased androgen action at the level of the hypothalamic-pituitary axis may result in greater feedback suppression of gonadotropin and, in turn T levels with aging.

As mentioned previously, testosterone affects sexual function. Davidson *et al.* [56], demonstrated that testosterone administration in hypogonadal men enhanced sexual activity. Schiavi *et al.* [57] demonstrated strong correlations between frequency of sexual thoughts and desire for sex, easiness of becoming aroused, degree of coital erections, and frequency of sleep erections and bioavailable testosterone in healthy aging men. Morales *et al.* [58] reported a 61% increase in sexual attitudes and performance in men receiving testosterone undeconoate orally for hypogonadal impotence. Hajjar *et al.* [59] demonstrated a marked increase in libido in older men receiving testosterone injections as compared with placebo. In conjunction with improvement in sexual function, it has been show that testosterone also increases strength [60]. Testosterone increases the fractional rate of protein synthesis in young [61] and older men [62]. With aging there is a clear decline in upper and lower muscle strength [63]. Testosterone has been reported to improve severe muscle weakness when replacement therapy was given [64]. Testosterone replacement in older men has been shown to increase grip strength in two studies [65, 66]. Urban *et al.* [67] reported an increase in leg muscle strength after 4 weeks of testosterone in men with a mean age of 67 years. Insulin Growth Factor I (IGF-1) increased in muscle in this study, suggesting a possible role for IGF-1 as the mediator of testosterone-induced muscle protein synthesis. In a cross sectional study strong correlation between a variety of memory tests and bioavailable testosterone was noted [68]. Bone mineral density [69], and coronary artery disease [70, 71] have also shown improvement. A psychiatric study showed that testosterone levels decrease in people who are depressed [72], and another showed success with testosterone when it was used as an adjuvant with antidepressants [73]. Barrett-Conner *et al.* [74] found the Beck Depression Inventory score to be significantly and inversely correlated with bioavailable testosterone.

Once a diagnosis of hypogonadism has been made, testosterone supplementation can be started and side effects can be monitored. Several testosterone supplementations are available, which include oral, parenteral, and transdermal preparations. The 2 oral forms of testosterone available in the United States (methyltestosterone, and fluoxymestrone) undergo first-pass inactivation in the liver and are the least efficient form of androgen replacement therapy. They may also present a significant risk for hepatotoxicity. Mesterelone and testosterone undecanoate are oral androgens available outside the United States. The later is taken up by the intestinal lymphatics, thus bypassing the liver. The parenteral forms include the testosterone esters (testosterone enanthate, and testosterone cypionate). They are more effective, safe, and least expensive and are the treatment of choice. Two-hundred milligrams are administered intramusculary every 2 weeks. Transdermal preparations have also avoided the firs-pass inactivation of the liver and have been shown to be effective. The therapeutic efficacy of androgen replacement is assessed primarily by monitoring the patient's clinical response. The ADAM questionnaire can be readministered. Adverse effects should also be monitored to monitor for polycythemia, symptoms of benign prostatic hypertrophy, fluid retention, rise in prostate specific antigen, sleep apnea, and gynecomastia in patients with hepatic cirrhosis. About 10% to 20% of men do not have any effects from testosterone supplementation. Future preparations include a long-acting 17-beta-hydroxyl ester of testosterone and a testosterone buciclate known as 20-AET-1 have been developed and tested by the World Health Organization. Intramuscular injection of 600 mg of the buciclate preparation have been tested and been shown to maintain serum testosterone levels in the low-normal to hypogonadal state for 12 weeks. It is not know whether this approach will provide adequate replacement for hypogondal men [75].

The aging of the male reproductive system is complex and intricate. Although some functionality is preserved, many changes occur effecting the well being of the individual. It is imperative that a good screening questionnaire be used followed by an appropriate laboratory measurement such as bioavailable testosterone to correctly diagnose hypogonadism in men. Subsequently, appropriate treatment should be instituted and certain clinical and laboratory parameters need to be followed to ensure the maximum benefit and augmentation of the quality of life of the individual.

References

1. Vermulen A (1991). Androgens in the aging male. *J Clin Endocrinol Metab.* 73: 221.
2. Bhasin S (1992). Clinical review 34: androgen treatment of hypogonadal men. *J Clin Endocrinol Metab.* 74: 1221.
3. Pirk KM, Doerr P (1970). Age related changes in free plasma testosterone, dihydrotestosterone and oestradiol. *Acta Endocrinol.* 80: 171.
4. Gray A, Feldman HA, Mckinlay JB, *et al.* (1991). Age, disease, and changing sex hormone levels in middle-aged men: results of the Massachusetts male aging study. *J Clin Endocrinol Metab.* 73: 1016.
5. Mitchell S, Harman E, Metter J, Tobin J, Pearson J, Blackman M (2001). Longitudinal effects of aging on serum total and free tesosterone levels in healthy men. *J Clin Endocrinol Metab.* 86(2): 724.
6. Wishar JM, Need AG, Horowitz M, Morris HA, Nordin BE (1995). Effect of age on bone density and bone turnover in men. *Clin Endocrinol.* 42: 141–46.
7. Vermeulen A, Deslypere JP (1985). Testicular endocrine function in the ageing male. *Maturitas* 7: 273–9.
8. Morley JE, Charlton E, Patrick P, *et al.* (1998). Validation of a screening questionnaire for androgen deficiency in aging males. Endocrine Society Abstracts.
9. Lafferty FW, Spencer GE, Pearson OH (1964). Effects of androgens,estrogens and high calcium intakes on bone formation and resorption in osteoporosis. *Am J Med.* 36: 514.
10. Baker HWG (1998). Reproductive effects of nontesticular illness. *Endocrinol Metab Clin N Am.* 27: 831–50.
11. Handelsman DJ (1994). Testicular dysfunction in systemic disease. *Endocrinol Metab Clin N Am.* 23: 839–56.

12. Morley JE, Melmed S (1979). Gonadal dysfunction in systemic disorders. *Metabolism* 28: 1051–73.
13. Gray A, Feldman HA, Mckinlay JB, *et al.* (1991). Age disease, and changing sex hormone levels in middle-aged men: results of the Massachusets Male aging Study. *J Clin Endocrinol Metab.* 3: 1016–25.
14. Haji M, Tanaka S, Nishi Y, *et al.* (1994). Sertoli cell function declines earlier than Leydig cell function in aging Japanese men. *Maturitas* 18: 143–53.
15. Kaiser FE, Vviosca SP, Morley JE, *et al.* (1988). Impotence and aging: clinical and hormonal factors. *J Am Geriatric Soc.* 36: 511–19.
16. Korenman SG, Morley JE, Mooradian AD, *et al.* (1990). Secondary hypogonadism in older men. Its relation to impotence. *J Clin Endocrinol Metab.* 71: 963–9.
17. Nankin HR, Calkins JH (1986). Decreased bioavailable testosterone in aging normal and impotent men. *J Clin Endocrinol Metab.* 63: 1418–20.
18. Orrell RW, Woodrow DF, Barrett MC, *et al.* (1995). Testosterone deficiency myopathy. *J Royal Soc Med.* 88: 454–6.
19. Bremmer WJ, Vitiello MV, Prinz PN (1983). Loss of circadian rhythmicity in blood testosterone levels with aging in normal men. *J Clin Endocrinol Metab.* 56: 1278.
20. Plymate Sr, Tenover JS, Bremner WJ (1989). Circadian variation in testosterone, sex hormone-binding globulin, and calculated non-sex hormone-binding globulin bound testosterone in healthy and elderly men. *J Androl.* 10: 366.
21. Werner AA (1946). The male climacteric: report of two hundred and seventy-three cases. *J Am Med Assoc.* 126: 472–7.
22. Morley JE, Charlton E, Patrick P, *et al.* (1998). Validation of a screening questionnaire for androgen deficiency in aging males. Endocrine Society Abstracts.
23. Morley JE, Patrick P, Perry HM (2002). Evaluation of assays available to measure free testosterone. *Metabolism* 51: 554–9.
24. Vermeulen A, Verdonck L, Kaufman JM (1999). A critical evaluation of simple methods for the estimation of free testosterone in serum. *J Clin Endocrinol Metab.* 84: 3666–72.
25. Morely JE, Kaiser FE (1989). Sexual function with advancing age. *Med Clin N Am.* 73: 1483–95.
26. Feldman HA, Goldstein I, Hatzichristou DG, Krane RJ, McKinlay JB (1994). Impotence and its medical and psychosocial correlates: results of the Massachusetts male aging study. *J Urol.* 151: 54–61.
27. Haji M, Tanaka S, Nishi Y, *et al.* (1994). Sertoli cell function declines earlier than Leydig cell function in aging Japanese men. *Maturitas* 18: 143–53.
28. Neaves WB, Johnson L, Porter JC, Parker CR Jr, Petty CS (1984). Leydig cell numbers, daily sperm production, and serum gonadotropin levels in aging men. *J Clin Endocrinol Metab.* 59: 756–63.
29. Kaler LW, Neaves WB (1978). Attrition of the human Leydig cell population with advancing age. *Anat Rec.* 192: 513–18.
30. Morley JE, Kaiser FE (1993). Impotence: the internist's approach to diagnosis and treatment. *Adv Intern Med.* 38: 151–68.
31. Neaves WWB, Jhonson L, Porter JC, Parker CR Jr, Petty CS (1984). Leydig cell numbers, daily sperm production, and serum gonadotropin levels in aging men. *J Clin Endocrinol Metab.* 59: 756–63.
32. Jhonson L, Grumbles JS, Bagheri A, Petty CS (1990). Increased germ cell degeneration during postprophase of meiosis is related to increased serum follicle-stimulating hormone concentrations and reduced daily sperm production in aged men. *Biol Reprod.* 42: 281–7.

33. Neaves WB, Johnson L, Petty CS (1987). Seminiferous tubules and daily sperm production in older adult men with varied numbers of Leydig cells. *Biol Reprod.* 36: 301–8.
34. Neischlag E, Lammers U, Freischem CW, Langer K, Wickings EJ (1982). Reproductive functions in young fathers and grandfathers. *J Clin Endocrinol Metab.* 55: 676–81.
35. Schwartz D, Mayaux MJ, Spira A, *et al.* (1983) Semen characteristics as a function of age in 833 fertile men. *Fertil Steril.* 39: 530–5.
36. Rolf C, Behre HM, Nieschlag E (1996). Reproductive parameters of older compared to younger men in infertile couples. *Int J Androl.* 19: 135–42.
37. Silber SJ (1991). Effects of age on male fertility. *Semin Reprod Endocrinol.* 9: 241–8.
38. Plas E, Berger P, Hermann M, Pfluger H (2000). Effects of aging on male fertility? *Exp Gerontol.* 35: 543–51.
39. Rolf C. Nieschlag E (2001). Reproductive functions, fertility and genetic risks of aging men. *Exp Clin Endocrinol Diabetes* 109: 68–74.
40. Grey A, Berlin JA, Mckinlay JB, Longcope C (1999). An examination of research design effects on the association of testosterone and male aging: results of a meta-analysis. *J Clin Epidemiol.* 44: 671–84.
41. Rubens R, Dhont M, Vermeulen A (1974). Further studies on Leydig cell function in old age. *J Clin Endocrinol Metab.* 39: 40.
42. Harman SM, Tsitouras PD (1980). Reproductive hormones in aging men. I. Measurement of sec steroids, basal leuteinizing hormone, and Leydig cell response to human chorionic gonadotropin. *J Clin Endocrinol Metab.* 51: 35.
43. Longscope E (1973). The effect of human chorionic gonadotropin on plasma steroid levels in young and old men. *Steroids* 21: 583.
44. Hammar M (1985). Impaired *in vitro* testicular endocrine function in elderly men. *Andrologia* 17: 444.
45. Vermeulen A, Reubens R, Verdonck L (1972). Testosterone secretion and metabolism in male senescence. *J Clin Endocrinol Metab.* 34: 730–5.
46. Morely JE, Perry HM (2000). Androgen deficiency in aging men: role of testosterone replacement therapy. *J Lab Clin Med.* 135: 370–8.
47. Winters SJ, Sherins RRJ, Troen P (1984). The gonadotropin-suppressive activity of androgen is increased in elderly men. *Metabolsim* 33: 1052–9.
48. Veldhuis JD, Urban RJ, Lizarralde G, *et al.* (1992). Attenuation of leuteinizing hormone secretory burst amplitude as a proximate basis for the hypoandrogenism of health aging men. *J Clin Endocrinol Metab.* 75: 707–13.
49. Morley JE, Baranetsky NG, Wingert TD, *et al.* (1980). Endocrine effects of naloxone-induced opiate receptor blockade. *J Clin Endocrinol Metab.* 50: 251–7.
50. Billngton CJ, Shafer RB, Morley JE (1990). Effects of opioid blockade with nalmefene in older impotent men. *Life Sci.* 47: 799–805.
51. Mikuma N, Kumamoto AM, Maruta H, *et al.* (1994) Role of the hypothalamic opioidergic system in the control of gonadotropin secretion in elderly men. *Andrologia* 26: 39–45.
52. Tenover JS, Matsumoto AM, Plymate SR, *et al.* (1987). The effects of aging in normal men on bioavailable testosterone and leuteinizing hormone secretion: response to clomiphene citrate. *J Clin Endocrinol Metab.* 65: 1118–26.
53. Lona P, Petraglia F, Concari M, *et al.* (1998). Influence of age and sex on serum hormone secretion concentrations of total dimeric activin A. *Eur J Endocrinol.* 139: 487–92.
54. Krithivas K, Yurgalevitch SM, Mohr BA, *et al.* (1999). Evidence that the CAG repeat in the androgen receptor gene is associated with the age realted decline in serum androgen levels in men. *J Endocrinol.* 162: 137–42.

55. Beilin J, Ball EEM, Favaloro JM, Zajac JD (2000). Effect of the androgen receptor CAG repeat polymorphism on transcriptiona; activity: specificity in prostate and non-prostate cell lines. *J Mol Endocrinol.* 25: 85–6.

56. Davidson JM, Camargo Ca, Smith ER (1979). Effects of androgen on sexual behavior in hypogonadal men. *J Clin Endocrinol Metab.* 48: 955–8.

57. Schiavi RC, Schreiner-Engel P, White D, *et al.* (1991). The relationship between pituitary-gonadal function and sexual behavior in healthy aging men. *Psychosom Med* 53: 363–74.

58. Morales A, Jhonston B, Heaton JP, Lundie M (1997). Testosterone supplementation for hypogonadal impotence: assesment of biochemical measures and therapeutic outcomes. *J Urol.* 157: 849–54.

59. Hajjar RR, Kaiser FE, Morely JE. Outcomes of long-term testosterone replacement in older hypogonadal males: a retrospective analysis. *J Clin Endocrinol Metab.* 82: 3793–6.

60. Bhasin S, Storer TW, Berman N, *et al.* (1996). The effects of supraphysiologic doses of testosterone on muscle mass and strength in normal men. *N Engl J Med.* 335: 1–7.

61. Brodsky IG, Balangopal P, Nair KS (1996). Effects of testosterone replacement on muscle mass and muscle protein synthesis in hypogonadal men: a clinical research center study. *J Clin Endocrinol Metab.* 81: 3469–75.

62. Urban RJ, Bodenburg YH, Gilikson C, *et al.* (1995). Testosterone administration to elderly men increases skeletal muscle strength and protein synthesis. *Am J Physiol* 269: E820–6.

63. Frischknecht R (1998). Effect of training on muscle strength and motor function in the elderly. *Reprod Nutr Dev.* 38: 167–74.

64. Orrell RW, Woodrow DF, Barrett MC, *et al.* (1995). Testosterone deficiency myopathy. *J Royal Soc Med* 88: 454–6.

65. Sih R, Morley JE, Kaiser FE *et al.* (1997). Testosterone replacement in older hypogonadal men: a 12 month randomized controlled trial. *J Clin Endocrinol Metab.* 82: 1661–7.

66. Morely JE, Perry HM III, Kaiser FE, *et al.* (1993). Effects of testosterone replacement in older hypogonadal men: a preliminary study. *J Am Geriatric Soc.* 41: 149–52.

67. Urban RJ, Bodenburg Yh, Gilikson C *et al.* (1995). Testosterone administration to elderly men increases skeletal muscle strength and protein synthesis. *Am J Physiol.* 269: E820–6.

68. Morley JE, Kaiser FE, Raum Wj, *et al.* (1997). Potentially predictive and manipulable blood serum correlates of aging in the healthy human male: progressive decreases in bioavailable testosterone, dehydroepiandrosterone sulfate, and the ratio of insulin-like growth factor 1 to growth hormone. *Proc Natl Acad Sci USA* 94: 7537–42.

69. Tenover JS (1998). Testosterone replacement therapy: a step closer. World Congress on Aging – Male. Geneva, abstract P2.

70. Hajjar RR, Kaiser FE, Morley JE (1997). Outcomes of long-term Testosterone replacement in older hypogonadal males: a retrospective analysis. *J Clin Endocrinol Metab.* 82: 3793–6.

71. Rosano GM, Leonardo F, Pagnotta P, *et al.* (1999). Acute anti-ischemic effect of testosterone in men with coronary artery disease. *Circulation* 99: 1666–70.

72. Schweiger U, Deuschle M, Weber B, *et al.* (1999). Testosterone, gonadotropin, and cortisol secretion in male patients with major depression. *Psychosom Med.* 61: 292–6.

73. Seidman SN, Rabkin JG (1998). Testosterone replacement therapy for hypogonadal men with SSRI-refractory depression. *J Affect Disord.* 48: 157–61.

74. Barrett-Connor E, von Muhlen DG, Krotz-Silverstein D (1999). Bioavailable testosterone and depressed mood in older men: the Rancho Bernardo study. *J Clin Endocrinol Metab.* 84: 573–7.

75. Bhasin S, Swerdloff RS, Steineer B, *et al.* (1992). A biodegradable testosterone micro-
    capsule formulation provides uniform eugonadal levels of testosterone for 10–11 weeks in
    hypogonadal men. *J Clin Endocrinol Metab.* 74: 75–83.

Index